The Elements of
Palaeontology

D0807867

The Elements of Palaeontology

Second edition

RHONA M. BLACK

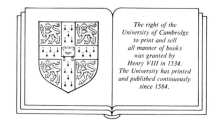

CAMBRIDGE UNIVERSITY PRESS

CAMBRIDGE   NEW YORK   NEW ROCHELLE

MELBOURNE   SYDNEY

Published by the Press Syndicate of the University of Cambridge
The Pitt Building, Trumpington Street, Cambridge CB2 1RP
32 East 57th Street, New York, NY 10022, USA
10 Stamford Road, Oakleigh, Melbourne 3166, Australia

© Cambridge University Press 1970, 1988

First published 1970
Reprinted 1972, 1973 (with additions) 1975, 1978, 1979
Second edition 1988

Printed in Great Britain at the University Press, Cambridge

British Library cataloguing in publication data
Black, Rhona M.
    The Elements of Palaeontology. – 2nd ed.
    1. Palaeontology
    I. Title
    560   QE711.2

Library of Congress cataloguing in publication data
Black, Rhona M.
    The Elements of Palaeontology.

Bibliography: p.
    Includes index.
    1. Palaeontology.   I. Title.
QE711.2.B5   1987   560   87–9353

ISBN 0 521 34346 1 hard covers
ISBN 0 521 34836 6 paperback

(First edition:
ISBN 0 521 07445 2 hard covers
ISBN 0 521 09615 4 paperback)

TM
560
B627.2
238247

CONTENTS

PREFACE

This book is written primarily for the use of students who require basic information on fossils: invertebrates, vertebrates, microfossils, plants and trace fossils. In each case, the morphology of hard and soft parts is described and illustrated by diagrams and photographs; biological affinities and mode of life are discussed; and the stratigraphic range is given. Relevant technical terms are defined and a glossary is included for each of the main invertebrate groups; but the use of jargon has been minimised. The introduction covers the necessary background concerning distribution and habitats of modern organisms, the process of fossilisation, the stratigraphic column and geological time scale. The final chapter summarises the fossil record and considers a few of the problems presented by its study.

In this second edition of the book a chapter on trace fossils has been added, the chapter on microfossils has been expanded and a section on pollen grains included with the plants. In each case there are additional illustrations including scanning electron micrographs. Much of the remainder has been rewritten to take account of advances in knowledge in recent years.

ACKNOWLEDGEMENTS

The author is deeply indebted to those who read and suggested amendments to the typescript of this new edition: Dr R. J. Aldridge (conodonts); Mr J.E. Almond (myriapods); Dr R.B. Rickards (graptolites); Professor H.B. Whittington (trilobites); and Dr W.W. Black for detailed discussion and help, particularly with the final chapter.

Grateful acknowledgement is made to all those, cited below, who so generously gave photographs:

Hunterian Museum: 3, 43, 59, 89, 99, 154, 173, 187, 190, 200, 204, 226.
Hunterian Museum and Dr W.D.I. Rolfe: 16, 45, 85a–c, 100, 133, 138, 144.
Hunterian Museum and Dr J.K. Ingham: 91, 174a, b, 186.
Hunterian Museum and Norman Brothers: 172.
(Except where otherwise stated in the captions, the above photographs are of
   specimens or exhibits in the Hunterian Museum, Glasgow.)
National Institute of Oceanography: 12.
Dr R.G. West: 14; a and c reproduced from West, *Pleistocene Geology and
   Biology* with permission from the Longman Group.
Dr P.M. Porter Kier: 74, 107, 113, 114, 129.
Dr T.D. Ford: 77a by permission of the Council of the Yorkshire Geological
   Society.
Dr G.B. Curry: 80.
Professor H.B. Whittington: 87, 88.
Professor Dr K.J. Müller: 96, 167.
Mr A.M. Honeyman: 98.
Mr J.E. Almond: 102.
Professor F.M. Carpenter: 103, 104, 105.
Dr R.B. Rickards: 139, 141, 142.
British Museum (Natural History): 149, 177a, 185, 194, 196, 197, 198.
Professor H.B. Whittington and Dr C.P. Hughes: 151.
Dr I.C. Harding: 155, 162d–h, 165c, d, 168g, h, 220i.
Dr M. Black: 157a, 158.
Dr P. Echlin and the Cambridge Scientific Instrument Co., Ltd: 157b.

Dr D. Gobbett: 165a.

Dr R.J. Aldridge: 170, 171c.

Professor A. Heintz: 177b, 178.

Dr N.C. Fraser: 206.

Dr J.H.J. Penny: 232, 233.

Sedgwick Museum: Photographs not mentioned above are of specimens in the Sedgwick Museum, Cambridge, and are the work of Dr C.P. Hughes, Mr A. Barlow and Mr D. Bursill; and also of Dr D. Price (69b, c, 70a, b, 71b, 146).

Diagrams of fossils were drawn directly from specimens where possible; otherwise they are modified after illustrations in Monographs of the Palaeontographical Society, Zittel (*Textbook of Palaeontology*) vols. I–III, various palaeontological journals, and also A.S. Romer (*Vertebrate Palaeontology*, with permission) and W.N. Stewart (*Palaeobotany and the Evolution of Plants*, extra diagrams for revised edition). The diagrams retained from the original edition include those redrawn for it by Mrs J. Friend: 82, 84; Dr J.K. Ingham: 86, 90, 93 including some of his own drawings, notably 86c, e; Dr R.B. Rickards: 134, 135, 137; Mr D. Batten: 32, 35, 37, 39, and, in part, 220, 221, 225, 227; Dr T. Kemp: most of 176, 179, 180, also 184, 192. New diagrams for this revision have been redrawn in the Artwork division of the publishers.

I

Introduction

Palaeontology is the study of ancient life-forms of which remains or other evidence are found in sedimentary rocks, and which are described as FOSSILS. It spans a period of some 3500 million years (my).

Fossils

Organic origin is implicit in the term 'fossil'. A fossil represents part of a once-living animal or plant, or has been formed by the action of a once-living organism. For example, the bones of a dinosaur are fossils, and so also are the footprints which a dinosaur made in wet sand. While the term 'fossil' implies great age, this does not mean that all fossils must represent extinct organisms; a wide variety of forms living today are known also from the fossil record. Many fossils, more or less similar to present day organisms, can be compared with these directly. Others belong to groups long since extinct but which show some (varying) degree of similarity to organisms alive today. The remains are usually of the hard parts only, and even these are often incomplete or fragmentary. Their interpretation requires a knowledge of the biology of living forms. Palaeontology as a subject, therefore, lies across the frontier of geology and biology; its unique contribution concerns the ancestry of modern faunas and floras.

Fossils have an intrinsic value, but their study is a fundamental part of geology. In economic geology they are important as a means of identifying the rock formations in which fuels like coal and oil occur. Their main importance to the geologist, however, lies in aspects of historical geology. This is the branch of geology which aims to reconstruct the record of events in earth's history, of past geography, climate and environments, using the evidence of sedimentary (stratified) rocks and their contained fossils. In this way palaeontology impinges closely on STRATIGRAPHY.

Stratigraphy

This subject deals with the nature and origin of stratified rocks; with their sequence in the earth's crust; and with their correlation, i.e. the identification of isolated outcrops of rocks with those of similar age elsewhere. The first practical step in stratigraphy is the grouping of strata into lithological units, e.g. limestones, clays, etc. The second step is to establish the sequence of these units in time. For this purpose the basis of all stratigraphical work is the principle of superposition of strata – simply, that in any normal undisturbed sequence of sedimentary rocks a stratum is younger than that on which it rests. This principle has been used to establish the correct sequence of strata in the stratigraphical column.

Early work in this field showed that the fossils which occur in any one part of the stratigraphical column are distinctive and different from those which occur at other levels. This made possible the delineation of groups of strata containing particular characteristic fossils (zonal fossils) and known as BIOZONES. Once the order of these biozones was established it became possible to refer any set of sedimentary rocks to its correct position in the stratigraphical column by examination of the fossil content.

Stratigraphical column

The complete succession of stratified rocks which contain fossils is sum-marised in the stratigraphical column in which the strata, combined into major groups and systems, are arranged in sequence with the oldest at the base and the youngest at the top (fig. 1). The earliest rocks in which fossils occur in appreciable numbers, the Cambrian, overlie still older rocks, the Precambrian, in which fossils are rare.

Classification of organisms

Organisms, living and fossil, are immensely varied, and some orderly arrangement (classification) is essential for their proper study. The basis of such a classification is the identification and naming of each organism which is recognisably different from all others. The system of naming in universal use is the BINOMIAL system, each different type of organism being given a name consisting of two parts, a SPECIFIC name and a GENERIC name. The specific name, always written with a small initial letter, denotes the SPECIES. A species is the smallest unit of division in common use and, as a biological concept, defines a group of similar organisms which can breed only within that group. A fossil species must obviously be identified by the characters of the hard parts and these may show variation. If sufficient individuals from the same bed are available to show that this variation is continuous, they may be referred to the same species without hesitation. With smaller numbers, however, the only practicable definition of a fossil species is 'a collection of individuals which show only minor differences'. To some extent the decision as to whether a fossil is a separate species is subjective.

The generic name, always written with a capital initial letter, denotes the

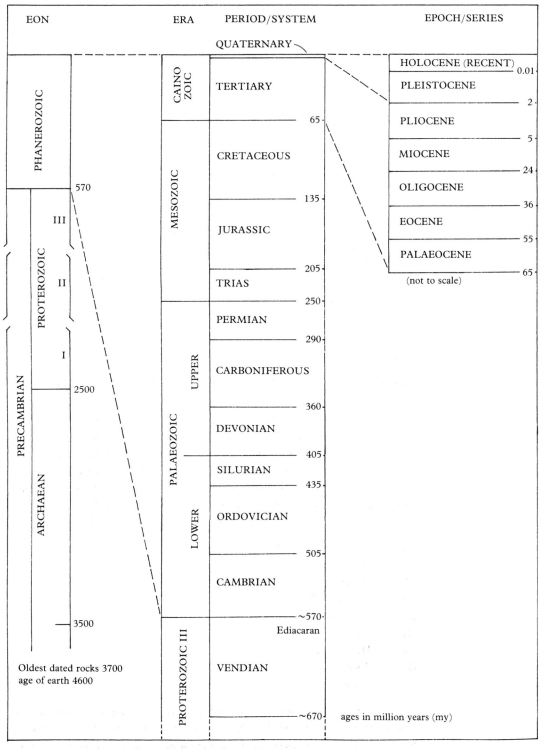

EON	ERA	PERIOD/SYSTEM	EPOCH/SERIES

QUATERNARY

PHANEROZOIC	CAINOZOIC	TERTIARY	HOLOCENE (RECENT) 0.01
			PLEISTOCENE 2
			PLIOCENE 5
			MIOCENE 24
			OLIGOCENE 36
			EOCENE 55
			PALAEOCENE 65

570

III

MESOZOIC — CRETACEOUS — 65

PROTEROZOIC — II — CRETACEOUS — 135

JURASSIC — 205

TRIAS — 250

(not to scale)

I — PERMIAN — 290

2500

UPPER — CARBONIFEROUS — 360

PRECAMBRIAN

ARCHAEAN

PALAEOZOIC — DEVONIAN — 405

SILURIAN — 435

LOWER — ORDOVICIAN — 505

CAMBRIAN

3500 — ~570

Ediacaran

PROTEROZOIC III — VENDIAN — ~670

Oldest dated rocks 3700
age of earth 4600

ages in million years (my)

1 Geological time scale.

GENUS to which the species belongs. A genus is an arbitrary unit consisting of a number of species which have similar features and are closely related.

The binomial system of classification is both comprehensive and flexible. It allows for modifications in the light of further information about particular organisms and their ancestry. Inevitably, the system is hedged in by conventions. These ensure, as nearly as possible, unambiguous identification of fossils. When a fossil is first named, a particular specimen is figured and described as the TYPE, and ideally is lodged in a museum. Thus, if doubt exists about the identity of a particular fossil it may be compared directly with the type of a species.

Confusion sometimes arises because a fossil is called by one generic name in an earlier textbook but by a different generic name in a later one. There are several reasons why a change of name may occur. It may simply reflect improved knowledge about the actual relationship of the fossil. Or a worker who named the fossil may have been unaware that it had already been described under a different name by an earlier worker. The earlier name has priority and the later name is invalid. Occasionally, a fossil is given a name which, unknown to the author, is already in use for another organism; when the mistake is discovered a new name must be chosen.

Higher categories of classification

The binomial system identifies individual species but these, living and fossil, are counted in millions. It is convenient, therefore, to have them arranged in a hierarchical system, each category of which contains groups with structures in common. The following are the main categories used in this higher classification of organisms. Just as related species comprise a genus, so related genera make up a FAMILY, related families an ORDER, orders a CLASS, classes a PHYLUM and phyla a KINGDOM. In cases where additional categories are required the prefixes 'sub' or 'super' may be added thus: subphylum; superkingdom. As an illustration, man is classified thus:

Superkingdom: Eucaryota
Kingdom: Animalia
Phylum: Chordata
Subphylum: Vertebrata
Class: Mammalia
Order: Primates
Family: Hominidae
Genus: *Homo*
Species: *sapiens*

The ultimate division of organisms is into two superkingdoms, Procaryota and Eucaryota, and is based on the nature of the CELL, i.e. the basic unit of the body. The cell consists of the living protoplasm enclosed by a semi-

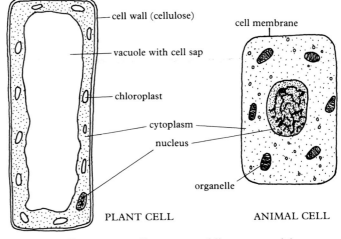

cell wall (cellulose)

cell membrane

vacuole with cell sap

chloroplast

cytoplasm

nucleus

organelle

PLANT CELL   ANIMAL CELL

2 Eucaryote cell structure (diagrammatic).

permeable membrane (fig. 2). The procaryote cell, the more primitive, lacks a nucleus, i.e. its genetic material (DNA) is dispersed in the protoplasm. The eucaryote cell has its genetic material organised in strands (chromosomes) of DNA and protein inside a membrane-bounded nucleus. In addition it has further membrane-bounded bodies (organelles) with special functions such as energy-production or, in plants, food manufacture using sunlight.

Principal groups of organisms of interest to palaeontologists

Category and geological range	Examples and summary of characteristics
Superkingdom Procaryota	
Kingdom Monera Precambrian (about 3500 my)–Recent	BACTERIA (ubiquitous) and CYANOBACTERIA (aquatic). Microscopic; one-celled, filamentous, or in colonies; often motile; feed on organic or inorganic source, some using light for food manufacture (photosynthesis). Fossils rare. Important role in geological processes.
Superkingdom Eucaryota	
Kingdom Protoctista Late Precambrian–Recent	Plant-like 'algae' (diatoms, seaweeds) and animal-like protozoans (foraminifera). Microscopic; one-celled (except seaweeds); aquatic; skeleton organic, calcareous or siliceous. Fossils abundant.
Kingdom Fungi ? Precambrian, Silurian–Recent	TOADSTOOLS. One-celled or many-celled body of filaments; plant-like but lack light-absorbing pigments for photosynthesis; usually land-based;

Category and geological range	*Examples and summary of characteristics*
	feed on plants, living or dead. Rare as fossils but important role in recycling plant debris.
Kingdom Plantae Silurian–Recent	PLANTS. Many-celled, non-motile; have light-absorbing pigments for photosynthesis; cell walls of cellulose; land-based and aquatic.
Phylum Bryophyta U Devonian–Recent	MOSSES. Simplest land plants; lack water/nutrient conducting tissue; reproduce by spores; damp-loving. Fossils rare.
Phylum Tracheophyta Silurian–Recent	VASCULAR PLANTS with organised water/nutrient conducting tissue. Plant body different-iated into leaves, stems and roots; mainly land-based; very small to forest tree size. Primitive plants reproduce by spores, e.g. LYCOPODS, HORSETAILS, FERNS. Advanced plants reproduce by seeds, e.g. GYMNOSPERMS (conifers) and ANGIOSPERMS (flowering plants). Fossils, including reproductive spores and pollen grains, very abundant.
Kingdom Animalia Later Precambrian–Recent	ANIMALS. Many-celled; motile or fixed; Typically with bilateral symmetry, but radial in some; have specialised nervous/sensory system; land-based and aquatic. Main phyla below.
Phylum Porifera M Cambrian–Recent	SPONGES. Sessile; aquatic, mainly marine; lack definite tissues and organs; may have internal skeleton of calcareous, siliceous or organic spicules. Fossils locally abundant.
Phylum Archaeocyatha L–M Cambrian	Extinct forms with combined sponge/coral features. Sessile; marine; skeleton inverted cone shape, calcareous. Fossils associated in reefs.
Phylum Cnidaria Late Precambrian–Recent	CORALS and JELLYFISH. Sessile or motile; single or colonial; mainly marine; have tissues but no true organs; radial symmetry; body cavity with single opening bordered by food-catching tentacles; stinging cells. Corals with calcareous skeleton common as fossils; jellyfish soft-bodied, rarely fossilised.
Phylum Bryozoa	SEA MOSSES. Sessile; mainly marine; colonial, individuals very small (1 mm); organs well developed; gut with mouth and anus; feeding tentacles filter tiny organisms; skeleton external, usually calcareous. Fossils abundant.
Phylum Brachiopoda Cambrian–Recent	LAMP SHELLS. Solitary; sessile; marine; filter-feeders; soft parts similar to bryozoans; external shell of two unlike valves; calcareous or horny phosphatic. Fossils abundant especially in Palaeozoic.
Phylum Mollusca Cambrian–Recent	AMMONITES, BELEMNITES, BIVALVES, GASTROPODS and others. Mainly vagrant; typically marine but

Category and geological range	*Examples and summary of characteristics*
	some fresh-water or land-living; highly organised with head and foot differentiated; plant, flesh or filter-feeders; typically external calcareous shell of one, two or more parts and varied in form; rarely, with internal or no shell. Fossils abundant.
Phylum Annelida Precambrian–Recent	SEGMENTED WORMS. Mobile or sessile; typically marine, also fresh-water and land-living; soft-bodied but may build organic or calcareous tubes; mainly filter or organic-debris-feeders. Fossils mainly tubes, burrows or trails.
Phylum Arthropoda Cambrian–Recent	TRILOBITES, CRUSTACEANS (crabs, etc.), ARACHNIDS (spiders), EURYPTERIDS.. Mobile, aquatic and land-living; body segmented, typically with head, body and tail regions; paired jointed limbs on each segment; plant, flesh or filter-feeders; external skeleton, mainly chitin, moulted periodically for further growth. Trilobites (Cambrian–Permian) most important as fossils.
Uniramian arthropods Silurian–Recent	INSECTS and MILLIPEDES Mainly land-living arthropods; either long-bodied with many segments, each with paired legs (millipedes); or short-bodied with fewer segments, three pairs of legs, and often with one or two pairs of wings (insects); varied plant or animal feeders. Fossilised in exceptional conditions.
Phylum Echinodermata Cambrian–Recent	ECHINOIDS (sea-urchins), CRINOIDS, STARFISH and many extinct groups. Mobile or sessile; typically five-rayed symmetry; water vascular system used in feeding, breathing and moving; algae, flesh or filter-feeders; internal skeleton of calcite plates. Fossils mainly crinoids and echinoids.
Phylum Hemichordata Cambrian–Recent	ACORN WORMS, PTEROBRANCHS; Class GRAPTOLITHINA probably belongs here. Sessile or free; marine; skeleton, if any, organic. GRAPTOLITE fossils abundant in lower Palaeozoic.

Phylum Chordata
?M, U Cambrian–Recent
	MAMMALS, BIRDS, REPTILES, AMPHIBIANS, FISH and minor groups. Mobile (usually); aquatic or land based; at some stage in development have:
	  (i) internal supporting rod (notochord), usually replaced by backbone, and
	  (ii) gill slits in throat (primitive function filtering food).
	Skeleton, usually present, internal, bony or cartilaginous. Fossils moderately common as fragments, rarely complete.

Use of informal terms

No attempt has been made to provide a detailed classification. At this level of treatment it is enough to recognise the main biological group to which a fossil belongs, e.g. that *Venus* belongs to the Class Bivalvia. It is convenient, however, to have a term which embraces a number of closely allied genera while, at the same time, clearly relating them to a genus already described. For this purpose the name of that genus, but with the termination 'id' (or in some instances 'oid') is used. Thus the term 'agnostid' is used to denote forms closely related to the trilobite genus *Agnostus*, and, similarly, the term 'cidaroid' embraces the close relatives of the echinoid *Cidaris*. In a similar way a generic name may be used in a broad sense as opposed to the strict definition. This is indicated by enclosing the name in inverted commas, e.g. '*Trigonia*', or by adding the letters *s.l.* after the name, meaning *sensu lato* (in the broad sense). The strict definition is indicated by adding the letters *s.s.*, meaning *sensu stricto* (in the strict sense). It should be noted that the stratigraphical range of the genus *s.l.* will be greater than that of the genus *s.s.*

Occurrence and preservation of fossils

Rock types in which fossils occur

Fossils are mainly found in sedimentary rocks such as limestones, mudstones and sandstones deposited in former seas, lakes, deltas or river flood-plains. They are most abundant in marine rocks deposited in relatively shallow seas where organisms are abundant and sediments accumulate rapidly. They are often sparse or lacking in continental rocks.

Limestones are often highly fossiliferous, sometimes consisting almost entirely of shells (fig. 3). These may be whole and uncrushed but more commonly are in fragments, and may have been rolled about so that delicate structures, like spines on brachiopods, are removed. Reefs built by corals, calcareous algae and other organisms are a feature of many limestones and often contain well-preserved fossils. Fine-grained rocks like clays and shales may also be very fossiliferous, with well-preserved fossils showing fine details. The fossils may, however, have been flattened as a result of vertical compaction of the rock. Clays and shales frequently contain nodules of limestone, ironstone, phosphate or flint. Often these nodules have formed around organic remains and the fossils they contain are usually well preserved and uncrushed (fig. 4). Coarser-grained rocks like sandstones are not commonly very fossiliferous apart from certain bands of sandstone, usually marine, which may be locally rich in fossils. These are usually uncrushed, but calcareous shells may be dissolved by permeating water, leaving moulds in the rock (fig. 5).

3 Shelly limestone with ammonites: *Asteroceras*, **Marston Limestone, L Jurassic, Yeovil, Somerset (× 1.3). The shell has flaked off in places, exposing internal moulds.**

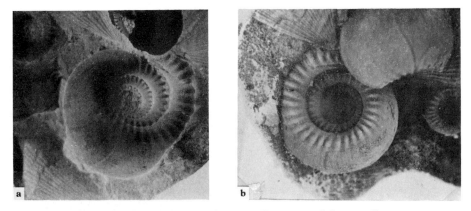

4   External mould and replica. Fossil extracted from a limestone nodule.

a, external moulds of a goniatite, *Gastrioceras*, U Carboniferous. b, a replica made by pouring latex solution over the mould; once set the replica was peeled off (× 1.5).

6 A derived fossil: an ammonite (*Mortoniceras*) from the Gault, L Cretaceous, of which internal moulds, in phosphate, occur in the Cambridge Greensand, U Cretaceous (× 2.2).

Transported and derived fossils

Fossils may be of organisms in position of life (*in situ*), e.g. corals in a reef, and may be well-preserved and entire specimens. In other cases, however, they are more or less fragmented, abraded, size-sorted and perhaps aligned – clear indications of transport by bottom currents. They may also be an assemblage of organisms from more than one environment, e.g. a mixture of pelagic and benthic shells (fig. 7); or land plants drifted into the sea and occurring with marine forms. A further complication is that of fossils eroded from an older formation and incorporated in younger sediments together with the younger fauna. Such fossils are said to be 'DERIVED' (fig. 6). Small forms, like foraminifera, are easily redistributed in this way.

Factors favourable for fossilisation

Factors favourable for fossilisation include an abundance of organisms, minimum physical disturbance, rapid entombment in sediment, and subsequent exclusion of oxygen and percolating water. Small size is also a

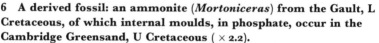

5 Internal moulds in Portland Limestone, U Jurassic.
The moulds are of (a) gastropod, and (b) bivalve shells which consisted of aragonite. Aragonite shells are more readily dissolved out of porous rock by percolating water than are calcite shells (× 0.8).

favourable factor: tiny organisms like foraminifera are more quickly buried than, say, the skeleton of a large plesiosaur and are therefore more likely to be preserved. These favourable conditions are most likely to be found in lakes, lagoons and seas; least likely on land. On land, fortuitous burial by flood, or entombment in bogs, tar-pools, etc. may preserve organic remains in good condition; but generally terrestrial animals are rare as fossils and seldom well preserved.

Preservation of fossils

Since soft tissues decay quickly after death, fossils are mainly the hard parts of organisms. The composition of these varies among the different groups and shows varying resistance to solution and decay.

Nature of the hard parts

Vertebrates. The skeleton, which is internal, consists of BONE or CARTILAGE or both. The matrix of bone is mainly collagen (a sclero-protein) hardened by mineral salts, largely calcium phosphate; it has a cellular structure ramified by channels containing blood vessels. Cartilage (gristle), a resilient, partly fibrous protein, is rarely preserved. TEETH form a minor part of the skeleton; like bone they also consist of calcium salts, but their structure is denser and they have a surface layer of ENAMEL, almost pure calcium phosphate and carbonate. They are relatively resistant to decay and are commonly fossilised. Superficial hard parts, e.g. horns and claws, contain KERATIN (a scleroprotein).

Invertebrates. The shell or skeleton may consist of organic matter only; or it may contain inorganic minerals. Common organic compounds include (i) CHITIN, a nitrogen-containing polysaccharide (carbohydrate) forming long fibrous molecules and found, e.g. in arthropods; and (ii) SCLEROPROTEINS (fibrous proteins) such as collagen, e.g. in graptolites, and conchiolin, e.g. in molluscs.

Common minerals include (i) CALCIUM CARBONATE forming a rigid intergrowth of crystals in an organic matrix; they may be in the form of ARAGONITE as in some molluscs, or CALCITE as in echinoderms. (ii) OPALINE SILICA, occurring as discrete parts, spicules, in some sponges, or as a coherent meshwork as in radiolarians.

Plants. The main structural substances in plants are (i) CELLULOSE, a fibrous polysaccharide forming the cell walls, and (ii) LIGNIN, a complex aromatic polymer binding the cellulose fibres.

Modes of preservation

The way in which organisms are preserved as fossils depends on their original composition and the physical and chemical conditions prevailing at death and during burial. Soft tissues decay quickly in the presence of bacteria, oxygen and water, the process being hastened by warmth. Further changes, after burial, may affect the entire skeleton or only the organic matrix.

Preservation of soft parts

Rarely, in unusual and transient circumstances, soft parts may be preserved entirely or to a limited extent. Examples follow:

Permafrost. Mammoths and other large vertebrates have been found completely preserved, some with stomach contents intact. They were frozen into surface sediments in the Siberian permafrost region.

Amber. Amber, fossilised resin which oozed from trees, mainly conifers, is found in some Mesozoic and Cainozoic rocks. The resin, sticky when fresh, often trapped and engulfed small creatures, especially insects, and in the resin the chitinous skeletons and some soft tissue have been preserved (fig. 104).

Mummification. Animals, mummified by dehydration in hot, dry climates (e.g. extinct sloths in New Mexico) are preserved only while these conditions persist. Eventually, with changing climate, decay sets in. In some cases the mummy may last long enough for its shape to be preserved as an impression in rocks. The nature of the skin in some dinosaurs has been preserved in this way.

Tar. Tar, the residue left on evaporation of oil seepages, has, when covered with water, acted as a trap for animals coming to drink. The best known instance, Rancho la Brea in California, contains complete skeletons of a range of extinct Pleistocene animals. Tar is antiseptic and the bones are little altered.

Peat. Peat, partly decomposed vegetable matter, forms in waterlogged conditions which exclude bacteria and oxygen. Animal bodies may be remarkably well preserved in this antiseptic medium. Preservation of soft tissues is due to their rapid tanning by humic acids. Bones, however, are decalcified by acidic water and become soft and flexible; but in alkaline water may be well preserved. Historic examples include the bodies of Iron Age bog people in Denmark, and 'Pete Marsh' from a peat deposit in Staffordshire.

With increasing depth of burial, peat is converted to lignite and then coal. A very wide range of animal remains are preserved in a 'brown coal' in the Eocene of East Germany. These show finely preserved details of soft tissue, such as bits of muscle, delicate frog skin, and cells with nuclei. In this case preservation is partly due to later penetration of the tissue by solutions introducing silica.

Impressions

Very delicate surface markings may be imprinted into a soft sediment surface, especially if this is fine-grained. Impressions of creatures resembling jellyfish, showing details of the body, are known from the Jurassic Solnhofen Limestone. This deposit also contains imprints of the feathers of the primitive bird, *Archaeopteryx*. Prompt and gentle burial must follow imprinting.

Traces of animal activity such as footprints come under this heading and are considered in Chapter 12.

Unusual preservations

Rarely, delicate features such as the appendages of trilobites may be preserved, as may remains of soft-bodied creatures such as worms, showing bristles and even traces of the gut (fig. 151). The Cambrian Burgess Shale contains an abundance of such rarities, more than half the genera being of soft-bodied animals not otherwise known in the geological record.

Changes affecting organic components

Carbonisation. An organic substance like scleroprotein may be quite resistant to decay if it is sealed rapidly in sediment, and little-altered graptolite skeletons have been dissolved out of limestone (fig. 136). More generally, organic materials like scleroprotein, chitin and the cellulose and lignin of plants are carbonised, i.e. the relative carbon content is increased by liberation of volatile constituents. The outline and sometimes details of the soft anatomy of an organism may be preserved as a carbonised residue, as in the outlines of the body and tail of occasional fossils of ichthyosaurs (fig. 200).

Pyritisation. After burial in mud, soft parts are occasionally replaced by pyrites formed under reducing conditions within the sediment. Traces of cephalopod arms, outlined by pyrites, have been found by X-ray probing of lower Devonian slate from West Germany.

Changes in the inorganic substance of hard parts

Fossils showing little change in the composition of the hard parts are likely to be of Cainozoic age. In these the organic components have decayed, leaving the mineral substance of the shell or bone unaltered. Such fossils are porous and rather fragile, as for instance the mollusc shells from the Pliocene Crag deposits in East Anglia.

Conversion of aragonite to calcite. Many shells consist of aragonite which is liable, in time, to convert to the more stable form, calcite. This recrystallisation destroys the fine structure of the shell while not affecting its overall shape. Aragonite shells are common in Cainozoic rocks, less common in Mesozoic and are hardly known in earlier rocks.

Petrifaction. Petrifaction (literally, 'turning into stone') involves the impregnation or replacement of the hard parts by minerals deposited from solution in waters percolating through permeable remains. IMPREGNATION is the infilling of the interstices left in a bone or shell on the decay of the organic matrix, by an inorganic compound. REPLACEMENT is the substitution of a different mineral for the original mineral matter of the shell or bone. The resulting fossil has the outward form of the original skeleton, but is heavier, may differ in chemical composition, and in some cases the finer details of the skeletal structure are destroyed.

The commonest petrifying minerals are calcite, silica, and iron compounds. Most frequently, petrifaction is due to the impregnation of calcareous shells by calcite, as in echinoderms where the meshwork of pores in each plate is filled by calcite resulting in a solidly crystalline plate which, if broken, shows cleavage faces. Silica may replace calcite, chitin or wood. Amorphous silica (opal), in particular, may preserve the original microstructures in almost perfect detail (fig. 219). Silicified fossils which are enclosed in limestone can be dissolved from their matrix by dilute acid and so retain surface details like delicate spines which would normally be broken during extraction of the fossil from its matrix. Compounds of iron include haematite and limonite (both oxides), siderite (carbonate), and pyrites and marcasite (both sulphides). For the most part they replace calcareous shells, and fossils tend to be poorly preserved; but pyritised graptolites and other forms may retain fine detail. Pyritised fossils, however, tend to oxidise and disintegrate once exposed to air.

Solution of hard parts. Acidified water percolating through a permeable rock will dissolve and remove calcareous shells. The space left, the EXTERNAL MOULD, bears the surface markings of the shell in reverse (fig. 4a). A latex (rubber) solution poured into the mould will provide a replica of the original shell (fig. 4b). Sometimes a shell may have been filled with matrix before solution occurred; this forms an INTERNAL MOULD (fig. 5).

Incompleteness of the fossil record

The fossil record can only be a partial catalogue of past life on earth. Knowledge of soft-bodied organisms, which today are overwhelmingly abundant, is minimal. Neither is there a complete record of skeletal animals: new forms are still being discovered. Because of the vagaries of the fossilisation processes there will always be 'gaps in the fossil record'; it will, nevertheless, continue to be improved as new discoveries are made.

Organisms and environment

Organisms live in harmony with their environment, and ecology is the study of the relationship between the two: how organisms are adapted to their environment. PALAEOECOLOGY is the corresponding study of fossil organisms.

The environment is a complex of variable factors, biological, physical and chemical, which determine the distribution of organisms, each species being adapted to a particular ecological niche. For example, the ecological niche of the common cockle is on the lower sea-shore lying just below the sediment surface where it feeds on microscopic organisms filtered from sea water drawn in through a short tube extended upwards when the tide is in (fig. 19b). It occurs as fossils in upper Cainozoic rocks and may be assumed to have lived then in similar circumstances.

Using the relationship of fossil forms with their present day relatives, palaeoenvironments can confidently be deduced for many of the younger rocks and their fossils. As we go further back in time, however, relationships with living forms become more remote and we find extinct groups about whose environment nothing is directly known. In such cases there may be associated fossils which do have living relatives and thus can be used to establish the general conditions in which the extinct forms lived.

There are two major environments, aquatic and dry land. Most fossils are found in sediments which formed under water, and the range of aquatic environments is considerable, from fresh-water rivers, lakes and swamps, to saline or brackish-water lakes, lagoons and estuaries, and the open sea. In each of these, conditions of salinity, temperature, light, depth, aeration, and food supply may differ and will control the particular organisms to be found there. A sudden change, albeit temporary, in these conditions may cause mass mortality of the fauna; for instance the sudden lowering of temperature in a coastal lagoon during a spell of unusually cold weather. Quite small variations may be important in determining the presence or absence of a particular organism, and where these occur as fossils the appropriate conditions may be safely inferred.

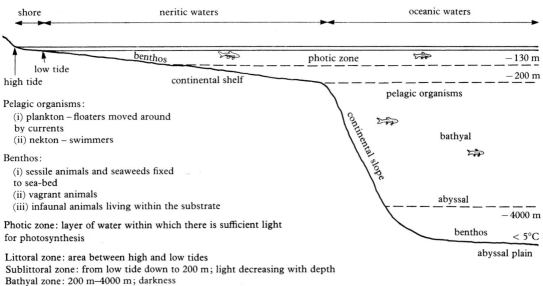

Pelagic organisms:
 (i) plankton – floaters moved around
 by currents
 (ii) nekton – swimmers

Benthos:
 (i) sessile animals and seaweeds fixed
 to sea-bed
 (ii) vagrant animals
 (iii) infaunal animals living within the substrate

Photic zone: layer of water within which there is sufficient light
for photosynthesis

Littoral zone: area between high and low tides
Sublittoral zone: from low tide down to 200 m; light decreasing with depth
Bathyal zone: 200 m–4000 m; darkness
Abyssal zone: below 4000 m

7 Cross-section showing major marine habitats.

Throughout geological time most of the earth's area, some 70%, has been covered by sea. In terms of living space, however, the sea provides perhaps 99% by volume. For this and other reasons most fossils are of marine origin. Accordingly, in the following sections attention is focussed on the marine realm (fig. 7).

The marine realm comprises two parts with contrasting types of populations: (i) the water mass with PELAGIC organisms which either drift passively with currents, PLANKTON, or swim freely, NEKTON; (ii) the sea-bed (substrate) with BENTHIC organisms which may be sessile or freely moving.

The plankton, mostly one-celled forms, consist of plant-like, photosynthetic PHYTOPLANKTON, which take their nutrients directly from solution; and animal ZOOPLANKTON which feed on phytoplankton or other animal plankton. The nekton includes fish, reptiles, mammals and invertebrate forms like the cephalopods. The benthos includes seaweeds and many different kinds of animals, mainly invertebrates.

The benthos may live upon the substrate, EPIFAUNAL, or within it, INFAUNAL, either by burrowing in soft sediment or boring into hard rock, Mobile benthos move around to find their food. They may be HERBIVORES which graze on seaweeds, CARNIVORES eating flesh, or SCAVENGERS eating dead matter. Some are DEPOSIT FEEDERS, ingesting soft sediment for its organic content. Sessile forms must collect food and get rid of wastes *in situ*; these are generally SUSPENSION (filter) FEEDERS which filter plankton and fine organic debris from water wafted towards the mouth by the beating action of tiny thread-like cilia.

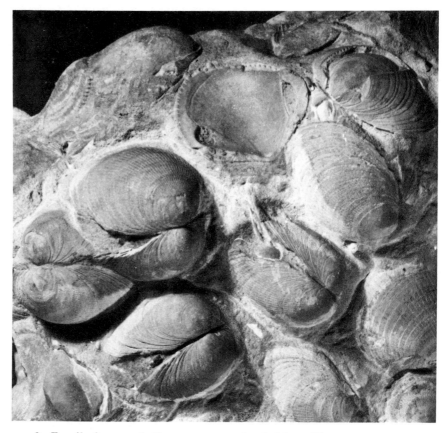

8 Fossils from a shallow-water marine habitat: *Glycymeris* **(bivalve), Eocene, Hampshire Basin** (× 0.7).

Salinity

Salinity refers to the amount of dissolved salts, principally sodium and chlorine ions, in water. It is fairly constant in the open seas, about 35‰, and is increased in closed seas and lagoons, often to over 40‰, where the rate of evaporation exceeds replenishment. It is lowered in estuaries by dilution from river water. The water in rivers and most lakes is fresh, containing about 1‰ or less of salts, mainly calcium and bicarbonate ions.

Aquatic organisms regulate the salt concentration in their body-fluids relative to that of the water they live in. Many marine organisms cannot adjust to variations in salinity, i.e. they are STENOHALINE. Groups which today live exclusively in normal marine waters include brachiopods, corals, echinoderms (fig. 107.) and cephalopods. It can be assumed that their ancestors also lived in a marine habitat. Associated with them in the older rocks are trilobites, an extinct group which, presumably, was also marine.

A variety of organisms can tolerate a range of salinities, i.e. they are EURYHALINE. Such forms may live in normal sea water and also in brackish

9   Brackish-water fossil: *Potamides vagus* (gastropod), Middle Headon Beds, Oligocene, Isle of Wight (× 3).

10   Non-marine fossils.
a, *Modiolopsis complanata* (bivalve); and b, *Platyschisma helicites* (gastropod); U Silurian, Old Red Sandstone facies (× 2).

(b)

(a)

11   Fresh-water fossil: *Viviparus* (gastropod), Eocene. Inset (× 1.8).

water. Certain bivalves, like *Mytilus*, gastropods and arthropods fall in this category. Others are restricted to either hypersaline (over 40‰), or brackish, or fresh water; and these as fossils may be useful indicators of such salinities, characteristic of lagoons and estuaries (fig. 9). 'Non-marine' is a useful term to use for fossils which were clearly not normal marine but cannot confidently be assigned to particular salinity conditions (fig. 10).

Living fresh-water organisms are mainly vertebrates, arthropods, a few genera of gastropods and bivalves. Similar forms occur in the fossil record, especially in the newer rocks (fig.11) and may be associated with plant remains.

Sunlight and the photic zone

Sunlight is absorbed as it passes through water, its intensity lessening with increasing depth until darkness prevails. The depth of the illuminated region, the PHOTIC zone (fig. 7), varies according to latitude, season and clarity of water. The shorter wavelengths (blue light) penetrate most deeply. Their maximum penetration is in clear tropical seas where about 1% remains at about 150 m; whereas in coastal, more turbulent waters this intensity of light is found at only 10–30 m.

Plants depend on light for photosynthesis, and in the sea they are therefore restricted to the photic zone, most occurring at shallow depths. Those plants which grow on the sea-floor are accordingly best developed in relatively shallow coastal waters. Their fossils, e.g. calcareous algae, give a general indication of the depth of the sea in which they grew.

Plants are the primary food producers and their distribution determines that of hordes of animals, including those which eat plants, e.g. herbivorous gastropods, the carnivores which eat the herbivores, the small fish and crustaceans which shelter among the seaweeds, and the reef-corals which have an intimate association with microscopic 'algae' (p. 116).

Many organisms live in complete darkness. Even in the deeper abyssal waters there is a relatively rich fauna (fig. 12). It includes both pelagic (e.g. fish, squids) and benthic (mainly invertebrates and bacteria) forms. Some feed on other organisms but there is a rich food source for deposit and suspension feeders in the rain of dead plankton from the surface waters.

12 Deep-sea environment: a view of the ocean floor taken at bathyal depth, 651 m.
The sea-bed consists of shell-sand and rock fragments. Many stemless crinoids are shown attached by their cirri to pieces of rock and bent over with the current. Galicia Bank, 42° 41′ N, 11° 35′ W; area shown is about 1 m square.

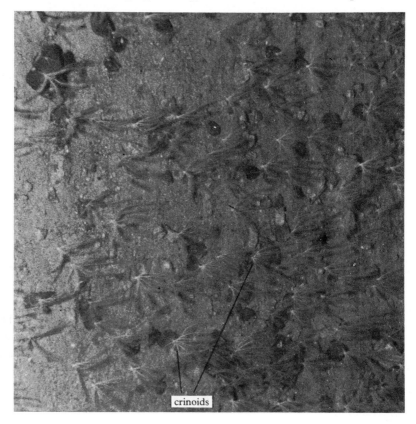

crinoids

Oxygenation

Oxygen is essential to the metabolism of the vast majority of organisms. In many parts of the sea enough oxygen is present to satisfy even a dense population of animals, and as it is used up it is constantly replaced. During the day oxygen is evolved as a byproduct of the photosynthesis of plants, and at all times it is absorbed from the air at the water surface, a process augmented by turbulence, e.g. by waves breaking. Thus, there is most oxygen near the sea-surface and especially in coastal areas, including coral reefs. In deeper waters the oxygen content is depleted by the respiration of animals, and by the bacterial decay of organic matter, which involves oxidation. So long as there is free circulation of water from the surface towards the sea-floor enough oxygen is available to sustain life. But if the circulation is very slow, as in the case of basins with limited inflow, oxygen in the deeper water is used up more quickly than it is replenished and in extreme cases is entirely removed with consequent anoxic conditions. Most organisms cannot exist in an anoxic environment. Anaerobic bacteria, however, do not need free oxygen because their energy may, for instance, be based on the reduction of dissolved sulphates to form sulphides, or on the breakdown of proteins in organic matter. Hydrogen sulphide is released in these processes. This will reduce iron compounds in sea-bed sediments to ferrous sulphide which, in due course, becomes pyrites or marcasite. A modern occurrence of anoxic conditions is in the Black Sea where there is virtually no circulation of the water. Black mud accumulates on parts of the sea-floor, the colour being due to finely divided ferrous sulphide and organic matter. The latter consists of the remains of plankton which have settled from the aerated surface water. It has been decomposed by ordinary bacteria until all the available oxygen has been exhausted, and at this point anaerobic bacteria continue the process of decay. A strong smell of hydrogen sulphide is characteristic of this phase. Anoxic environments of this nature are devoid of a bottom-living fauna, including scavengers, and conditions may be favourable for the preservation of skeletons of pelagic animals. Instances of dark shales which are believed to have formed in poorly oxygenated or anoxic conditions include the graptolitic shales of the lower Palaeozoic and goniatite shales of the Carboniferous. In both cases the fossils may sometimes be pyritised.

Nutrients and the food chain

In the oceans the phytoplankton is the basis of the food chain, being eaten by herbivorous zooplankton which in turn are eaten by carnivorous zoo-plankton and vertebrate predators. The food substances needed by plants for growth are contained in sea water in the form of mineral salts, dissolved gases (e.g. carbon dioxide) and soluble organic matter. The limiting factor in growth is the availability of nitrate and phosphate and as these are depleted growth slows down. Nutrients are replenished by recycling. Animals feeding in the food chain respire carbon dioxide, excrete nitrogenous waste, and produce faeces containing phosphate. They also contribute further on death and decay. Bacteria play a major role in the recycling process by breaking down organic matter and are themselves recycled as food for animals. Some, e.g. cyanobacteria, fix nitrogen dissolved in the water.

Nutrient supply varies greatly, being low in many central oceanic areas and greatest in areas of upwelling cold water which lie along certain continental margins and in latitudinal belts in equatorial and subpolar regions. The plankton population is very large in these nutrient-rich waters and includes siliceous diatoms and radiolarians, and calcareous coccoliths and foraminifera (chapter 11). Skeletal remains of these may form extensive deposits on the sea-floor, but much is dissolved and recycled.

In the case of calcareous skeletons no solution occurs in the upper layers of the ocean which are generally saturated with respect to calcium carbonate. The deeper waters, however, are at lower temperatures and higher pressures, both of which increase the solubility of calcium carbonate which is, for this reason, absent from large areas of the deep ocean floor. The depth below which calcareous sediments do not accumulate is known as the carbonate compensation depth (CCD) and is variable both between major oceans and also within them. In the Pacific at the present day it lies at about 5000 m in equatorial regions but is less deep towards the poles.

Sea water at all levels is markedly undersaturated with respect to opaline silica which forms the skeletons of diatoms and radiolarians, most of which dissolve soon after death. It is estimated that only about 5% survive, being protected from solution by a resistant organic sheath, or having a low surface area per unit volume so that the solution rate is reduced. It follows that highly siliceous deposits are confined to areas of very high abundance.

Climate and temperature

Climate is a decisive factor in the distribution of most living plants and animals. The information about ancient climates which may be deduced from fossils is in some cases direct but generally is circumstantial.

Temperature varies according to latitude and season, and fluctuates more extremely on land than in the sea. At present, sea temperatures at the surface range from about 30 °C in low latitudes to − 2 °C in high latitudes. In depth it falls quickly through the photic zone, then more slowly to about 1000 m, below which the range is 5–0.5 °C. Most types of invertebrates today tolerate a wide range of temperature: molluscs, for instance, range from tropical to arctic waters. However, particular species of molluscs are virtually restricted to either warm or cold water and their fossils may be used to demonstrate fluctuating water temperatures during glacial and interglacial periods of the

13 Molluscs from the Pleistocene Crags, East Anglia.
a, *Boreoscala greenlandicum* (gastropod); and b, *Macoma calcarea* (bivalve) both
of which live today in arctic waters. c, d, *Acila cobboldiae* (bivalve, external and
internal views, L Pleistocene only) which lives today in Japanese waters. (All
× 1.8.)

Pleistocene (fig. 13). Modern reef-forming corals indicate a precisely defined
marine environment (p. 116). Their preferred temperature range is
25–29 °C and they are therefore confined to tropical and subtropical waters.
The reef-corals found in Cainozoic rocks are closely related to modern forms
and may, by analogy, be used to establish the location of the warmer waters
during that period. Further back in time, relationships with modern forms
become more remote and deductions more tentative. It is likely, however,
that massive carbonate deposits with organic reef structures were formed in
similar circumstances to present day reefs. It may be noted that sea water in
tropical areas is usually slightly supersaturated with respect to calcium
carbonate, favouring its direct or organically controlled precipitation.

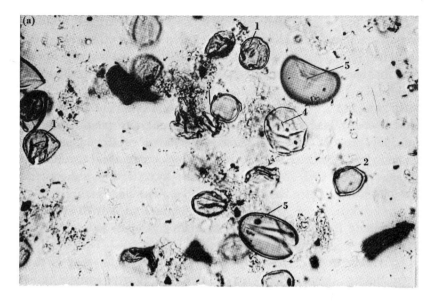

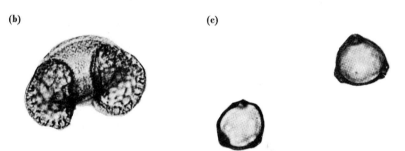

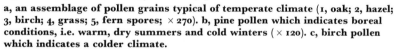

14 Pollen grains from Quaternary peats.
a, an assemblage of pollen grains typical of temperate climate (1, oak; 2, hazel; 3, birch; 4, grass; 5, fern spores; × 270). b, pine pollen which indicates boreal conditions, i.e. warm, dry summers and cold winters (× 120). c, birch pollen which indicates a colder climate.

On land the well-defined climatic zones have characteristic faunas and floras but their adaptations to climate are physiological in the main and do not affect the hard parts. However, in the younger rocks, fossils of modern plants, including their pollen grains, can be good climatic indicators since the plants grow in conditions of quite narrow climatic range. Pleistocene interglacial deposits, for example, show a sequence of climatic change from cold and dry when an arctic flora (including dwarf birch, fig. 14c) colonised the area from which the ice had retreated, to a milder and wetter climate when a mixed deciduous forest (e.g. oak, hazel, fig. 14a) became established. A very different flora in the Eocene London Clay includes the palm *Nipa* and its fruit *Nipadites*, found today in the Indo-Malay region and indicating low-lying tropical conditions. Further back in time more reliance must be placed on modifications in plant structure which are related to temperature and humidity. For instance, in some plants adapted to a dry habitat the leaves have a tough leathery texture with a heavy cuticle and sunken stomata (p. 351), features which help to conserve water. Woody trees living in a region with seasonal changes in temperature and humidity show growth rings, formed in alternating periods of fast and slow growth. By contrast, in an equable climate, both warm and humid, trees grow steadily and growth rings are absent, as is generally the case in fossil trees of the Carboniferous Coal Measures. This accords with other evidence of the warm, humid and equable nature of the Coal Measure climate. It may be noted here that most of the plants in the Coal Measures reproduced by spores, which require continued humidity for germination.

Among land vertebrates, the reptiles, whose body temperature is roughly that of their surroundings, are today restricted mainly to the warmer regions. This was probably true of most reptiles during the Mesozoic when they were widely distributed throughout the world, including regions (like the British Isles) which are now relatively cold. That the climate at this time was humid as well as warm is suggested by the abundance, and often great size, of plant-eating reptiles, implying plentiful vegetation. Indirect evidence of climate comes from fossil horses found in upper Miocene and Pliocene rocks. Their tooth structure shows that, like modern horses, they grazed on grass. The distribution of their remains suggests that wide stretches of grassland existed in North America and Eurasia at that time. The main grasslands today are in regions with a temperate climate of low rainfall.

Fossil fish might seem an unlikely source of information on climate. However, one group, the lungfish (p. 298), are air-breathing forms which live in regions liable to seasonal drought. During the dry season they survive by burrowing in the mud of the river-bed. Remains of lungfish in mud-filled vertical burrows in Permian rocks suggest similar conditions at that time.

2

Mollusca – introduction and Class Bivalvia

Introduction

The molluscs are animals with a complex soft body, and most secrete a hard external shell. They are a major group, enormously varied, and include familiar living forms like the whelks and limpets, cockles and mussels, the octopus and the cuttlefish; but many are known only as fossils.

Early in their history the basic body-plan was modifed by adaptation to disparate life-styles. Each of these imposed particular and distinctive characters which are the basis for subdivision into classes. Three classes are important as fossils: the Bivalvia, Gastropoda and Cephalopoda. Four other classes with a minor fossil record may be mentioned: the Monoplacophora, Amphineura, Scaphopoda and the extinct Rostroconchia.

Most molluscs are mobile creatures living in the sea, but some are found in fresh water and some gastropods occur on land. While externally the molluscan shell is very varied in form, internally the body largely conforms to a basic plan of organisation. In general they are bilaterally symmetrical except for the gastropods in which the body is twisted. There are four main regions: the head, the visceral mass, the mantle and the foot. The VISCERAL MASS, containing the internal organs (heart, liver, etc.) is dorsal (uppermost) in position. On the ventral (lower) side is usually a muscular organ used for moving and hence known as the FOOT (except in the cephalopods in which it is modified). The body is covered by the MANTLE, two folds of tissue which hang down freely on either side and enclose a cavity in which lie the gills. The HEAD with the mouth is at the anterior end; it leads into the digestive tract which in turn leads to the anus at the posterior end. The nervous and circulatory systems are highly organised throughout the phylum. The SHELL, secreted by the mantle, is partly calcareous and partly organic in composition.

Modern marine molluscs occur world-wide from abyssal depths to the relatively shallow water where most live. They may be benthic or nektonic or planktonic. They range in size from almost microscopic to gigantic; the giant squid, the largest known invertebrate, may measure nearly 20 m.

The fossil record of the Mollusca dates from the earliest Cambrian.

Class Bivalvia

(Alternative names: Lamellibranchia, Pelecypoda)
The bivalves are distinguished from other molluscs by their laterally compressed body enclosed between two calcareous valves which are united on the dorsal side by an elastic horny ligament. In most cases the animal is bilaterally symmetrical and the plane of symmetry is the plane along which the two valves meet. Thus the valves are almost perfect mirror images of each other. The group is entirely aquatic (marine and fresh water), the vast majority living in quite shallow water, though some, like the recently discovered giant 'clams' associated with tube-worms around sea-floor hydrothermal vents, live in deeper water, and a few in abyssal depths. Most lead a relatively sedentary life, not moving far or quickly, and following one of a limited number of modes of life. Modern examples of bivalves include cockles, mussels and oysters. Most measure a few centimetres across; the largest, *Tridacna*, which lives embedded in coral reefs, may exceed 1.3 m.

Morphology

Usually, the bivalve shell completely encloses the soft body. It serves as a support, as a surface for muscle attachment, and as a protection against sediment and predators. It consists of two valves united by an elastic ligament and is secreted by the mantle. The latter is a fleshy tissue which hangs down as two folds, one on the right side of the body and one on the left (fig. 15b). The valves are described as RIGHT and LEFT. They are united on the DORSAL surface of the body by the LIGAMENT, and they separate along the other margins; these are distinguished as the ANTERIOR (where the mouth is situated), the VENTRAL (opposite the dorsal), and the POSTERIOR (where the anus opens).

Soft parts

The head is rudimentary, and parts like the radula (toothed tongue), typical of some other molluscs, are absent. The main mass of the body, including the viscera and organs such as the heart, lies in the dorsal part of the shell (fig. 15). The foot, lying between the mantle lobes on the ventral side, is a muscular wedge-like organ which can be extended outside the shell and, by alternately lengthening and contracting, can pull the animal through soft sediment. The gills, also within the mantle cavity, typically have a dual role: respiration and food-gathering. The gill filaments bear tiny hair-like cilia

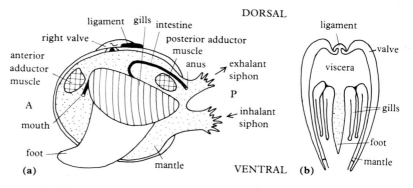

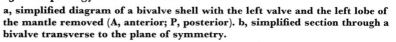

15 Morphology of the bivalves.
a, simplified diagram of a bivalve shell with the left valve and the left lobe of
the mantle removed (A, anterior; P, posterior). b, simplified section through a
bivalve transverse to the plane of symmetry.

which, by beating to and fro, draw water through the mantle cavity.
Incoming water is separated from the outgoing water because the mantle
margins, which may be fused locally, form two open folds or tubes, SIPHONS,
at the posterior end. Water enters via the lower, INHALANT siphon and
particles of food (microscopic plants and animals) are sieved out by the gills
and passed forwards in strings of mucus to the mouth at the anterior end.
The outgoing current with waste products is passed out through the
EXHALANT siphon. Most bivalves use this system of suspension feeding but a
few are deposit-feeders.

The muscle systems, alone of the soft parts, leave imprints on the inner
surface of the shell. There are muscles which attach the mantle to the valves,
others which operate the foot, and, the largest, the ADDUCTOR muscles
which contract to close the valves. The adductors are of two sorts: 'catch'
muscles which can close and lock the shell for long periods; and 'quick'
muscles which snap the valves shut rapidly but have no staying power.

Shell

The substance of the shell is partly calcareous, partly organic (conchiolin, a
proteinaceous material); a thin-section shows it to consist of three layers
(fig. 18e). On the outside is a thin horny layer, the PERIOSTRACUM. The
inner layers consist of crystals, usually of aragonite, sometimes of calcite or of
both. Muscle attachments are underlain by aragonite. The structure of these
inner layers varies in different bivalves and is used in classification. A
common structure found in the middle layer consists of prisms lying
transverse to the periostracum. The innermost layer in many shells is
made of NACRE, otherwise known as mother-of-pearl. Nacre is made of
alternate very thin lamellae of aragonite and conchiolin lying parallel to the
outer layers. A structure seen in many other shells consists of lath-shaped

body cavity

16 Section of a bivalve shell, *Gryphaea*, L Jurassic (× 2).
a, the outer, relatively thin layer of shell secreted during growth by the edge of
the mantle. b, the inner, thick layer of shell laid down as a series of laminae by
the whole surface of the mantle.

lamellae arranged in a regular alternating series; these shells have a dull porcellanous appearance. The outer two layers are secreted by the edge of the mantle; the innermost layer is laid down by the whole surface of the mantle and continues to grow during the life of the animal so that it may be quite thick compared with the outer two layers (fig. 16).

The shape of the bivalve shell may vary considerably according to its mode of life but its basic design is simple. The initial part of each valve is the BEAK. With growth of the shell the beak forms part of a prominent convex area, the UMBO (pl. umbones), projecting above the dorsal margin. Typically, it lies in front of the ligament (fig. 18a). As the animal grows, further layers of shell are laid down by the mantle: (i) along the margins (except near the ligament) which extend the shell; and (ii) on the inner surface, which thicken it. The marginal increments are seen on the outer surface as growth lines which merge towards the hinge line (fig. 18b).

The surface of the shell may be relatively smooth or it may be sculptured by ORNAMENT consisting of radial or concentric markings. The latter vary from fine growth lines to quite coarse lamellae of regular or, sometimes, quite irregular appearance. Radial ornament varies from fine lines to coarse ribs and grooves. Sometimes the radial and concentric elements are combined to produce a reticulate pattern; occasionally, spines (fig. 17) or tubercles are present.

17   A spinose bivalve: *Spondylus,* **Cretaceous** (× 1.5).

On the INNER surface of the valves is the HINGE PLATE, which is a thickening of the dorsal margin just below the umbo (fig. 18a). On each plate are projections, TEETH, which fit into SOCKETS in the opposite hinge plate. The teeth under the umbo are the CARDINAL teeth, and those beyond it are the LATERAL teeth. The teeth and sockets together make up the DENTITION and form a simple interlocking mechanism which ensures correct alignment of the valves as they open and shut. Some of the commoner types of dentition are defined on page 50.

The LIGAMENT, which connects the two valves dorsally, consists of resilient conchiolin (p. 29). It may be EXTERNAL, lying above the hinge plate, or INTERNAL, lying between the hinge plates. In general, an external ligament lies behind the umbo, a condition described as OPISTHODETIC (fig. 18a): in some forms it extends in front of as well as behind the umbo and this condition is described as AMPHIDETIC (fig. 30f). In those forms with an

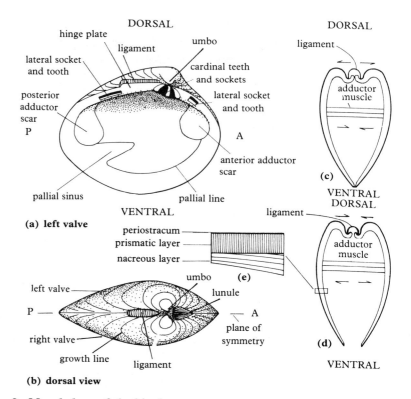

18   Morphology of the bivalves.

a, interior view of the left valve of an equivalve and inequilateral shell. In this
and later diagrams which show dentition, the sockets are black and the teeth
unshaded. b, dorsal view of the shell. c, transverse section of a closed shell
showing the adductor muscles contracted and the ligament stretched; and d, of
an open shell with the adductor muscles and ligament relaxed. e, diagrammatic
transverse section of a shell fragment, enlarged, showing a layered structure
commonly found in bivalves.

internal ligament it may be situated in one pit or in a series of pits along the
hinge plate (fig. 27i). The ligament is rarely preserved.

The hinge plate, with its dentition and ligament, is only a part of the
mechanism by which the valves are opened and shut. The resilient ligament
by itself would keep the valves gaping; they are closed by the contraction
of ADDUCTOR muscles (fig. 18c). When these muscles relax, the external
ligament, which has been under tension, PULLS the valves apart (fig. 18d).
In forms with an internal ligament, the ligament is under compression and
thus PUSHES the valves open when the muscles relax.

The areas where the adductor muscles are attached to the valves are
marked by SCARS which can be clearly seen on the inner surface of each
valve; typically there is one at the anterior end (ANTERIOR ADDUCTOR),
and one at the posterior end (POSTERIOR ADDUCTOR) of each valve. Shells
with scars of roughly equal size are described as ISOMYARIAN (fig. 18a)
while those in which the anterior one is smaller are ANISOMYARIAN

(fig. 23d); shells lacking the anterior scar are MONOMYARIAN and in these the posterior scar is enlarged (fig. 23c, f). A faint groove runs parallel to the ventral margin from the anterior muscle scar to the posterior muscle scar; this is the PALLIAL LINE along which the retractor muscles of the mantle are inserted (pallium is another word for mantle) (fig. 18a). Bivalves which burrow in sediment have elongated tubular siphons which are extended from the burrow during feeding, and can be retracted into the shell quickly if the animal is disturbed. The pallial line in such forms is not entire but shows an embayment, the PALLIAL SINUS, at the posterior end (fig. 18a). The depth of the sinus gives some indication of the length of the siphons and thus of the depth of burrowing.

When the valves are closed, the margins are, typically, pressed tightly together. Bivalves which live in burrows, however, may have a permanent opening, a GAPE, at the posterior end for the siphons (fig. 21a), and there may be a similar gape at the anterior end for the foot. The valve margins are usually quite smooth, but accurate closing is aided in some forms by the development of small crenulations of the margin (fig. 19a).

Most bivalves are bilaterally symmetrical about a plane passing between the two valves and these shells are said to be EQUIVALVE (fig. 18b). Each valve is, however, usually asymmetrical about a line from the umbo to the ventral margin and is said to be INEQUILATERAL (fig. 18a). In the majority of bivalves the umbo is nearer to the anterior end than to the posterior end. Thus the right and left valves can be readily distinguished: if the shell is held dorsal surface up with the anterior end pointing away from you, the right valve is then on your right and the left valve on your left. There are, however, exceptions to this rule, e.g. *Nucula*.

Classification

Classification of bivalves is under constant reappraisal and no attempt is made here to provide a systematic treatment. However, some mention must be made of the characters which have been found of use. Many features of the soft body, such as gills, used in classifying living forms are unknown in most fossils, in which shell characteristics and the limited knowledge of soft parts gained from internal markings, have to suffice. The main characters used are the nature of the dentition; the ligament; muscle scars and pallial line; shape; and shell mineralogy and microstructure. These provide the bases for identifying some six subclasses which are further subdivided into many orders and superfamilies.

Shell form and mode of life of some common bivalves

Bivalves are adapted to a wide range of habitat: epifaunal, including forms fixed by byssus or cementation, and others which are freely moving; and

infaunal, including a variety of burrowers. The associated modifications of
the basic anatomy are reflected in the form of the shell. Shell shape may,
therefore, be a clue to mode of life, as illustrated by the following examples
chosen from the more usual life-styles of modern bivalves having a fossil
record. Similar modifications can be found in forms known only as fossils.

Marine bivalves

Cerastoderma (fig. 19a, b).   Rather globular shell with rounded outline;
equivalve; slightly inequilateral; prominent umbones with beak directed
forwards; coarse ornament of radial ribs. Ligament opisthodetic; heterodont
dentition (p. 50); isomyarian; pallial line entire; ventral margin crenulate.
U Oligocene–Recent
Many bivalves move sluggishly at or near the surface of soft sediments.
Cerastoderma edule, the cockle, is an example found on the lower shore of many
sandy beaches where, for much of the time, it lies just covered in sand. It

19   Shallow-burrowing bivalves.
**a, b, *Cerastoderma*: a, interior of right valve; b, in feeding position in sandy
sediment. c, d, *Venus*. c, in feeding position; d, interior of left valve. Arrows
indicate the direction of the incurrent and excurrent flow of water.**

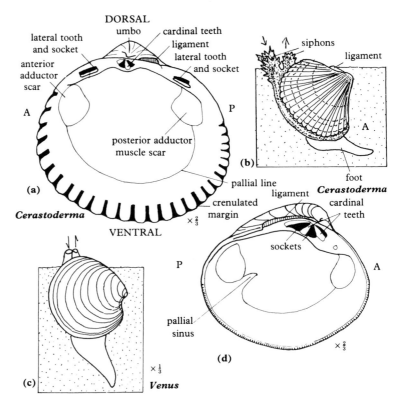

feeds when the tide is in by extending its siphons into the water. Small organisms such as diatoms are sieved out from the inhalant current. When moving, the shell opens slightly and the foot is extended as a narrow probe which elongates as it is thrust into the sand. The tip of the foot dilates, forming an anchor, as the rest of the foot contracts to pull the shell through the sand. The process is made easier by water expelled from the mantle cavity which loosens the surrounding sediment. The cockle lives in a turbulent, unstable habitat and may be washed out of the sand, but its robust globose shell allows it to be rolled about on the beach without damage. The coarse ornament is typical of such bivalves living in this niche, being a strengthening feature of the shell. Unlike most bivalves, the cockle, when stranded, can leap along the sand in a series of hops. First, the foot is bent almost double under the shell and then, with the tip pressed against the sand, it is straightened suddenly, jerking the shell forwards.

Venus (fig. 19c, d). Shell compressed, more or less oval in outline; equivalve; inequilateral with umbones directed forwards; lunule present (oval depressed area, see fig. 18b); ornament of concentric lamellae. Ligament opisthodetic; dentition heterodont with three cardinal teeth in each valve; isomyarian; small pallial sinus; crenulate margin. Oligocene–Recent

20 Deep-burrowing bivalves.
a–c, *Mya*: a, interior of left valve; b, the animal in its burrow in muddy sand with the siphons extended; c, transverse section through the ligament and chondrophores. d, e, *Solen*: d, interior of right valve; e, the animal in its feeding position at the top of its burrow in sand.

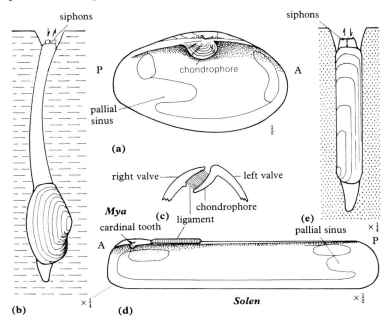

Venus is found from the extreme lower shore down to about 100 m or more, burrowing actively in sand or gravel. It moves up near the surface to feed by extending its siphons into the water, then down to a lower level to rest. The shell, more compressed than that of the cockle, is moved more easily and quickly through the sand. Related forms live at varying depths. *Dosinia* (Oligocene–Recent), for example, living at 5 cm from the surface, has a smoother more compressed shell and deeper pallial sinus. The latter is a guide to the length of the siphons.

Solen (fig. 20d, e).   Smooth, thin, compressed shell elongated parallel to hinge line (may be eight times as long as broad); equivalve; very inequilateral with umbones at truncate anterior margin; gapes at anterior and posterior ends. Ligament opisthodetic; heterodont dentition with one small tooth in each valve; elongate anterior adductor; shallow pallial sinus. Eocene–Recent

Solen, one of the 'razor' shells, burrows in sand on the extreme lower shore and in shallow water. The shell gapes at each end thus allowing the foot and siphons to protrude from the otherwise closed shell. The siphons are short and *Solen* feeds at the top of its burrow (fig. 20e). When the tide goes out, or if disturbed, it digs downwards with its large and powerful foot. The smooth compressed shell facilitates very fast movement through the sand.

Mya (fig. 20a–c).   Shell oblong; umbones almost central; posterior end truncated and with wide gape. Ligament mainly internal, supported by chondrophore (p. 50); no teeth; isomyarian; deep pallial sinus. Oligocene–Recent

21   Bivalves which bore into hard material.
a, dorsal view of *Pholas* in its cavity in rock. b, *Teredo* in its tunnel in wood.

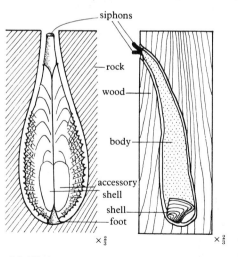

(a) *Pholas*   (b) *Teredo*

Mya lives in a burrow, about 30 cm deep, in sand or muddy sand from the middle shore down to about 70 m. It remains at the bottom of its burrow. Its siphons are long (they may be twice or three times the length of the shell) and are enclosed by a leathery sheath. They are not completely retracted, and are accommodated by the large posterior gape.

Pholas (fig. 21a). Elongated, almost cylindrical; gapes at both ends; inequilateral with umbones anterior; surface with rows of rasp-like spikes, roughest towards anterior end. Ligament reduced; no teeth; shelly deposit reflected over umbones which are smooth and rounded; isomyarian with anterior adductor on antero-dorsal margin; deep pallial sinus; short internal projection under each umbo for insertion of foot muscles.
Cretaceous–Recent

22 Internal mould of a cavity bored in coral by a bivalve, *Lithophaga, Coral Rag, Jurassic, Upware, Cambridgeshire* (× 4).
The cavity was drilled in the skeleton of a coral. After the death of the bivalve, the cavity and the interseptal spaces of the coral were infilled with calcareous sediment. Later, the coral skeleton, composed of aragonite, was dissolved away, leaving the space around the mould of the cavity.

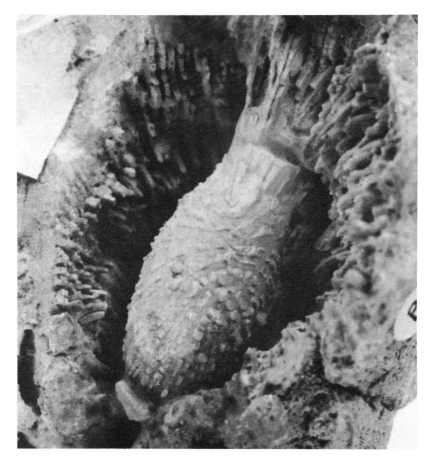

Pholas, the piddock, bores into soft rock (slate, sandstone, chalk) on the lower shore and in shallow water. It forms a symmetrical flask-shaped cavity, wider at the inner end and constricted at the entrance, so that the growing shell becomes imprisoned. The sucker-like foot grips the inner end of the cavity, and the rounded hinge area forms a fulcrum on which the valves see-saw as the adductor muscles, each in turn, alternately contract and relax. This action rasps the spiky ornament against the rock and, as it rasps, the foot changes position and rotates the shell. Internal moulds of such cavities made by fossil forms similar to *Pholas* are known from rocks of various ages (fig. 22) and may be associated with unconformities. *Teredo* (Eocene – Recent, fig. 21b), the ship-worm, is a related form which has adapted to drilling into submerged wood. The body is worm-like and the shell greatly reduced. It is used as a cutting tool to scrape out a long tunnel which may reach 20 cm in length and has a calcareous lining. *Teredo* feeds largely on wood scrapings, broken down by bacteria which also fix nitrogen. Fossil wood, riddled by similar tunnels, is found from the late Cretaceous onwards.

23 Anisomyarian and monomyarian bivalves.
a–c, *Ostrea*: a, external view of the left valve; b, a shell attached to rock; c, interior of the left valve. d, e, *Mytilus*: d, interior of the right valve; e, a shell attached to rock by its byssus. f, *Pecten*: interior of left valve.

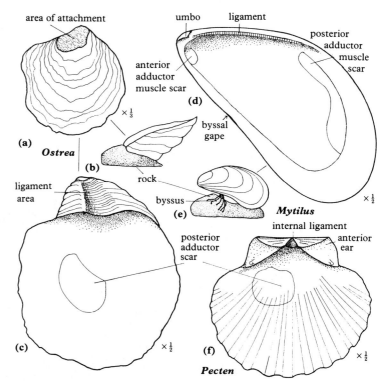

Mytilus (fig. 23d, e).   Very inequilateral with umbones near the pointed anterior end; obliquely elongated with broadly rounded postero–ventral margin. Ligament opisthodetic, long; dentition weakly developed or absent; posterior muscle scar larger than anterior; pallial line entire.

U Jurassic–Recent

Mytilus edule, the common mussel, is a euryhaline form living attached to rocks on the mid and lower regions of rocky shores and estuaries. It withstands exposure to rough waves and strong currents by means of an anchoring BYSSUS, accommodated by a slight marginal gape. This consists of horny threads (conchiolin) formed from a sticky fluid poured from a gland in the foot onto a rock. The fluid hardens quickly into a thread and a succession of such threads is spun until the shell is secured. Byssally attached forms exploit a well-oxygenated niche where food particles are constantly replenished.

Ostrea (fig. 23a–c).   Thick shell, irregularly circular in shape but variable; inequivalve with convex left and flattish right valve; surface with rough sculpture of radial ribs and frilled concentric lamellae. Short hinge; amphidetic ligament in triangular pit under the umbo; no teeth; mono-myarian with large posterior muscle scar almost centrally placed; pallial line absent; non-siphonate.

Cretaceous–Recent

Ostrea edulis, the common oyster, is an attached form living on a shelly or gravel substrate in shallow water down to about 80 m. The larva settles on a clean stone or shell (fig. 24) to which the left valve becomes fixed by means of

24   Area of attachment of a fossil 'oyster' which settled on an ammonite shell.

The area of attachment reflects the pattern of the ammonite in reverse. Exogyra, Cambridge Greensand, U Cretaceous (× 1.7).

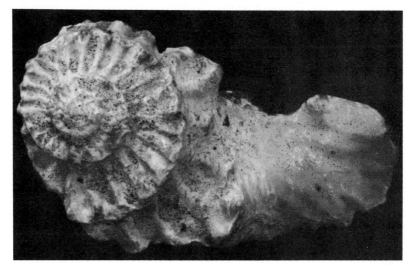

a calcareous cement poured out from the modified byssus gland. It avoids being silted over by rapid snapping of the right valve to flush sediment away, the 'quick' muscle being strongly developed for this purpose.

Pecten (fig. 23f).   Outline subcircular; almost equilateral; inequivalve with right valve rounded and left valve flat; surface with strong radial ribs. Hinge line short, straight, with equal sized ears at either end; amphidetic ligament with, internally, a triangular ligament cushion sited in a pit in hinge plate; no teeth; monomyarian with large subcentral muscle scar. Many allied genera range from the Trias.
Eocene–Recent
Pecten maximus, the large scallop, lives offshore in moderately deep water on a sand or gravel substrate. It may reach about 15 cm in diameter. It is partly sedentary and partly free-swimming. The very young *Pecten* is attached by a byssus. The growing *Pecten* discards the byssus and lies free on its right valve. It can swim for short distances, albeit somewhat inelegantly, by clapping its valves together rapidly to expel jets of water. Swimming is enabled by the combined design of ligament and adductor muscle working in opposition. The springy ligament, being under compression, causes the valves to gape widely, about 30°, as the muscle relaxes; then as the muscle contracts quickly to close the valves, water in the mantle cavity is ejected, thrusting the shell in the opposite direction. It alters direction by manipulating the mantle edges to control the outflow of water.

Fresh-water bivalve

Unio (fig. 27c).   Elongated oval shell; inequilateral with umbones towards anterior end; equivalve; smooth except for concentric growth lines; thick periostracum. Ligament opisthodetic; two strong cardinal teeth in left valve, one in right valve; isomyarian; pallial line entire.
Trias–Recent
Unio lives in rivers. It is a suspension feeder which burrows, only just covered by sediment. The periostracum and organic layers in the shell are exceptionally thick; even so, in waters of low pH the umbones are often eroded. The larva is, for a time, parasitic on fish.

Summary

The form of the bivalve shell is, to a considerable extent, a guide to the mode of life. Thus, bivalves which move through the sediments at or near the surface of the sea-floor have a shell which is equivalve and inequilateral; in addition, the shell is isomyarian and has an entire pallial line. Fully infaunal bivalves which burrow in soft sediment are characterised by a pallial sinus, and typically the shell is elongated and compressed; in many forms the shell gapes and the dentition may be poorly developed. Forms which bore into rock have a shell resembling that of the burrowing type except that it is more or less cylindrical. Bivalves which are attached by a byssus usually have a very inequilateral shell with the umbones at or near the anterior end, and the posterior end enlarged; the shell is anisomyarian. Bivalves which are cemented to the sea-floor have an inequivalve shell with the fixed valve the larger, and the free valve flat and lid-like; the shell is generally monomyarian.

Additional genera

Palaeozoic bivalves

Ctenodonta (Fig. 25b).   Almost equilateral oval outline; equivalve; surface smooth. Ligament external, opisthodetic; taxodont dentition (p. 50); pallial line entire.
Ordovician

Modiolus (fig. 25a).   Somewhat inflated; inequilateral with umbones near rounded anterior; equivalve; growth lines; long straight hinge line. Ligament external, opisthodetic; no teeth; anisomyarian; pallial line entire. Byssally attached with anterior end partly buried in sediment; subtidal to about 200 m. Related forms include *Mytilus* (fig. 23d, e) and *Lithophaga* (?Carboniferous, Jurassic–Recent, fig. 22).
Carboniferous–Recent

Dunbarella (fig. 26a).   *Pecten*-like shell; almost semicircular with straight hinge line ending in indistinct ears; inequivalve; surface with fine bifurcating radial ribs and growth lines. Found in black marine shales associated with goniatites.
U Carboniferous

Carbonicola (fig. 26b).   Roughly oval outline; inequilateral; equivalve.

25 Palaeozoic bivalves.

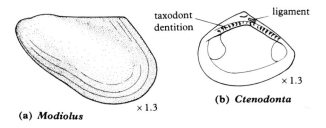

taxodont dentition   ligament

× 1.3

(b) *Ctenodonta*

(a) *Modiolus*   × 1.3

26   Upper Carboniferous bivalves.

a, marine bivalve, *Dunbarella* (× 1.5). b–d, non-marine bivalves: b, *Carbonicola*;
c, *Naiadites*; d, *Anthracosia*. (All × 1.2.)

Ligament external, opisthodetic; one, or two, or no teeth; isomyarian; pallial
line entire. Occurs in 'mussel' beds (shales) often overlying coal seams in
Coal Measures.
U Carboniferous

Mesozoic bivalves

Nucula (fig. 27d, e).   Oval to subtriangular; inequilateral; umbones point
backwards; surface may have fine radial ribs. Ligament internal, in
triangular pit below beak; taxodont dentition; isomyarian; pallial line
entire.
Cretaceous–Recent
Nucula lives on a richly organic substrate from offshore to abyssal depths. It is
one of a primitive, stable group, persisting since the Ordovician, which creep
over the substrate on a flat-soled foot. It is a deposit feeder, lacking siphons.
Organic detritus is picked up by grooved processes, near the mouth, which

27   Mesozoic bivalves.

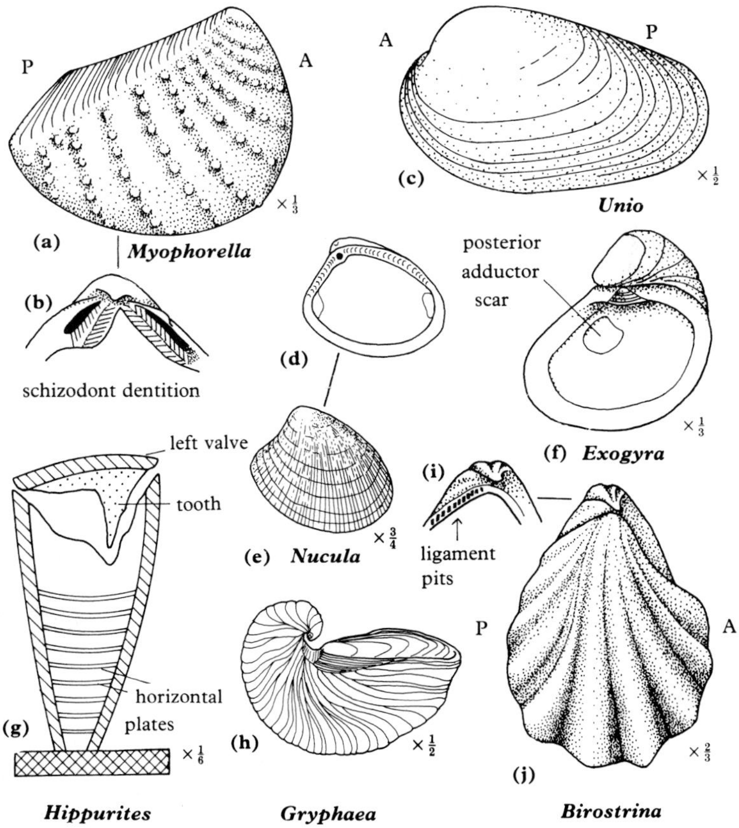

(a) | *Myophorella*

(b)

schizodont dentition

(c) $\times \frac{1}{2}$

Unio

(d)

posterior
adductor
scar

(f) *Exogyra* $\times \frac{1}{3}$

left valve

tooth

(e) *Nucula* $\times \frac{3}{4}$

(i)

ligament
pits

horizontal
plates

(g) $\times \frac{1}{6}$

(h) $\times \frac{1}{2}$

(j) $\times \frac{2}{3}$

Hippurites   **Gryphaea**   **Birostrina**

are extended from the anterior end. *Acila* (Cretaceous–Recent, fig. 13c, d) is a related form.

Birostrina (fig. 27i, j). Subtriangular; inequilateral; inequivalve, larger convex left valve with beak inclined forwards. Hinge with many small ligament pits; no teeth; may have been attached by byssus. Related forms common, and of some importance stratigraphically, in the Cretaceous, e.g. *Inoceramus* (L Jurassic–U Cretaceous) with longer hinge line and thick prismatic layer of calcite which provides much debris in the Chalk.
L Cretaceous

Gryphaea (fig. 27h). Related to *Ostrea* (p. 39). Very inequivalve, with much thickened left valve having umbo strongly incurved, and with radial groove on posterior flank; right valve flat. Attached by left umbo in early life; adult lying free; attachment area usually very small.
U Trias–Jurassic
Such thick-shelled ostreaceans are widespread and often common in shallow-water shales and calcareous mudstones of the Jurassic and Cretaceous (fig. 29). *Exogyra* (Cretaceous, fig. 27f) is another distinctive form with umbones spirally twisted towards the posterior; it remained cemented.

Myophorella (fig. 27a, b). Stout, thick-shelled; subtriangular; inequilateral; equivalve; surface with rows of tubercles oblique to margin, except on posterior slope where growth lines only; short curved hinge. Ligament

28 Mesozoic bivalves.
a, *Inoceramus*: the valves are slightly displaced, exposing the ligament pits along the hinge. U Cretaceous (× 1.1). **b,** *Hippurites,* U Cretaceous (× 0.4).

(a)   (b)

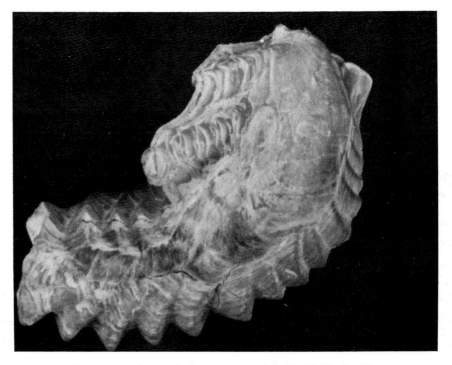

29 A Cretaceous 'oyster': *Arctostrea colubrina*, **L Chalk, Cherry Hinton, Cambridgeshire** (× 1.7).

The radial folding of the valve margins produces a sharp zigzag suture between the valves. This would have allowed an increased flow of water while restricting the gape of the shell, thus aiding efficiency of feeding.

external, opisthodetic; schizodont dentition (p. 50) with ridged teeth; isomyarian; pallial line entire.

U Trias–U Jurassic

Related forms are important and widespread in shallow-water deposits in the Mesozoic, e.g. *Trigonia* (Trias–Cretaceous) with ornament of stout concentric ridges. *Neotrigonia* (Oligocene–Recent) is the only surviving form; it is a non-siphonate shallow burrower living in muddy sands in shallow waters off Australia.

Hippurites (figs. 27g, 28b).   Much modified, inequivalve; right valve up to 30 cm high, fixed by apex, and conical to cylindrical in shape; lower part of valve cut off by horizontal plates, leaving a quite small living space above. Left valve lid-like, with two peg-like teeth projecting into pits in fixed valve.

U Cretaceous

Hippurites is an example of the Mesozoic rudists, aberrant sessile bivalves. They lived, often in large numbers closely associated, in the warm clear shallow waters of tropical and subtropical seas mainly in the region of Tethys, the ancient ocean in which the marine rocks of the

Alpine–Himalayan mountain chains were deposited. Earlier forms had spirally coiled umbones; in later members, like *Hippurites*, the fixed valve superficially resembles a rugose coral.

Pholadomya (fig. 31b). Elongated, inequilateral; somewhat inflated with wide posterior gape and slight anterior gape; surface with radiating ribs and concentric folds. Ligament opisthodetic; teeth weak; deep pallial sinus.
U Trias–Recent
This deep-burrowing, inactive form was widespread in shallow water in the Jurassic. It is now restricted to seas off the West Indies and Japan.

Cainozoic bivalves

(The genera described under 'shell form and mode of life' are all typical as Cainozoic fossils.)

Glycymeris (figs. 8, 30f). Thick-shelled; almost equilateral with subcircular outline; equivalve; surface with radial striations and growth lines. Ligament external, amphidetic, inserted in grooves of inverted V-shape; taxodont dentition, teeth slope inwards; isomyarian; pallial line entire; ventral margin crenulated. *Glycymeris* is a suspension feeder, actively burrowing just below the surface in sand or gravel, usually in shallow subtidal water.
L Cretaceous–Recent

Cardites (fig. 30a, b). Thick-shelled; subquadrate; inequilateral with umbones inclined forwards; equivalve; surface with strong radial ribs. Ligament opisthodetic; heterodont dentition with strong cardinal teeth; isomyarian; pallial line entire.
Eocene–Recent

Crassatella (fig. 30c, d). Inequilateral with umbones inclined forwards and posterior end truncated; equivalve; surface with concentric ribbing. Prominent internal triangular pit for ligament in each valve; heterodont dentition; isomyarian; pallial line entire.
U Cretaceous–Miocene

Macoma (figs. 13b, 30g). Suboval, almost equilateral with low umbones; slightly inequivalve; slight posterior gape; smooth surface with faint growth lines. Ligament opisthodetic, external; heterodont, but lateral teeth reduced; isomyarian; deep pallial sinus.
Eocene–Recent
Macoma balthica is a brackish-water form found in large numbers in sublittoral muds and muddy sands in coastal waters, e.g. in the Baltic. It is a

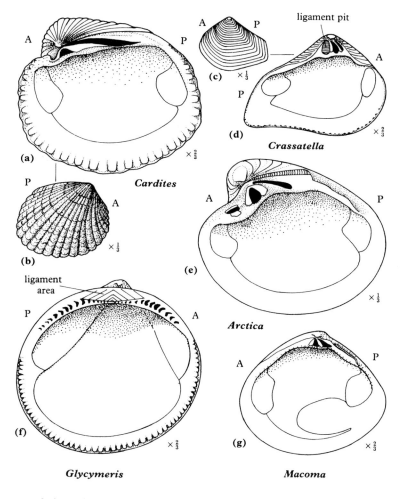

30 Cainozoic bivalves.
The dentition is heterodont in each case except (f) where it is taxodont.

burrower with separate siphons; the inhalant siphon, long and mobile, sucks up organic-rich detritus from the surface of the sediment.

Arctica (fig. 30e). Large, up to about 12 cm; thick-shelled, roughly oval and inequilateral with prominent umbones; equivalve; surface smooth except for growth lines; thick periostracum. Ligament opisthodetic; heterodont dentition, strongly developed; isomyarian; pallial line entire. Lives in sand and mud from low-tide level downwards.
L Cretaceous–Recent

Geological history

L Cambrian–Recent

Microscopic bivalves are recorded from the lower Cambrian. Macroscopic forms appeared in the lower Ordovician and numbers gradually increased until the end of the Palaeozoic, occurring in abundance in some upper Palaeozoic horizons. Many genera disappeared at the close of the Permian and further extinctions occurred in the late Trias. A great diversity of new genera appeared during the Mesozoic and the class reached its acme in the Tertiary. It remains an important group today.

The earliest bivalves so far identified are extremely small, 1.2–5 mm, and include two from the mid lower Cambrian, *Pojetaia* from Australia, and *Fordilla* from North America, Europe and Siberia; and one from the late middle Cambrian of New Zealand, *Tuarangia*. The group was, therefore, already widespread. These early forms are preserved as internal moulds showing traces of dentition and muscle scars. They are thought to represent separate stocks.

In the Ordovician, larger-sized forms (about 1 cm or more across) are found, a few genera at first. By mid-Ordovician time most of the main stocks were established. They include forms like *Ctenodonta* and related deposit feeders; and an array of suspension feeders, some byssally attached, others exploiting the infaunal niche by burrowing. Generally, however, bivalves are not really abundant in the lower Palaeozoic and the shell, originally aragonite, is often poorly preserved, or dissolved leaving moulds.

Bivalves are more varied and better preserved in the upper Palaeozoic rocks and at some horizons are common. Fresh-water forms (e.g. *Archanodon*) appeared in the Devonian. More varied byssally attached and burrowing forms occur in the marine facies. *Modiolis* is typical of lagoonal limestones and shales, and *Dunbarella* (fig. 26a) occurs in black marine shales in the mid-Carboniferous, often in association with goniatites. Great numbers of bivalves occur in some Coal Measure shales known as 'mussel' bands. They are referred to as 'non-marine' since it is uncertain whether they lived in fresh or brackish water. They include shallow-burrowing forms like *Carbonicola* (fig. 26b) and *Anthracosia* (fig. 26d); and byssally attached forms like *Naiadites* (fig. 26c). Species of these forms are of quite limited vertical range and of wide lateral distribution so they have stratigraphical value.

Bivalves were less affected than many other groups by the extinctions which occurred in the later Permian and at the end of the Trias. While many genera disappeared, the main stocks survived. Renewed radiation in the course of the Mesozoic brought new and widely distributed genera in increasing numbers. So abundant are bivalves at some horizons that whole or broken shells make up a high proportion of the rock; and the state of preservation is often good. Bivalves were faced during this time by increasingly efficient predators like fish and crabs – a spur to continued

(a)   **(b)**

31 Warm-water bivalves.
Two genera which disappeared from the rocks in England after the Cretaceous
(b) or early Tertiary (a). In warm-water regions, however, they are still present
today. a, *Pterotrigonia*, U Greensand, U Cretaceous, Blackdown. b, *Pholadomya*,
London Clay, Eocene. The specimen is tilted to show the posterior gape. (Both
× 0.9.)

diversification. Among the diversified bivalves were forms cemented by one
valve, an innovation by 'oysters' in the late Trias and followed by, for
instance, *Gryphaea* in the Jurassic, and *Exogyra* and *Ostrea* in the Cretaceous.
A different and very specialised style of cementation is shown by the rudists,
which were especially characteristic of the later Cretaceous, and were largely
confined to warm Tethyan waters (p. 45). Varied byssally-attached forms
include *Pecten*-like shells in the Jurassic and Cretaceous; and *Inoceramus* and
related forms like *Birostrina* in the Cretaceous. Infaunal types include active,
non-siphonate burrowers of the *Trigonia* group like *Myophorella*, common in
the Jurassic and lower Cretaceous; and siphonate, deep burrowers, like
Pholadomya, found in some Jurassic limestones and clays. Siphonate forms
with heterodont dentition were prominent during this time and subsequently in the Cainozoic.

In the Cainozoic, bivalves were abundant, sharing with gastropods a
dominant position in the marine shallow-water benthic faunas. They were
also widespread in fresh-water lakes and rivers. They include many familiar
genera still living. The majority have heterodont dentition. In the Eocene of
England there are a number of genera which today are restricted to warmer
waters, e.g. *Pholadomya* (fig. 31b). Later, in the Pliocene, bivalves which still
live in British waters appeared in increasing numbers. They were joined in
the Pleistocene by others which today only occur in colder northern seas, e.g.
Macoma calcarea (fig. 13b).

Technical terms

ADDUCTOR MUSCLE SCARS two scars, one anterior and one posterior,
    impressed in each valve by the muscles which close the valves. The scars
    are equal in size in ISOMYARIAN shells (fig. 18a); the posterior scar is

larger than the anterior scar in ANISOMYARIAN shells (fig. 23d); there is only one scar, the posterior scar in MONOMYARIAN shells (fig. 23c).

BYSSUS horny, fibrous outgrowth from the body by which the shell, in some forms, is secured to a firm surface (fig. 23e). A gape or notch may be present in the valve margin near the anterior to accommodate the byssus when the valves are closed.

CHONDROPHORE a special device found in *Mya* (fig. 20a, c) to accommodate the internal ligament. In the left valve it is a spoon-shaped process projecting from the hinge line at about 90°; in the right valve it is a shallow pit under the umbo.

CLAM An American term for a bivalve. In Britain, usually refers to the scallop (*Pecten*).

DENTITION the collective term for the teeth and sockets of the hinge plate. The principal types of dentition are:

(i) TAXODONT many small similar teeth and sockets all along the hinge plate (fig. 25b); (ii) HETERODONT a few teeth varying in size and shape, distinguished as the cardinal teeth which radiate from the umbo, and the lateral teeth which lie obliquely along the hinge plate on both anterior and posterior sides of the umbo (fig. 18a); (iii) SCHIZODONT two or three teeth, rather thick and sometimes grooved, lying under the umbo (fig. 27b).

EQUILATERAL the shell is more or less symmetrical about a line from the umbo to the ventral margin (fig. 30f). A shell in which the umbones are nearer the anterior end than the posterior end is INEQUILATERAL (fig. 18a).

EQUIVALVE the valves are mirror images of each other except for the minor differences on the hinge line due to the dentition (fig. 18b). A shell in which the valves are dissimilar in size and shape is INEQUIVALVE (fig. 27g).

GAPE a permanent opening between the valve margins; a posterior gape occurs in some bivalves with a large siphon; an anterior gape through which the foot extends may also occur in some burrowers, e.g. *Pholas* (fig. 21a).

HINGE PLATE a thickening of the dorsal margin, either straight or curved, along which the valves articulate. It supports the dentition and the ligament (fig. 18a).

LIGAMENT a resilient horny substance (formed from an organic material, conchiolin) which unites the two valves along the dorsal margin. It may lie posterior to the umbones, OPISTHODETIC (fig. 18a), or on both sides of the umbones, AMPHIDETIC (fig. 30f). It may be external or internal. An external ligament is under tension when the valves are shut and, being elastic, causes the valves to open when the adductor muscles are relaxed (fig. 18c, d). An internal ligament usually occurs in a pit (fig. 23f), or a series of pits (fig. 27i) in the hinge line. It is compressed when the valves are closed; when the adductor muscles relax, it pushes the valves open.

ORNAMENT surface markings on the outside of the shell which may be concentric about the umbo, or radial from the umbo, or both concentric and radial. The ornament may consist of striae, ribs or, occasionally, of knobs or spines.

PALLIAL LINE a line running between the anterior and posterior muscle scars, parallel to the ventral margin of the valve. It marks the attachment of the mantle retractor muscles to the valve and it may be entire (fig. 19a), or deflected by the pallial sinus (fig. 18a).

PALLIAL SINUS a bending inwards of the pallial line at the posterior end of the shell; it is found in burrowers (fig. 18a).

UMBO (pl. umbones) the earliest formed part of the valve, usually a rounded boss which projects above the hinge line (fig. 18a).

3

Mollusca–Class Gastropoda plus minor mollusc classes

The distinguishing feature of the gastropods is the twisting of the viscera (internal organs). The body is asymmetric; there is a head at the anterior end, and on the ventral surface is a muscular creeping foot. In most forms the body is protected by a single (univalve) shell (forms lacking a shell are not considered here). Typically the shell is a tapering tube, coiled in a right-handed spiral.

Gastropods are more abundant than any other group of molluscs at the present day; they also occupy a greater range of habitat. The majority are aquatic, and of these most live in shallow seas; they are also widespread in fresh water and on dry land. Modern examples include the marine limpets, winkles and whelks, and the terrestrial snails and slugs.

Gastropods are classified by features of the soft parts, but these leave few clues in the empty shell. Accordingly, fossil shells are classified by comparing them with similar modern shells. There are three subclasses: Prosobranchia, Opisthobranchia and Pulmonata. Only the prosobranchs are important as fossils and are dealt with in more detail here.

Morphology

Soft body

The shell is a refuge into which the entire body can be retracted. When the animal is moving, the head and foot are extended outside the shell but the soft visceral hump remains within it. The head bears sensory tentacles and eyes, and below these is the mouth in which is a food-rasping tool, the RADULA. This is a flexible horny ribbon bearing rows of teeth (its detailed structure is useful in classifying modern forms). The foot (fig. 32b) is an elongate organ with a flat sole on which the animal glides over the sea-floor by continuous waves of muscular contraction. Terrestrial forms lubricate the dry surface over which they move by exuding mucus.

The visceral hump lies on the dorsal surface of the body. It contains the digestive system and other organs. Much of it is spirally coiled within the

POSTERIOR

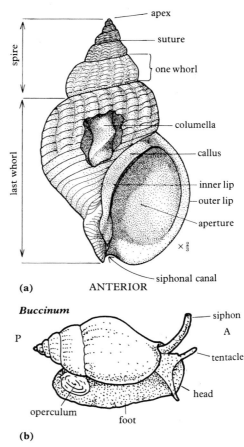

apex

suture

one whorl

spire

columella

callus

last whorl

inner lip

outer lip

aperture

$\times \frac{2}{3}$

siphonal canal

(a)   ANTERIOR

Buccinum

siphon

P   A

tentacle

head

operculum

foot

(b)

32   Morphology of gastropods based on *Buccinum*.
a, shell with part of last whorl broken to expose the columella. b, gastropod
crawling with head and foot extended.

shell. It is covered by the mantle which extends towards the head to form a
space, the mantle cavity, which is thus anterior in position, in contrast to its
normal position in other molluscs. Other parts of the body, including the
anus, are also anterior in position because during the development of the
larva the mantle and viscera are twisted (torted) through 180° relative to the
head and foot. The effect of torsion is to juxtapose the clean water intake and
the waste outlet. In most gastropods the arrangement is modified to avoid
possible pollution and, in many, a degree of detorsion has occurred.

In aquatic forms the mantle cavity generally contains feathery gills. These
extract oxygen from water circulated through the cavity by ciliary action. In
some forms the anterior margin of the mantle may form a tubular extension,
the INHALANT SIPHON (fig. 32b) to direct water to the gills. Gills are

lacking in terrestrial forms in which the mantle cavity is modified to act as a lung.

Shell

The shell consists mainly (96%) of crystals of calcium carbonate, usually aragonite but sometimes calcite, intermeshed with organic matter; it is covered with a layer of periostracum. There are usually two layers in the shell, each with a distinct crystal arrangement which resembles the microstructure found in bivalve shells (p. 29), for instance, nacreous and crossed lamellar structures.

The shell is basically a conical tube, closed at the pointed end, the APEX, and open at the wide end, the APERTURE (fig. 32a). The form of the shell results from variations on a very few themes which, combined in different ways, produce distinctive shells for each of hundreds of species of gastropods. The themes include coiling, the rate of increase in diameter of the shell, the shape of the cross-section, form of aperture, and ornament.

Coiling. The shell, secreted by the mantle, grows by increments to the margins of the aperture; in most cases the increments are greater along one margin so that the shell coils about an axis. Coiling is typically in a helical spiral, like the thread of a screw, descending from apex to aperture. This type of shell is described as CONISPIRAL; planispiral coiling is exceptional. In some gastropods coiling is confined to the embryonic shell, (protoconch) at the apex, and the fully grown shell is a conical 'cap' shape. Coiling of the shell is, of course, independent of the torsion of the body which occurs in the larval stage (p. 53).

Each complete coil of a shell is a WHORL; the line along which successive whorls meet is the SUTURE. The ultimate whorl is the LAST WHORL, and the earlier whorls together form the SPIRE (fig. 32a). The shell may be coiled tightly about its axis so that a solid central pillar, the COLUMELLA (fig. 32a, 33), is formed. In loosely coiled shells there is an axial space in the columella; this shows as the UMBILICUS (fig. 35d) where it opens at the base of the last whorl.

The last whorl may be little larger than the previous whorl if the diameter of the shell increases slowly (fig. 35f); if the diameter of the shell increases rapidly the last whorl may be much larger than the spire (fig. 35d). The spire may be high, pointed and consist of many whorls (fig. 35f); it may be short, and have few whorls (fig. 35d); it may be depressed (fig. 37c), or occasionally it may be concealed by the last whorl (fig. 37a).

The aperture. The aperture may be rounded, oval, or long and narrow. The margin nearest the apex is termed the POSTERIOR margin; the opposite side is the ANTERIOR margin, where the head emerges. The margin in contact

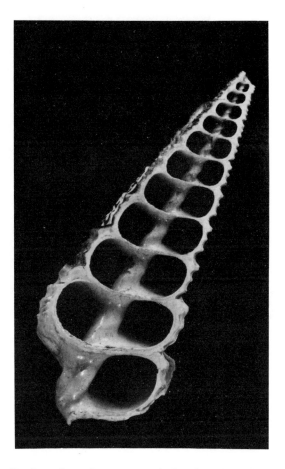

33 **Longitudinal section of a gastropod showing the columella** (× 2).

with the previous whorl is the INNER lip; the free margin is the OUTER lip (fig. 32a).

The aperture may be entire (fig. 35d), notched or extended at the anterior margin by a SIPHONAL CANAL (fig. 39), or cut by an EXHALANT SLIT (fig. 37). The siphonal canal ranges from a short deflection of the anterior margin to a long narrow canal like a split tube. It supports the inhalant siphon (p. 53). The exhalant slit is a narrow slit in the outer lip at right-angles to the edge. The earlier part is filled in by shell during growth; the resultant trace on the shell is called a SELENIZONE (fig. 37). An exhalant slit is found in living gastropods with two gills; it occurs in many fossils, especially Palaeozoic forms. In some genera a subsequent layer of shell, CALLUS (fig. 35d), is deposited by the mantle on the inner lip and adjacent part of the whorl. In many marine gastropods a horny lid, the OPERCULUM (fig. 32b), closes the aperture when the body is withdrawn into the shell. It is borne on the posterior end of the foot; it is rarely preserved in fossils.

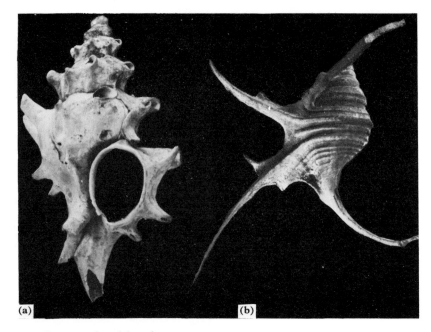

34 Gastropods with spinose processes.
a, *Typhis*, Barton Beds, Eocene (× 4), the spines are hollow tubes some of which are former siphonal canals. b, *Tessarolax*, L Cretaceous (× 1.5), the spines are extensions of the outer lip which were resorbed during shell growth so that only the last-formed remain.

Ornament.   The surface of the shell may be smooth, or bear fine or coarse markings arranged transversely or spirally; tubercles and spinose projections sometimes occur (fig. 34).

Internal markings are confined to scars left by the muscles which attach the animal to its shell. These are not usually seen except in cap-shaped shells (fig. 35c); in coiled shells they occur on the columella.

Orientation.   Most gastropods are asymmetric. The shell is conventionally drawn with the aperture facing you, and the apex of the spire pointing upwards. Most genera are coiled in a clockwise direction so that the aperture is on your right, dextral coiling (fig. 32). Occasionally, shells are coiled in a left-handed direction, sinistral coiling (fig. 39a), and the aperture is then on your left.

Classification of Subclass Prosobranchia

The prosobranchs are fully torted and are divided into three orders:

1. Archaeogastropoda (Cambrian–Recent), one or two-gilled forms usually with low-spired or cap-shaped shell; inner layer of nacre; may have exhalant slit; eggs and sperm discharged into sea for external fertilisation.

2. Mesogastropoda (M Ordovician–Recent), one-gilled forms with typically conispiral shell; often with siphonal notch; may have operculum;
fertilisation internal.

3. Neogastropoda (Cretaceous–Recent), differ from mesogastropods in soft
part features; conispiral shell with siphonal notch or canal.

Shell form and mode of life of some Recent gastropods

Archaeogastropods

(marine; most are herbivores)

Patella. (fig. 35a–c).   Thick shell, widely conical cap-shape; aperture oval;
ornament of prominent ribs radiating from subcentral apex; inside, a U-
shaped muscle scar opens towards the anterior.
Cretaceous–Recent

35 Gastropods with an entire aperture.

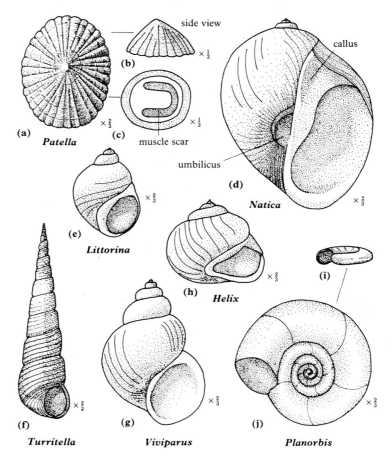

Patella vulgata, the common limpet, is specialised to live on exposed rocky shores. At rest, it clings tenaciously to its rock with the sole of its rounded foot so that the roughest waves cannot dislodge it. Its broad-based cap-shape offers minimal resistance to wave drag, and when the tide is out this habit reduces risk of desiccation. At high tide, *Patella* crawls a short distance in search of encrusting algae on which it browses; it then returns 'home' to its oval-shaped depression. Cap-shaped shells are common in littoral deposits. They include *Fissurella* (Eocene–Recent) with an exhalant opening near the apex; and *Emarginula* (Trias–Recent) with a marginal slit.

Mesogastropods

(marine, fresh-water, terrestrial; feeding habits varied).

Littorina (fig. 35e).   Thick shell, up to about 2.5 cm high; low spire; last whorl large; aperture entire with sharp outer lip; surface smooth or with faint spiral striae.
Pliocene–Recent
Littorina littorea is the largest of several species of periwinkle found on rocky shores around the British coasts. They are vagrant forms, each restricted to a preferred habitat on either upper, or middle, or lower shore. Thus at low tide some are out of water for many hours each day. They avoid desiccation because of their tightly fitting operculum. The thick shell is damage-resistant however much it is rolled around by turbulent water. They are herbivorous, feeding on seaweeds. One species, *Littorina neritoides*, when adult, may live above high-water mark for as long as five months at a time and can breathe air direct. This suggests one way in which gastropods may have adapted to life on dry land. Some mesogastropods are terrestrial.

Natica (fig. 35d).   Thick shell, up to about 1.5 cm high; almost globular, with low spire and very large last whorl; aperture entire with callus partly obscuring umbilicus.
Miocene–Recent
Natica alderi is an infaunal form with a large foot (partly reflected over the head) with which it ploughs through sand just below the surface seeking prey (e.g. bivalves). It grips its prey in the foot and, using the radula plus a chemical secretion, drills a neat funnel-shaped hole through which its proboscis (mouth part) scoops out the soft body. Similar holes are often seen in fossil shells (fig. 36).

Turritella (fig. 35f).   Shell high-spired, about 5 cm high; many whorls; ornament of spiral ribbing; aperture entire, subquadrate; this distinguishes

36   A hole drilled by a carnivorous gastropod, probably *Natica*, in a bivalve shell (*Dosinia*) from the Red Crag, Pleistocene, East Anglia (× 2).

Turritella from otherwise similar shells like *Cerithium* (U Cretaceous–Recent), which have a siphonal canal.
Cretaceous–Recent
Turritella communis lives gregariously in shallow water from about 10 to 100 m deep, where it burrows, spire downwards, to lie at a low angle in muddy gravel or sand. It feeds by filtering water drawn into the mantle cavity by ciliary action; food particles, collected by the gills, are embedded in mucus and passed to the mouth.

Viviparus (fig. 35g).   Thin shell with thick periostracum; spire moderately high; about 2.5–3.5 cm high; last whorl large; aperture entire.
Cretaceous–Recent
Viviparus viviparus, unlike the previous gastropods, lives in *fresh* water (in slow-moving streams). It is a filter-feeder. Fossils may occur in sufficiently large numbers to form shell-beds as in the Sussex 'marble' (L Cretaceous) of Britain. Most fresh-water gastropods, however, are pulmonates (p. 64).

Neogastropod

(marine; active carnivores or scavengers)

Buccinum (fig. 32a).   Thick shell; spire moderately high; last whorl large; aperture wide, with short siphonal canal; callus on inner lip; ornament of spiral and transverse ribs.
Oligocene–Recent
Buccinum undatum, the common whelk, occurs from the lower shore down to about 200 m. It is often large, ranging from about 5 cm high inshore, to about 15 cm in deeper water. It lives on a soft substrate, ploughing through mud or sand with its long inhalant siphon extending upwards to draw clear water through the mantle cavity. It is a flesh-eater, and rasps the soft body of its prey, live or dead, with its radula.

Summary

The form of the gastropod shell is not a reliable guide to the habitat of the animal, but some broad generalisations can be made:

  (i) Gastropods are typically benthic, though pelagic forms do occur.
 (ii) Those living in the littoral zone usually have a thick shell, which may be cap-shaped as in the limpets, or rounded with a short spire as in the winkles.
(iii) Forms in which the shell has an entire aperture are often herbivorous, and usually live on a hard substrate (but *Natica* and *Turritella* are notable exceptions).
 (iv) Forms with a siphonal canal are often found on soft sediment and are carnivorous.
  (v) Fresh-water gastropods for the most part have thin shells with a thick periostracum.

Additional common fossils

Archaeogastropods

Bellerophon (figs. 37a, 40).   Shell smooth; unusually for a gastropod it is bilaterally symmetrical and planispiral, each whorl enveloping the previous one so that only the last whorl is visible; exhalant slit cuts the outer lip in the median plane, its earlier trace forming a well-marked selenizone.
Silurian–Trias

Euomphalus (fig. 37c).   Shell smooth, planispiral with depressed spire and wide umbilicus; aperture subpentagonal; outer lip with notch; selenizone forms spiral ridge on shoulder.
Silurian–Permian

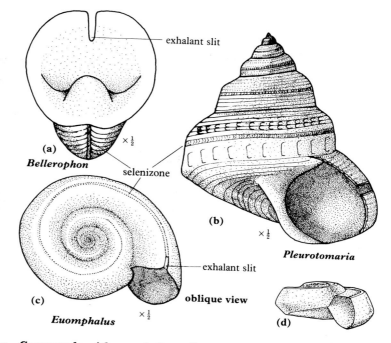

exhalant slit

(a)
Bellerophon   × ½
                    selenizone

(b)
                           × ½
                    Pleurotomaria

exhalant slit

(c)   × ½   oblique view
Euomphalus   (d)

37   Gastropods with an exhalant slit.

Pleurotomaria (fig. 37b).   Spire moderately high; umbilicus in some; exhalant slit long; broad selenizone; ornament of spiral striae and tubercles; represents an ancient family now restricted to quite deep water off Japan and in the Caribbean.
L Jurassic–L Cretaceous

Mesogastropods

Purpuroidea (fig. 38a).   Thick robust shell; spire moderately low; last whorl large; wide aperture; ornament of spirally arranged tubercles on shoulder.
Jurassic–Cretaceous
Tessarolax (fig. 34b).   Spindle-shaped with long narrow aperture; short siphonal canal; outer lip wing-like with long spinose processes. *Aporrhais* (U Cretaceous–Recent, fig. 38b) is a related form.
Jurassic–Cretaceous

Neogastropods

Fusinus (fig. 39c).   Long slender shell; high turreted spire; very long narrow siphonal canal; ornament of intersecting spiral and transverse ribs.
U Cretaceous–Recent

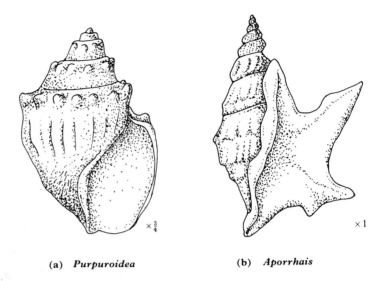

(a) *Purpuroidea*   (b) *Aporrhais*

38 Mesogastropods.

39 Gastropods with a siphonal canal.

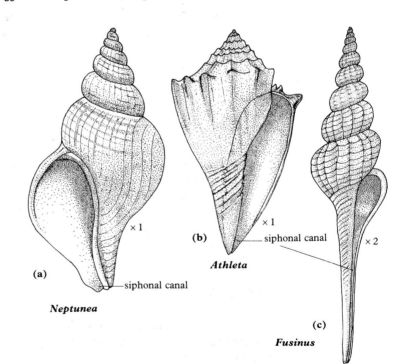

(a)

Neptunea

(b)

Athleta

siphonal canal

siphonal canal

(c)

Fusinus

Athleta (fig. 39b).   Thick shell with low spire and large last whorl; aperture elongate, relatively narrow, passing into short siphonal canal; tubercles prominent on shoulder of last whorl.
Cretaceous–Recent

Neptunea (Fig. 39a).   Similar to *Buccinum* (p. 60) but with longer spire and wider siphonal canal. Shell smooth or with spiral ribbing. Some species are sinistrally coiled.
Eocene–Recent

Geological history

Basal Cambrian–Recent
Gastropods are first recorded in basal Cambrian rocks. Many families appeared in the course of the Palaeozoic and Mesozoic; few of these became extinct and by the Tertiary gastropods were numerous and highly diversified. Today, gastropods far outnumber any other class of mollusc.

Palaeozoic gastropods are mainly archaeogastropods. They include cap-shaped and spirally coiled forms. The latter were generally low-spired or planispiral. In many, the aperture was entire but commonly there was an

40 Palaeozoic gastropods.
a, *Bellerophon*, L Carboniferous, showing the selenizone (× 2). b, *Poleumita*, Wenlock Limestone, Silurian (× 1); a form related to *Euomphalus* but differing in its relatively pronounced ornament.

exhalant slit as in *Bellerophon* (fig. 37a) and *Euomphalus* (fig. 37c). They are found in various types of deposits but are, however, commoner in limestones such as the mid-Silurian Wenlock limestone and the reef-limestones of the lower Carboniferous.

Mesogastropods, which appeared in the Ordovician, became prominent in the Mesozoic. In most the aperture is entire but forms with a siphonal canal began to appear during the Mesozoic and were quite common in the Cretaceous. Some Palaeozoic families persisted. *Pleurotomaria* (fig. 37b) represents an ancient group which appeared in the upper Cambrian; it survives today in warm seas. Fresh-water gastropods, both prosobranchs and pulmonates, also occur in the Mesozoic.

The Cainozoic marks the acme of gastropod diversity with an abundance of mesogastropods and neogastropods. The latter, with siphonal notch or canals, had appeared in the Cretaceous and had become established as prime carnivores. Many of those common as fossils in the Eocene and Oligocene of Britain persist in tropical seas today: for instance *Fusinus* (fig. 39c); and the specialised *Conus* (U Cretaceous–Recent), which injects its prey with poison from a tooth. Fresh-water and land gastropods are common in non-marine horizons, e.g. the Oligocene Bembridge limestone in England. Later, in the Plio-Pleistocene crags of East Anglia, gastropods are again prominent. The earlier crags contain 'warm'-water species which are later replaced by 'cold'-water forms. Some, like *Buccinum*, still occur around British coasts; others, like the related *Neptunea*, are now confined to northern waters (see also fig. 13a).

Pulmonata

?U Carboniferous, Jurassic–Recent

Gills are lacking in pulmonate gastropods and the mantle cavity acts as an air-breathing lung. The shell is usually coiled, but is vestigial in slugs; it is thin with a smooth surface and no operculum. Most live in fresh water or on land.

Among fresh-water pulmonates the shell ranges from high-spired, e.g. *Galba* (U Jurassic–Recent), to almost planispiral, e.g. *Planorbis* (Oligocene–Recent, fig. 35j), the latter with sinistral coiling. A few are cap-shaped, e.g. *Ancylus* (Pleistocene–Recent). They feed on algae and plants and are widespread throughout the world in running or still (including stagnant) water.

Terrestrial forms are independent of water for breeding. Shell form ranges from high-spired to almost planispiral. In *Helix* (Miocene–Recent, fig. 35h) the shell is almost globular with few whorls. Terrestrial forms are found in a wide variety of habitats throughout the world; they are especially common in areas with a humid climate and with calcareous soils.

Opisthobranchia

?U Carboniferous–Recent
Opisthobranchs are highly varied gastropods which have undergone
detorsion and in which the shell is frequently reduced or lacking. Shelled
forms are generally uncommon as fossils apart from the microscopic, pelagic
pteropods (see p. 266).

Technical terms

APERTURE the opening of the shell through which the head and foot extend
(fig. 32a).

APEX the first-formed part of the shell at the tip of the spire (fig. 32a).

CALLUS subsequent layer of shell over the inner lip and adjoining part of the
last whorl (fig. 32a).

COLUMELLA the solid axis of the spire formed by the inner wall of the coiled
shell (figs. 32a, 33).

CONISPIRAL cone-shaped spire (fig. 32a).

EXHALANT SLIT a narrow fissure cutting the outer lip through which faeces
are passed away from the aperture (fig. 37).

OPERCULUM the lid, horny or calcareous, which closes the aperture when
the head and foot are withdrawn into the shell (fig. 32b).

OUTER LIP the outer edge of the aperture; the inner margin is the INNER lip
(fig. 32a).

SELENIZONE the spiral trace of the exhalant slit, filled in by shell during
growth (fig. 37).

SIPHONAL CANAL the gutter-like extension of the anterior end of the
aperture which is moulded around the inhalant siphon (fig. 39).

SPIRE all the whorls except the last whorl (fig. 32a).

SUTURE the spiral line along which the whorls are contiguous (fig. 32a).

UMBILICUS the cavity around the axis of coiling found in loosely coiled shells
(fig. 35d).

WHORL a complete turn of a coiled shell; the ultimate one is the last whorl
(fig. 32a).

Minor mollusc classes

Amphineura–Subclass Polyplacophora

U Cambrian–Recent
Polyplacophorans, or chitons, have a dorsal skeleton consisting of eight
overlapping plates. These are surrounded by a muscular girdle (fig. 41a).
Chitons often live in high-energy areas of the sea, and cling to rocks with

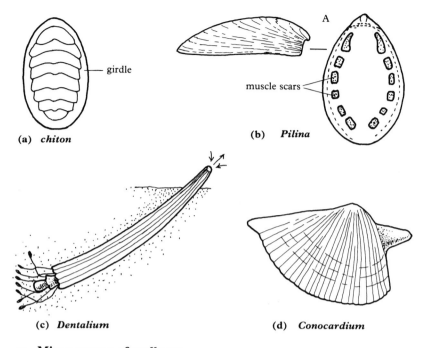

girdle

muscle scars

(a) *chiton*

A

(b) *Pilina*

(c) *Dentalium*

(d) *Conocardium*

41   Minor groups of molluscs.
a, polyplacophoran. b, monoplacophoran: above, profile of shell; right, internal mould. c, scaphopod. d, rostroconch. (a–c × 1, d × 1.5).

their large sucker-like foot, the girdle pressed tightly against the surface. They have a radula used for grazing on algae. Fossils are not common and are usually disarticulated.

Monoplacophora

L. Cambrian–Devonian, Recent

Monoplacophorans have a univalved, cap-shaped shell with the apex subcentral or curved forwards; and may be planispiral. They have a large foot and multiple paired muscles which leave discrete muscle scars (fig. 41b). They graze on algae. They are common as fossils in shallow-water rocks from Cambrian to Silurian but have not been found as fossils after the Devonian. At the present day they live at depths between 175 and 6500 m but mainly in the deeper waters where they are widespread. Some bellerophont shells (p. 60) are referred to this group.

Scaphopoda

M. Ordovician–Recent

Scaphopodans ('tusk' shells) have a curved, tapering, tube-like shell which is open at either end (fig. 41c). The foot, at the anterior end, is used to burrow obliquely into soft sediment. Water currents pass in and out via the narrower posterior end which projects above the sediment surface. They have ciliated tentacles which collect detritus and small organisms (e.g. foraminifera) from the sediment. Fossils are rare.

Rostroconchia

Cambrian–U Permian

Rostroconchs are extinct molluscs once classed with bivalves. They differ from these, however, in their one-piece shell which is continuous over the dorsal margin but is open along the other margins. Neither ligament nor muscle scars are present. In early forms a dorsal transverse plate lies across the shell towards the anterior end. Later forms have a characteristic tubular projection at the posterior end, and the shell over the dorsal surface may be organic-rich (fig. 41d).

4

Mollusca–Class Cephalopoda

The cephalopods are nektonic molluscs distinguished by the eight, ten or many prehensile arms, or tentacles, which surround the head. The foot is closely associated with the head and is modified to form the FUNNEL, a swimming organ powered by water ejected from the mantle cavity. One living, and most fossil cephalopods possess an external chambered univalve shell, designed as a buoyant device. Basically it is a 'cone' which may be straight, curved, or coiled in a planispiral. It is divided by transverse partitions, SEPTA (sing. septum); the septa are perforated by a tube, the SIPHUNCLE, which extends from the mantle to the apex of the shell. In most living and some fossil cephalopods the shell is either a modified internal structure, or it is absent (as in the octopus). In this respect then, living cephalopods are untypical of the fossil record of the class. Fossil forms were jet-propelled 'mini-submarines', their history reflecting diverse experiments in techniques of adjusting shell buoyancy.

Living cephalopods are the most highly organised molluscs and include the largest and fastest-moving invertebrates. Numerically, they are well past their zenith. Their habitat is exclusively marine, ranging from shallow to deep water, some living at abyssal depths. They are predatory creatures of diverse habit and include: the torpedo-like, fast-moving squids, some reputed to exceed 60 kph and able to glide for short distances out of water; the less actively swimming cuttlefish; the relatively sedentary octopus which lurks under rocks; and *Nautilus*, which alone has an external shell and which swims slowly near the sea-floor.

Cephalopods have an extensive fossil record from late Cambrian onwards. The main subclasses are:

Nautiloidea, upper Cambrian–Recent
Ammonoidea, lower Devonian–end Cretaceous
Coleoidea, lower Devonian–Recent

Nautiloidea

The nautiloid shell is divided into CHAMBERS by saucer-like SEPTA with circular or gently fluted edges. Each septum has a more or less central

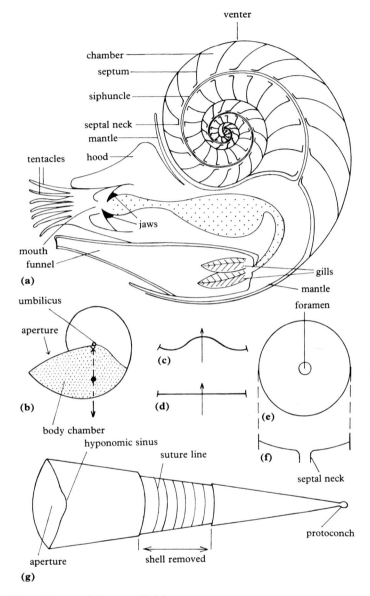

42   Morphology of the nautiloids.

a–c, *Nautilus*: a, a simplified median section of the shell to show the
arrangement of the soft parts and internal structures; b, attitude of the shell
when floating in water (the cross marks the approximate position of the centre
of buoyancy, and the filled dot the centre of gravity); c, suture line (the arrow is
pointing in the direction of the aperture). d–g, *Orthoceras*: d, suture line; e,
anterior view of a septum; f, transverse section of a septum; g, idealised view of
a shell showing the main features.

perforation for the siphuncle and this is surrounded on the posterior (convex) face of the septum by a short tube, the SEPTAL NECK.

The structure of fossil nautiloids is interpreted by reference to the single living genus, *Nautilus* (fig. 42). *Nautilus* lives in the last-formed chamber, the BODY CHAMBER, at the anterior end of its shell. The body can be retracted within this chamber and the aperture closed by a muscular hood. The shell is thus a support and a refuge; it is also, by virtue of gas contained in the chambered part, a hydrostatic organ which controls the buoyancy and facilitates a nektonic mode of life.

Morphology

Soft body

Nautilus has a distinct head with two highly developed eyes and, around the mouth, many retractable tentacles; these lack hooks or suckers that may occur in other cephalopods. The mouth is furnished with two jaws, like a parrot's beak, made of chitin with calcified tips. Between them lies the radula, bearing transverse rows of teeth, which functions as an aid to swallowing food. The visceral mass is covered by a thin mantle which secretes the shell and, ventrally, encloses the mantle cavity containing two pairs of gills; other living cephalopods have only one pair of gills. The mantle is prolonged as a fleshy cord enclosed in a tubular sheath, the SIPHUNCLE, which passes back through the chambers to the PROTOCONCH (initial shell) at the apex of the shell. The siphuncle, partly horny and partly a porous chalky substance, plays a key role in buoyancy control.

The funnel is a flexible tube-like structure (fig. 42a). It opens just below the mouth, and its other end leads into the mantle cavity. Oxygenated water drawn in around the edges of the mantle bathes the gills and is passed out via the funnel; when this water is ejected with force it propels the animal backwards. *Nautilus* alters course by bending its funnel.

Shell

The shell consists of aragonite with an organic substance (conchiolin). There are two layers, an outer opaque layer with minute grains of aragonite in an organic matrix, and an inner layer composed of nacre (p. 29).

In its simplest form the nautiloid shell is a cone, closed at the apical end and open at the other end, the APERTURE. This form of shell is described as an ORTHOCONE (fig. 42g). More generally the shell is either curved (fig. 43), or is coiled planispirally, with the ventral side, the VENTER, forming the circumference. As the angle of coiling increases, successive complete (360°) coils, or WHORLS, come into contact. Since the shell widens with growth, the last whorl encloses a depression, the UMBILICUS, on each

43 A nautiloid with loosely coiled shell: *Halloceras*, **M Devonian, Rhineland** (× 0.6).

side of the shell, centred on the axis of coiling (fig. 42b). The umbilicus is wide in a loosely coiled, or EVOLUTE shell. In a tightly coiled, INVOLUTE shell, each whorl embraces the previous one closely so that the umbilicus is small.

The chambered part of the shell is called the PHRAGMOCONE. In *Nautilus* the chambers contain a gas, mainly nitrogen, at almost atmospheric pressure; later chambers contain a little fluid. The gas in the chambers increases the buoyancy of the animal relative to sea water. The septa are concave forwards (fig. 42a), and they unite with the shell wall so that their edges are seen only when the latter is removed. In fossils the septal edges (seen only on internal moulds) form lines, the SUTURE LINES, which are straight in orthocones (fig. 42d, g) and gently undulating in coiled shells (fig. 42b, c); (contrast with ammonite suture lines, p. 78). Each septum is pierced in the centre by an opening, the FORAMEN, through which the siphuncle passes; a short tubular extension of the septum, the SEPTAL NECK, encircles the siphuncle on the posterior side (fig. 42f). The last chamber, the body chamber, is relatively large in *Nautilus*, occupying about one-third of a whorl. The inner surface is smooth apart from some broad muscle scars. The APERTURE varies in shape; it may be round or oval, but sometimes is restricted by inward growth of the margin. In coiled shells there is a re-entrant on the inner (dorsal) side of the aperture due to overlap on the previous whorl; this re-entrant is deeper in involute shells. Most nautiloids have an embayment, the HYPONOMIC sinus, on the ventral margin of the aperture to accommodate the funnel (fig. 42g).

Shell form and mode of life of nautiloids

Nautiloids are abundant and highly varied as fossils, especially in Palaeozoic rocks. Speculation about their behaviour is based largely on a study of living *Nautilus*.

Nautilus (figs. 42a–c, 44).   Quite large, smooth, involute shell, about 15 cm or more in diameter; umbilicus small or hidden by callus; aperture oval with deep dorsal re-entrant; hyponomic sinus. Septa concave forwards with broadly folded suture lines; siphuncle subcentral; short, backwards-projecting septal neck.
Oligocene–Recent
Nautilus, initially widespread, is now confined to the warm coastal waters of Indo-Pacific islands with a depth range of 5 to 500 m (common from 150 to 300 m). The shell wall collapses under pressure at depths below 800 m. *Nautilus* swims over the substrate (nektobenthic) preying on small animals, e.g. crustaceans, which it catches with its tentacles and cuts up with its horny jaws. The shell, well camouflaged with cream and brown markings, is almost invisible in water. It floats in a quite stable position with the centre of gravity lying vertically below the centre of buoyancy from which it is well separated (fig. 42b). More or less neutral buoyancy is maintained by adjusting the amount of gas in the chambers. During growth the body chamber is enlarged and the body moved forwards leaving a space between it and the previous septum which is filled by fluid. A new septum is now secreted by the posterior end of the mantle. Fluid is then pumped out by the siphuncle and replaced by gas which diffuses in. Buoyancy is adjusted by the distribution and ratio of gas to fluid in the chambers, the amount of gas increasing progressively in earlier chambers. In this way a distribution of weight is maintained which ensures stability of the shell. *Nautilus* swims by ejecting jets of water via the funnel. Its normal speed is leisurely, either backwards or forwards, but it can move quickly in short bursts, useful as an escape mechanism. Compared with the squid, for example, it is a poor swimmer because of the drag from the

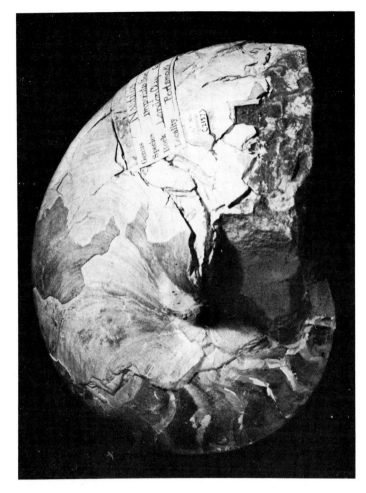

44 *'Nautilus' imperialis*, **London Clay, Eocene (× 1).**

large buoyant shell. As each jet is emitted the shell rotates forwards and upwards through a small angle, then swings back like a pendulum before the next jet occurs. Its horizontal motion is thus accompanied by a rocking action through an angle of about 20°.

Orthoceras (fig. 42d–g). Orthocone with adult part almost cylindrical; cross-section circular; septa concave forwards; siphuncle subcentral; suture straight; hyponomic sinus.
L Ordovician–U Trias
Orthoceras may reach several metres in length and the larger shells must have been very vulnerable to damage in shallow turbulent water; certainly, fossils are often incomplete. It is probable that *Orthoceras* and similar forms swam in typical cephalopod fashion by jet propulsion; this is implied by their having a hyponomic sinus. Because of its buoyancy, the shell must have tended to float

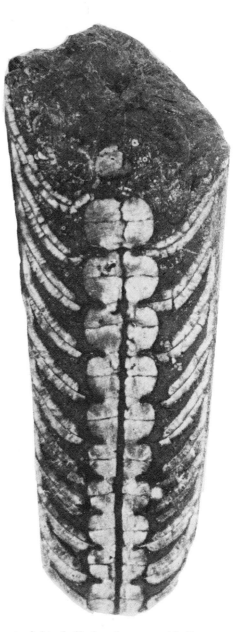

45   Section of a straight-shelled actinoceratoid: *Rayonnoceras*, L Carboniferous (× 0.9).

The section shows 'string-of-beads' swelling of the siphuncle formed by calcareous deposits between septa. There is further calcareous tissue on the septa. It is estimated that individuals of this form may have reached a length of 6 m.

in a vertical position with the aperture facing downwards, but this is an impractical attitude. Instead, the shell was probably held horizontally, an attitude achieved (by analogy with *Nautilus*) by the presence of fluid in the chambers to adjust the buoyancy. Camouflage markings, preserved in rare cases, occur on only one side of the shell (the dorsal), and lend support to such an attitude. In some fossils, shelly deposits in the chambers provided ballast; and others overcame the problem by shedding the early part of the phragmocone in later life.

Further solutions to adjusting buoyancy are shown by other cephalopods, formerly classed with nautiloids but now referred to separate subclasses, the Endoceratoidea and the Actinoceratoidea.

Endoceratoids (L Ordovician–Silurian) include medium to very large (up to 9 m) orthocones and gently curved shells with a large siphuncle filled by a series of closely spaced cone-in-cone calcareous sheaths.

Actinoceratoids (L Ordovician–U Carboniferous) are typically ortho- cones with a large siphuncle swollen between the septa to give a string-of- beads effect, and also with shelly deposits in the chambers, fig. 45. Length up to 6m.

Geological history

U Cambrian–Recent
Nautiloids, the earliest cephalopods, appeared in late Cambrian times. They were abundant and diverse throughout the Palaeozoic but nearly became extinct at the end of the Trias. They made a modest recovery later in the Mesozoic but declined gradually during the Tertiary to leave only *Nautilus* (five species) today.

Early nautiloids were small with slightly curved shells, close-set septa and marginal siphuncles. They diversified markedly in the Ordovician ranging from orthocones to all grades of coiling up to evolute planispiral; and also short-coned forms with restricted slit-like apertures. Orthocones disap- peared at the end of the Trias leaving only planispiral forms which range from evolute to involute, showing a range in shape of aperture. The more involute forms have quite strongly zigzag suture lines. Fossil jaws of *Nautilus* type are found from mid-Trias to Recent.

Ammonoidea

Ammonoids are typically distinguished by a coiled, usually planispiral shell with folded and often complex suture lines; forwards projecting septal necks in the adult shell; a slender siphuncle lying close to the ventral margin; and an ovate embryonic shell (protoconch). The shell consists of three parts, the protoconch, the phragmocone and the body chamber; it is a record of successive growth stages.

Ammonoids, extinct since the end of the Cretaceous, have left few fossil clues as to the nature of the soft parts. Thus *Nautilus*, the sole surviving cephalopod with an external chambered shell, is an important point of reference. However, there are marked differences between the ammonoid and nautiloid shells; and probably, the same was true of the soft body.

Morphology

The ammonoid shell is made of aragonite; it resembles that of *Nautilus* in structure but is thinner.

Typically, the shell is planispirally coiled (figs. 46a, 49); less commonly the shell may be straight, curved or combine initial coiling with a distal straight or hooked section; a helicoid spiral shell is exceptional. Planispiral shells may be loosely coiled, EVOLUTE; or tightly coiled, INVOLUTE. The shape is also influenced by the form of the cross-section or whorl-section

46   Morphology of the ammonoids.
a–c, *Dactylioceras*: a, diagram of the shell in its presumed position of life (body chamber ornamented; ornament omitted on phragmocone; cross marks centre of buoyancy; filled dot marks centre of gravity); b, front view showing the aperture and shape of the whorl section; c, median section through the early whorls of a shell, much enlarged. d–f, ammonoid suture lines (L, lobe; S, saddle).

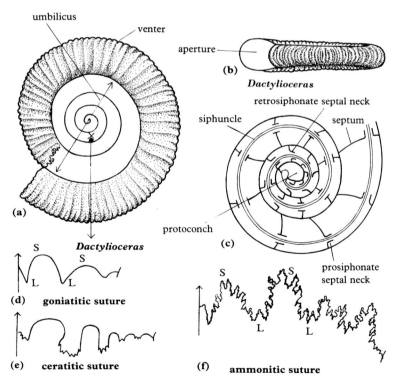

which may be round, oval, quadrate, compressed, triangular or depressed (fig. 47). The whorl section in involute shells is modified by the re-entrant on the dorsal side which is caused by the outer whorl impinging on the inner whorl (fig. 46b). Certain combinations of degree of coiling and shape of

47 An involute, highly globose ammonite with depressed whorl section: *Cadoceras*, Kellaways Rock, U Jurassic (× 1.1).
a, front view. b, lateral view.

whorl section are common, and have been given descriptive terms, e.g. oxycone, planulate, serpenticone, sphaerocone; these terms are defined in the 'technical terms' section (p. 98). The size range of shells is about 1 cm to 3 m in diameter.

The embryonic shell, or PROTOCONCH, is seen only in a median section of the shell. It is an ovate chamber cut off from the phragmocone by the first septum (fig. 46c).

The PHRAGMOCONE is divided into chambers by septa which typically are much more complex structures than in the nautiloids (fig. 46). In all ammonoids the septa are folded where they meet the shell wall, and the basic plan of their SUTURE (p. 71) is a zigzag line on which the forward pointing folds are called SADDLES and the backward folds are called LOBES (fig. 46d). The folding of the suture line is relatively shallow in primitive ammonoids and in the early septa of individuals, but it may be intensely crimped in the adults of the advanced ammonoids (fig. 48), a feature, developed, perhaps, to withstand pressure in deep water. Suture patterns and their development are of great value in taxonomy, but in practice their identification is a matter for the specialist. It is sufficient at this level to refer a suture line to one of three categories: (i) GONIATITIC suture (fig. 46d) in which both lobes and saddles are entire; typical of the goniatitids (Palaeozoic); (ii) CERATITIC suture (fig. 46e) in which the saddles are entire but the lobes are toothed; typical of the ceratitids (Permian and Trias); and (iii) AMMONITIC suture (fig. 46f) in which both lobes and saddles are intricately frilled; typical of the ammonitids (Permian and Mesozoic). Only a part of the suture is usually shown in diagrams, starting

48   Contrasts in ammonitic suture lines.
a, *Strigoceras*, M Jurassic, with a highly intricate suture line. b, *Hildoceras*, L Jurassic, with a relatively simple suture line. (Both × 0.7). The area between several suture lines has been painted to show their course more clearly.

from the venter (the position of which is indicated by an arrow pointing towards the aperture), and ending at the margin of the umbilicus.

The SIPHUNCLE is a slender tube which starts blindly in the centre of the first septum (fig. 46c). In most ammonoids it shifts outwards and lies on the margin immediately under the venter in later whorls; in the Devonian clymeniids, however, it lies on the dorsal, inner margin. Short septal necks encircle the siphuncle where it passes through the septa; the septal necks project forwards (PROSIPHONATE) in the later-formed septa of Mesozoic ammonitids (figs. 46c, 50) but in the earliest ammonoids (as also in the nautiloids) they project backwards (RETROSIPHONATE) (fig. 46c, 54e).

49 Section through an ammonite shell, approximately in the median plane.
This shows how the shell is preserved with rock matrix infilling later chambers, and crystalline calcite the earlier chambers. The latter contain cavities into which well-formed calcite crystals have grown. A part of this section is shown in fig. 50. *Parkinsonia*, Inferior Oolite, Jurassic; diameter of shell, 280 mm.

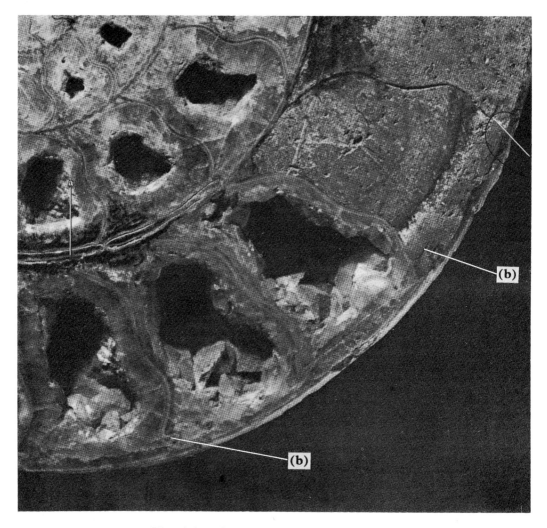

50   The siphuncle seen at a, and septal necks at b, in part of the section shown in fig. 49 (× 1).

The BODY CHAMBER varies in length; it may occupy only half a whorl in stout shells, or more than one whorl (360°) in slender shells. It is not always preserved; it lacks the support afforded the phragmocone by the 'fan-vaulting' of the septa and was, perhaps, the more easily broken.

The aperture is, in general, of similar shape to the whorl section; in some forms, however, it is constricted or modified by projections of shell. Lateral projections (one on each side) are LAPPETS (fig. 59), and a ventral projection is a ROSTRUM.

Where the body chamber is well preserved, parts of the jaws may be found inside it. The lower jaw in some forms is a horny plate made of conchiolin,

51 Aptychi, Kimmeridge Clay, U Jurassic (× 1.5).

the ANAPTYCHUS; in other forms it consists of a pair of calcite plates, the APTYCHI, lying side by side (fig. 51). The aptychi are thought to represent a calcareous thickening of the anaptychus. They usually separated from the shell on death and are well known as discrete and flattened fossils. In life the lower jaw lies in the body chamber, towards the aperture, associated with a smaller upper horny jaw and, in some cases, with the remains of the radula lying between them. Together they form jaws (fig. 52) comparable with those of, for instance, the octopus. The two aptychi may, where the fit in the aperture is good, have had a secondary function as an operculum or lid, an interpretation extended in the past to all these plates. Jaws have been found in a variety of ammonoids, anaptychi from Devonian to Cretaceous; aptychi appearing in the lower Jurassic. Those found in the Jurassic and Cretaceous may be more specialised, the lower jaw forming a longer scoop-like structure. The diversity now known suggests differences in feeding strategies. Evidence of diet comes from the contents of food crops, occasionally preserved, which include foraminifera, ostracods and bits of little ammonites, i.e. remains of small sedentary or slow-moving creatures. The rarity of jaws occurring *in situ* is a consequence of their not being articulated but operated by muscles which ordinarily would decay on death allowing the plates to drop out of the shell.

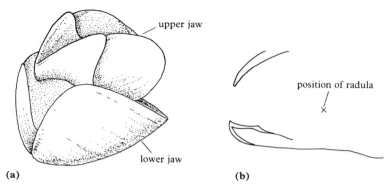

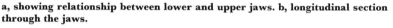

52 Model of ammonite jaws, anaptychus type (modified from Lehmann).
a, showing relationship between lower and upper jaws. b, longitudinal section through the jaws.

The radula is of a type (with seven teeth in each transverse row) found also in the octopus but differing from that of *Nautilus* (in which there are 13 teeth per row).

Ornament.   The shell may be smooth or, especially in Mesozoic forms, may be ornamented by fine lines (striae), ribs, tubercles or spines. Ornament may be transverse, spiral or both. It may be confined to the sides, or may also occur on the venter. Some forms may have a groove (SULCUS) (fig. 56b) on the venter, or a ridge (KEEL) (fig. 55b). The ornament may be a localised thickening of the shell, but more generally it is a folding of the shell so that it is seen also on internal moulds.

Dimorphism.   The sexes are separate and may be unlike in living ceph-alopods. This may also be the case with mature shells of certain ammonite species which occur in paired groups in the same bed, one of the pair being large, MACROCONCH, the other small, MICROCONCH. If this view is correct, such pairs actually represent a single species with distinct female and male shells. The macroconch is interpreted as the female on the grounds that the extra space was needed to contain eggs, and also by analogy with living coleoid cephalopods. In some cases lappets distinguish the aperture of the microconch (fig. 59).

Orientation.   Ammonoids are, with few exceptions, bilaterally sym-metrical; the plane of symmetry lies at right angles to the axis of coiling. The shell is usually figured in two positions, one showing the side view, and the other the view of the aperture. The periphery is the VENTRAL side.

Mode of life of ammonoids

The buoyant shell is a feature of all ammonoids, but in other respects they show much variety which suggests a corresponding variety in life-style. Evolute shells have potentially greater buoyancy and may have been less active swimmers than involute forms. The latter may be compressed and streamlined, not dissimilar to *Nautilus* and with the same jet-propelled locomotion hampered by drag from the buoyant shell (p. 72). There is no evidence to suggest that any ammonoid was a significantly better swimmer than *Nautilus*; some may even have been passive floaters (planktonic). Given an ability to adjust buoyancy and, therefore, position in the water column, food resources at any level in the sea could be exploited. Fragile hollow spines found on some forms may have been an aid to stability in floating. All forms would have been vulnerable to predators such as fish and marine reptiles, making camouflage highly desirable. Some well-preserved ammonoids retain original colouring which may have served this purpose (cf. *Nautilus*). Others have strongly developed ornament and sculpture in the form of ribs and tubercles which break up the outline of the shell as viewed by a predator. Highly compressed shells, viewed end-on, would be relatively inconspicuous.

In summary, it may be conjectured that the ammonoids ranged in habit from nektonic to planktonic, living at various levels in the seas, feeding on small organisms, and relying mainly on camouflage to evade predators.

Three examples

Oxynoticeras (fig. 53a, c). Shell involute, compressed, with small shallow umbilicus (oxycone); whorl section with sharply pointed venter and deep dorsal re-entrant; gently curved ribbing; ammonitic suture; centre of gravity well spaced from centre of buoyancy.
L Jurassic (Sinemurian)
In shape *Oxynoticeras* is not unlike *Nautilus* except that it is more compressed and streamlined with a sharp venter. Centres of buoyancy and gravity for the two are similarly placed, suggesting that *Oxynoticeras* could have swum like *Nautilus* but with less drag and therefore a little faster. Its slim end-on aspect would have been an aid to camouflage. A nektobenthic habit seems probable.

Dactylioceras (fig. 46a, b). Shell evolute with wide shallow umbilicus; whorl section oval (serpenticone); narrow, closely spaced ribs, most bifurcating on shoulder and extending across venter; ammonitic suture; lower jaw of anaptychus type, scoop-like, with radula (fig. 52); body chamber long,

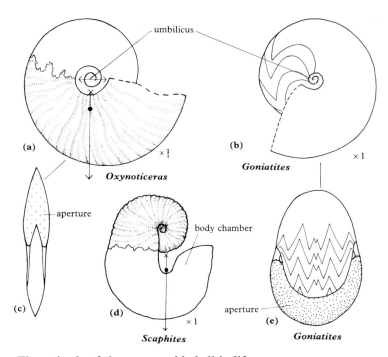

53   The attitude of the ammonoid shell in life:
a, c, *Oxynoticeras*: a, an oxycone shell in its presumed attitude in life (body chamber ornamented); c, front view showing the shape of the whorl section. b, e, *Goniatites*: b, lateral view with part of the shell removed to show the suture lines: e, front view showing the depressed whorl section. d, *Scaphites* in its presumed attitude in life.

about one whorl (360°); centres of gravity and buoyancy almost coincide. L Jurassic (Pliensbachian–Toarcian)

The large, very buoyant, evolute shell, and near coincidence of centres of gravity and buoyancy, suggest that *Dactylioceras* must have been a very poor swimmer with considerable rotation of the shell at each jet pulse. Perhaps it used some other mechanism, e.g. manipulation of the tentacles, to minimise the tendency to spin. Alternatively, if it was feeding within a cloud of small organisms a rotary motion may have been advantageous. Whatever its precise life-style may have been, it was clearly very successful since the genus had world-wide distribution during a time span of several million years.

Scaphites (fig. 53d).   Shell moderately inflated; early whorls involute and tightly coiled; body chamber in adult a short straight shaft bent into a hook with aperture facing the early whorls; ribbed, with tubercles in some; suture ammonitic; jaws of aptychus type; centre of gravity well-spaced from centre of buoyancy.
Cretaceous

The change in shape of the shell with growth indicates that *Scaphites* may have altered its mode of life when mature and ready for breeding, possibly collecting in shoals for that purpose as do some living cephalopods. In both its phases the centres of gravity and buoyancy were well-spaced giving the shell stability. The attitude of the mature shell, suspended in the water with the aperture facing upwards, is consistent with passive floating (planktonic) rather than active swimming. If, like various living cephalopods, it was capable of adjusting the shell buoyancy then it may have had a diurnal rhythm, moving up and down in the water, following the diurnal migration of the plankton on which it perhaps fed.

Scaphites is one of a diversity of ammonitids in which the shell, partly or wholly, was uncoiled or loosely coiled. They are known as HETEROMORPHS They occurred at various times but notably in the Cretaceous. Here they were common, some spread world-wide, and some were long-ranged. They represent very successful life-styles. They are, however, in marked contrast with the fully coiled ammonitids with which they co-existed and which for much of ammonoid history were the dominant designs.

Additional genera

Palaeozoic

Bactrites (fig. 54e).   Orthoconic with whorl section round to oval; suture slightly sinuous with small median ventral lobe; siphuncle ventral but septal necks project backwards (retrosiphonate).
?U Silurian, L Devonian–U Permian

54   Palaeozoic ammonoids.
a,b, *Gastrioceras*: a, lateral view; b, front view showing the shape of the whorl section. c, d, *Ceratites*: c, whorl section; d, lateral view. e, f, *Bactrites*: e, part of phragmocone, ventral view; f, shell with protoconch.

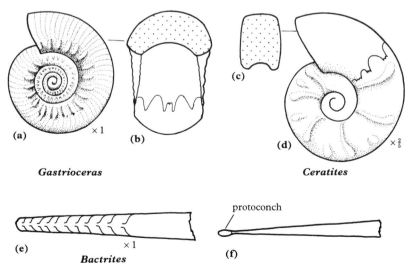

Gastrioceras

Ceratites

Bactrites

Goniatites (fig. 53b, e). Globose involute shell; umbilicus very small; whorl section depressed with rounded venter; ornament of faint striae; suture goniatitic.
L Carboniferous

Gastrioceras (figs. 4, 54a, b). Subglobular with moderately large open umbilicus; whorl section wide, depressed; fine ribs cross venter with stout tubercles on umbilical shoulder.
U Carboniferous

Goniatites and *Gastrioceras* are two of a number of goniatite genera which occur in marine shales in the Carboniferous and, being short-ranged in time, are valuable zonal forms. Another is *Reticuloceras* (fig. 57). They are relatively involute forms distinguished from each other mainly by their ornament.

Mesozoic

Ceratites (figs. 54, 58). Moderately evolute, with shallow umbilicus; whorl section relatively compressed; broad ribs with tubercles on shoulder; ceratitic suture.
M Trias

The remaining genera described here have an AMMONITIC suture.

Promicroceras (fig. 55f). Small, evolute with wide umbilicus; whorl section rounded; strong radial ribs cross venter.
L Jurassic (Sinemurian)

Amaltheus (fig. 55c). Compressed, moderately involute with shallow umbilicus (oxycone); whorl section with keeled venter; ribs bend forwards on shoulder and cross venter giving a 'corded' effect.
L Jurassic (Pliensbachian)

Hildoceras (fig. 55b) Relatively evolute with compressed quadrate whorl section; venter with keel, defined on each side by sulcus; sickle-shaped ribs on flanks.
L Jurassic (Toarcian)

Parkinsonia (fig. 55e). Evolute, compressed (planulate); sharp ribs bifurcate towards venter; ribs interrupted on venter by sulcus.
M Jurassic (Bajocian)

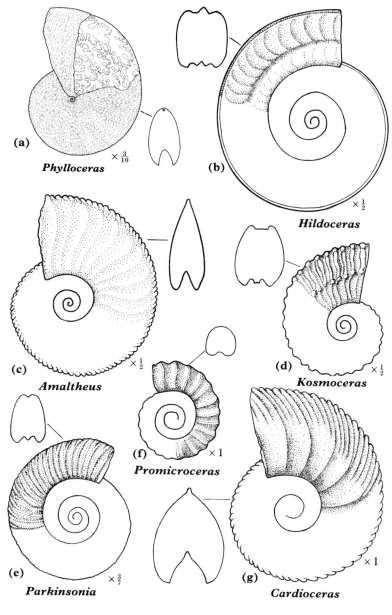

(a) × $\frac{3}{10}$ *Phylloceras*

(b) × $\frac{1}{2}$ *Hildoceras*

(c) × $\frac{1}{2}$ *Amaltheus*

(d) × $\frac{1}{2}$ *Kosmoceras*

(f) × 1 *Promicroceras*

(e) × $\frac{3}{7}$ *Parkinsonia*

(g) × 1 *Cardioceras*

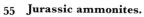

55 **Jurassic ammonites.**

Kosmoceras (fig. 55d). Moderately involute, compressed with flattened venter; strong ribs bifurcate with tubercles where ribs split and on margins of venter; shows dimorphism with lappets on microconchs (fig. 59).
M Jurassic (Callovian)

Cardioceras (fig. 55g). Moderately involute, compressed, with keeled venter; strong curved ribs with shorter ribs intercalated, extend across venter; keel corded.
U Jurassic (Oxfordian)

Euhoplites (fig. 56b). Moderately evolute, compressed with deep sulcus on venter; prominent ribs rising from tubercles on umbilical margin and curving forwards to stronger tubercles on shoulders.
L Cretaceous (Albian)

Schloenbachia (fig. 56a). Moderately involute, compressed with keeled venter; ribs bifurcating, slightly curved forwards with tubercles on umbilical margin and on shoulders.
L–U Cretaceous (U Albian–Cenomanian)

Hamites (fig. 56d). Heteromorph; early part coiled, later part bent to form three subparallel shafts; whorl section oval; strong closely spaced ribs.
L Cretaceous (Albian)

Turrilites (fig. 56c). Coiled in helical spiral; ribs and some tubercles.
U Cretaceous (Cenomanian)

Baculites (fig. 56e). Heteromorph; shell straight except for tiny initial coiled part (rarely seen); section oval with venter sharper.
U Cretaceous (Turonian–Maastrichtian)

Ammonoids and biostratigraphy

The details of ammonoid morphology changed quickly with time and individual species were relatively short-lived, a few hundred thousand years in some cases. They were also widely distributed because of their nektonic or planktonic habit, and their fossils are found in a wide variety of rock types. Some were truly cosmopolitan though others were more restricted in distribution, e.g. mainly in tropical and subtropical waters, or in the cooler seas of higher latitudes. Their numbers were great, so their fossils are common; they are, in short, extremely valuable as zonal fossils.

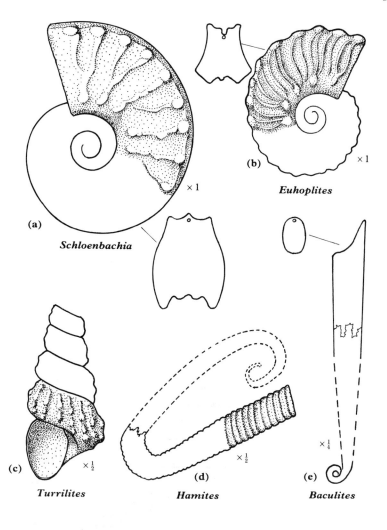

(a) *Schloenbachia* ×1

(b) *Euhoplites* ×1

(c) *Turrilites* ×½

(d) *Hamites* ×½

(e) *Baculites* ×¼

56 Cretaceous ammonites.

Geological history

L Devonian–end Cretaceous

A few small ammonoids, appearing towards the end of the lower Devonian, began a series of striking radiations interrupted by episodes of multiple extinctions (fig. 235). The first of the latter occurred in the late Devonian, others in late Permian and late Trias, while the final extinction of the group took place at the end of the Cretaceous. The major radiations were of goniatitids and clymeniids (later Devonian); new families of goniatitids (Carboniferous–Permian); ceratitids (Trias); and ammonitids (Jurassic–Cretaceous).

The basic ammonoid stock, the anarcestids, which appeared in the later part of the lower Devonian, were small forms ranging from orthocones, e.g. *Bactrites*, through curved shells to tightly coiled forms. *Bactrites* (L Devonian–U Permian) is unlike the orthocerids from which the ammonoids are thought to have arisen in, for instance, having a ventral siphuncle and gently sinuous suture with small ventral lobe. During the Devonian there was a marked diversification of shell form, and increased folding of the suture line. Most Devonian forms were goniatitids (M Devonian–U Permian) which were derived from the anarcestids and became the major Palaeozoic ammonoids. In late Devonian times they were joined by a small but important group, the clymeniids, distinguished by having the siphuncle

57 Goniatite from marine shales in the Millstone Grit facies, U Carboniferous: *Reticuloceras* (× 8).

placed dorsally. They were widespread but short-lived, becoming extinct in the late Devonian, a time when ammonoid diversity was at a low point.

Many new families appeared in a new radiation in the early Carboniferous dominated by the goniatitids which comprise the great majority of Carboniferous and Permian ammonoids. (The name 'goniatite' refers to the characteristic angular folding of the suture.) Carboniferous forms were typically small, more-or-less involute and smooth shelled, a strong ornament being unusual (fig. 57). The goniatitids remained important in the Permian but here changes in the suture line, begun in the Carboniferous, are found, with the addition of extra lobes and, in some, ammonitic frilling. However, diversity declined markedly in the late Permian, most of the Palaeozoic families becoming extinct. This decline was followed by an

58 Triassic ammonite showing ceratitic suture: *Ceratites*, M Trias (× 1.3).

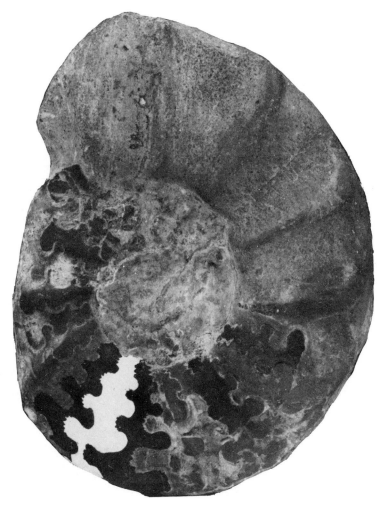

equally marked increase during the Trias in the form of a dramatic radiation of the ceratitids (M Permian–U Trias, fig. 58). Their origin lies in an early offshoot from goniatitid stock, the prolecanitids (L Carboniferous–L Trias). In these the suture line developed extra lobes and, in some, one or more lobes were toothed. The ceratitids showed great diversity of shell form, ornament and suture, the latter reaching a pinnacle of intricate frilling in some forms. During the relatively short time span of the Trias they were numerically abundant and generally widespread in North America and Europe (though not found in Britain where fully marine Triassic rocks do not occur).

The late Trias saw yet another striking fall in diversity, only one small

59 Jurassic ammonite with lappet: *Kosmoceras*, Oxford Clay, U Jurassic (× 1.7).

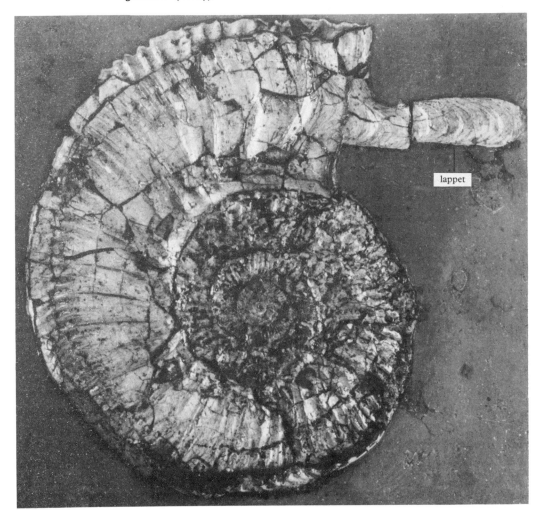

group, the phylloceratids (M Trias–U Cretaceous), surviving into the Jurassic. *Phylloceras* (fig. 55a) is a member of this group. They were an offshoot from early ceratitids and, in the Jurassic, were joined by two new groups, the lytoceratids (Jurassic–Cretaceous) and ammonitids (Jurassic–Cretaceous) whose origin is uncertain. In all these groups the suture is ammonitic. The phylloceratids (involute forms) and lytoceratids (evolute forms) were stable long-ranged groups which were widespread in deeper waters, e.g. in the Tethys region. They occur only rarely in Britain.

The ammonitids were the major group of the Jurassic (fig. 55) and Cretaceous; they represent the last great evolutionary radiation of the ammonoids which reached a peak at the beginning of the late Cretaceous. Some were widespread, others geographically confined but regionally important for correlation purposes. The planispiral forms characteristic of the Jurassic were joined in the Cretaceous by an assortment of hetero-morphs. Towards the end of the Cretaceous diversity declined sharply and only a few widespread genera are found in the uppermost beds, these finally disappearing in the end-Cretaceous extinctions.

Coleoidea–Order Belemnitida

Belemnites are an extinct order of the Subclass Coleoidea which includes living cephalopods other than *Nautilus*, and one other order not dealt with here (Aulococeratida, Carboniferous–Jurassic). In contrast with *Nautilus* and the ammonoids, the coleoid shell is internal and, in living forms, is modified or rudimentary. The fossil record of living forms, i.e. squids (Teuthida, fig. 60g), cuttlefish (Sepiida, fig. 60f) and octopuses (Octo-podida) is meagre. Belemnites, however, with more substantial shell, are common fossils especially in the Jurassic and Cretaceous.

Belemnite morphology

The belemnite skeleton consists of a PHRAGMOCONE which was partly contained in a cavity in one end of a solid calcitic GUARD.

The guard is normally the only part preserved of the belemnite shell. It is a cigar-shaped solid structure, tapered at the posterior end, and with a deep conical cavity, the ALVEOLUS, at its anterior end (fig. 60b, e). In most genera the guard is not a large structure, falling within a range of 2–20 cm in length. In transverse section the guard shows a fibrous structure with tiny calcite prisms radiating from an eccentric point (fig. 60c); rings concentric about this point denote growth stages.

The phragmocone (figs. 60a, 61) is a conical chambered shell with thin delicate walls and is less commonly preserved. The posterior end, with the oval protoconch at its tip, lies within the alveolus of the guard (fig. 60a). There is no body chamber, and the anterior end is drawn out dorsally as a

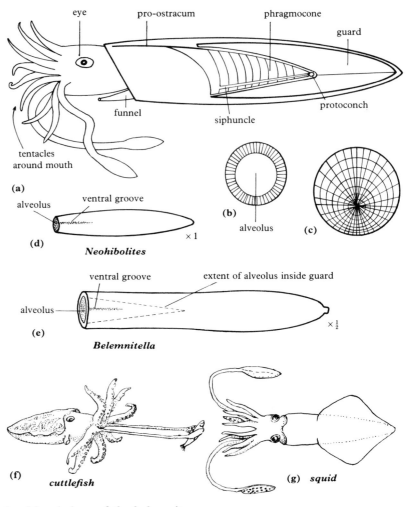

60 Morphology of the belemnites.
a, simplified section of the skeleton of a belemnite showing a possible reconstruction of soft parts. b, c, transverse sections across the guard in the region of the alveolus (b), and below the alveolus (c). d, e, examples of belemnites. f, *Sepia*. g, squid.

fragile horny process, the PRO-OSTRACUM. The septa are saucer-like with unfolded edges, and the septal necks project backwards. The siphuncle lies near the ventral margin of the phragmocone (fig. 60a).

An example

Neohibolites (fig. 60d, 61b). Small guard, length about 2–8 cm; more or less spindle-shaped, tapering to the pointed apical (posterior) end; alveolus not usually preserved but when present has a short ventral groove.
L–U Cretaceous (Aptian–Cenomanian)

61 Belemnites with the phragmocone preserved.
a, part of a guard, fractured lengthwise along the median plane, with part of the uncrushed phragmocone lying *in situ* in the alveolus; *Belemnites oweni*, Oxford Clay, Jurassic (× 2). b, guard with anterior part of phragmocone (crushed); *Neohibolites minimus*, Gault Clay, L Cretaceous (× 2).

Mode of life of belemnites

Belemnites are interpreted with reference to modern coleoids like cuttlefish and squids. These are streamlined, predatory forms, with ten arms and lateral fins. They swim with the body held horizontally, changing direction

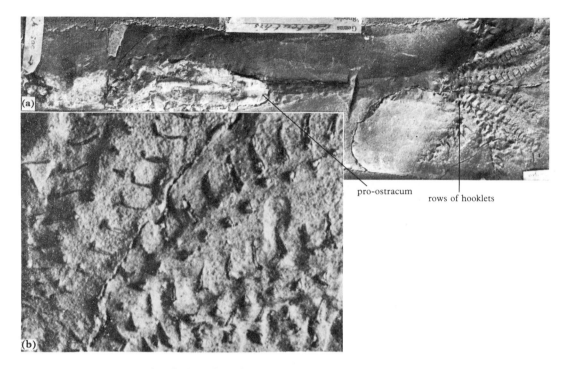

pro-ostracum   rows of hooklets

62 Belemnite with traces of arms.
a, the paired rows of hooklets indicate the traces of the arms (× 0.7). b, hooklets
(× 1.3). 'Belemnoteuthis', Lias, Jurassic, Lyme Regis, Dorset.

by varying the attitude of the funnel. They possess an ink-sac containing a pigment (sepia). In well-lit waters, the pigment is ejected via the funnel to form a dark cloud which hides them from predators. Cuttlefish are benthic forms able to hover or swim at leisurely speed, buoyancy being provided by the cuttle 'bone'. This is a modified phragmocone, a long oval and flattened structure with obliquely overlapping septa, which has a tiny remnant of guard at its posterior tip. Squids, torpedo-shaped with stabilising fins, are fast-moving, varied, and live in open waters, each species with its preferred depth. They lack the phragmocone, their internal support consisting of a horny pro-ostracum ('pen').

Rarely, fossil belemnites may show traces of the soft parts including triangular fins, arms with double rows of hooks (fig. 62), and an ink-sac. Presumably they swam rather like squids with the body held horizontally. The distribution of buoyant phragmocone in the middle, counterbalanced fore and aft by the denser body and guard, suggests this posture. Belemnite fossils show little variation in form and perhaps their life-style, too, was quite uniform. It is probable that they preyed on small animals, e.g. fish, and in turn were preyed on by larger animals, e.g. ichthyosaurs.

Geological history

L Carboniferous–Eocene

Belemnites are occasional fossils until the Mesozoic at which time they became abundant. By the end of the Cretaceous, however, they had almost disappeared and are last recorded from the Eocene.

Coleoids are considered an offshoot from forms similar to *Bactrites* (p. 85). Little is known of their early radiation apart from a fossil squid (*Eoteuthis*) recently found in the Hunsrück Shale (L Devonian) in West Germany. Traces of the soft parts preserved by pyritisation in this fossil indicate that squids have changed little since that time, implying an early diversification of coleoid stock. Belemnites are known from the later part of the lower Carboniferous (possibly also from the Devonian) but they were generally rare until the Mesozoic when they became abundant, especially in more northern waters. They also occur in the Tethyan belt and in the southern hemisphere. A very few forms continued into the Eocene.

A series of fossil cuttlefish from Tertiary rocks show stages in modification of guard and phragmocone in sepiids.

Technical terms

ALVEOLUS the cavity in the anterior end of the guard within which the phragmocone is located in belemnites (fig. 60b, e).

ANAPTYCHUS a single, V-shaped, horny plate sometimes found in the ammonoid body chamber and interpreted as the lower jaw (fig. 52).

APERTURE open end of the body chamber (fig. 46b).

APTYCHI (sing. aptychus) a pair of calcite plates interpreted as the calcified equivalent of the anaptychus (fig. 51).

BODY CHAMBER the last-formed chamber which contained the body.

CHAMBER (or camera) space enclosed by two adjacent septa (fig. 42a).

EVOLUTE term describing a loosely coiled shell in which the outer whorls enclose the inner whorls to only a small extent; evolute shells have a wide umbilicus (fig. 46a).

FORAMEN the hole in the septum through which the siphuncle passes (fig. 42e).

FUNNEL the muscular tube or swimming organ through which water is ejected from the mantle cavity (fig. 42a).

GUARD cigar-shaped structure enclosing the posterior end of the phragmocone in belemnites (fig. 60a).

HETEROMORPH refers to shells partly or wholly uncoiled or with helical coiling (figs. 53d, 56e).

HYPONOMIC SINUS curved re-entrant on the ventral margin of the aperture for the passage of the funnel (or hyponome) in nautiloids (fig. 42g).

INVOLUTE refers to a tightly coiled shell in which the inner whorls are largely concealed by the last whorl; the umbilicus is small in involute shells (fig. 53a).

KEEL a ridge on the venter (fig. 55b).

LAPPET a forward projection of the shell on each side of the aperture in some ammonites (fig. 59).

ORTHOCONE a 'straight' nautiloid shell (fig. 42g).

OXYCONE an involute shell with a sharp venter (fig. 53a, c).

PHRAGMOCONE the chambered part of the shell (fig. 46a).

PLANULATE describes an evolute shell with a more or less oval whorl section (fig. 55e).

PRO-OSTRACUM a dorsal projection of the phragmocone in belemnites (fig. 60a).

PROTOCONCH the initial part of the shell secreted by the embryo (fig. 46c).

ROSTRUM a forward projection of shell on the ventral side of the aperture in some ammonites.

SEPTA (sing. septum) transverse partitions which divide the phragmocone into chambers (figs. 42d, 46c).

SEPTAL NECK a funnel-like projection of the septum around the siphuncle (fig. 42a, f).

SERPENTICONE describes an evolute shell with many slender whorls which barely overlap (fig. 46a, b).

SIPHUNCLE a long slender tube extending from the body chamber back to the protoconch (figs. 42a, 46c).

SPHAEROCONE an almost spherical, very involute shell with a very small umbilicus and a depressed whorl section (fig. 53b, e).

SULCUS a groove on the venter (fig. 56b).

SUTURE LINE the trace of the edge of a septum exposed by removal of the shell wall; it is seen only in internal moulds (fig. 48). The suture line may be simple or folded; a fold convex towards the aperture is called a saddle, and a fold concave towards the aperture is a lobe.

TYPES OF SUTURE LINE (i) SIMPLE a straight line when projected on paper, e.g. *Orthoceras* (fig. 42d); (ii) GONIATITIC the suture line is bent into strong lobes and saddles, e.g. *Goniatites* (fig. 46d); (iii) CERATITIC the saddles are rounded and the lobes are toothed, e.g. *Ceratites* (fig. 46e); (iv) AMMONITIC both saddles and lobes are strongly frilled, e.g. most Mesozoic ammonites (figs. 46f, 48).

UMBILICUS the depression enclosed by the last complete whorl on each side of a planispiral shell (fig. 46a).

VENTER the ventral wall of a coiled shell, usually on the outside (fig. 42a).

WHORL one complete coil of a shell (360°).

5

Cnidaria

The Cnidaria (formerly called Coelenterata) include the simplest of the many-celled animals (Metazoa) in which definite tissues are developed. Examples include the jellyfish and sea-anemones which are soft bodied, and the reef-building corals which secrete an external calcareous skeleton; there are also a variety of extinct corals.

Cnidarians are exclusively aquatic and the majority are marine. They may be sessile (sea-anemone) or free-swimming (jellyfish). The sessile form, or POLYP, may occur as a single individual (SOLITARY) or be united with others to form a colony (COMPOUND). Basically it is sac-shaped and is attached at its base to the sea-floor with its one opening, the mouth, at its upper free end. The free-swimming form, or MEDUSA, resembles an umbrella with tentacles hanging down round the margin. Some cnidarians, like corals, exist only as polyps, while in others the life history involves an alternation of polyp and medusoid generations, the one giving rise to the other.

The cnidarian body is of simple structure consisting of an outer layer, ECTODERM, and an inner layer, ENDODERM, which in more-advanced members like the corals are separated by a jelly-like supportive substance, MESOGLOEA. The body wall encloses a central cavity, the gut or ENTERON, and this may be divided by radial partitions, the MESENTERIES, which aid in digestion and absorption of food. The mouth serves for both intake of food, and discharge of waste and larvae. It is surrounded by retractile tentacles which are armed with NEMATOCYSTS, stinging organs almost confined to cnidarians. There is no blood system, and the nervous system is a diffuse network of cells. Reproduction is sexual or asexual. Typically the body shows radial symmetry, the parts of the body being repeated around the mouth, but some forms also show bilateral symmetry. Cnidarians are common in warm shallow seas, but some forms live at depths down to 6000 m, and in temperatures as low as 1 °C. *238247*

Fossil cnidarians, first found in the late Precambrian (Vendian), are rare until the Ordovician when lime-secreting corals appeared. They are assigned to three classes; Hydrozoa (hydroids); Scyphozoa (jellyfish); and

Anthozoa (anemones and corals). Only the anthozoan corals are common as fossils and need be considered in detail here.

Class Anthozoa (corals)

Anthozoans exist only as polyps, either solitary or as colonies. The enteron is divided by six, eight or more mesenteries which may be paired. Bilateral symmetry is shown by the slit-like mouth which leads by a constricted gullet to the enteron. They are all marine.

Anthozoans with a fossil record include the following subclasses:

Rugosa
Tabulata
Zoantharia–Order Scleractinia (Hexacorallia)
Octocorallia

Rugosa

The rugose corals are solitary or compound Palaeozoic corals with a calcareous skeleton, the CORALLUM, divided by vertical radial partitions, the SEPTA. These include six main, PRIMARY septa together with a number of later-formed septa which, during growth, were inserted at *four* points in the corallum.

Rugose corals are common fossils in Britain, especially in lower Carboniferous limestones where they are of considerable stratigraphic value. Their structure is relatively simple and can be examined in detail either by grinding down the specimen with an abrasive, or by cutting through it at various points with a rock-saw.

Morphology

The corallum is bounded on the outside by a wall which may be transversely wrinkled, a feature which gives rise to the term 'rugose'. Its upper surface is a cup-shaped hollow, the CALICE (figs. 64, 65g), which by analogy with modern corals is the surface on which the rugose polyp 'sat'. The skeleton consists for the most part of vertical elements, the SEPTA; smaller convex plates, the DISSEPIMENTS, which lie between the septa; and transverse elements, the TABULAE, which are flattish plates. In solitary rugose corals the corallum is basically conical in shape (the tip being the first-formed part) and may be straight or curved (horn-shaped, fig. 63c); the later-formed part may be cylindrical (fig. 63g). In compound forms the corallum is built up of a number of individuals, each of which is referred to as a CORALLITE, and is bounded by a thin wall. The corallites may remain free, FASCICULATE (fig. 63e, f) or may be in contact, MASSIVE (fig. 63d). A fasciculate corallum

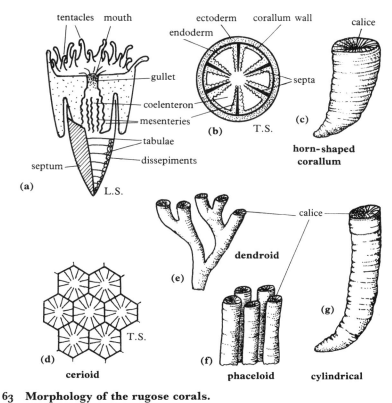

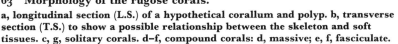

63 Morphology of the rugose corals.
a, longitudinal section (L.S.) of a hypothetical corallum and polyp. b, transverse
section (T.S.) to show a possible relationship between the skeleton and soft
tissues. c, g, solitary corals. d–f, compound corals: d, massive; e, f, fasciculate.

is further described as DENDROID if the corallites branch irregularly
(fig. 63e), and as PHACELOID if the corallites are more or less parallel
(fig. 63f). A massive corallum is described as CERIOID if the corallites are
polygonal in shape (transverse section) and are united by their walls
(fig. 63d), or as ASTRAEOID if the corallite walls are lacking (fig. 68d).

The calice (figs. 64, 65g) may be a shallow or relatively deep depression;
its centre is the AXIAL REGION. In a small number of forms it could be closed
by a lid or OPERCULUM (fig. 68c). The septa project from the floor of the
calice, and presumably they extended between pairs of mesenteries (p. 99) in
the living polyp.

The growth of the septa can be traced by cutting the corallite transversely
at intervals along its length. The first-formed are six PRIMARY septa; these
are named as the CARDINAL, COUNTER, two ALAR and two COUNTER-
LATERAL septa (fig. 65a). Subsequent septa are added at four points,
i.e. on each side of the cardinal septum, and on the counter side of the
alar septa (fig. 65b). In some Rugosa there are relatively conspicuous spaces,
FOSSULAE, at the points where new septa are inserted. The most obvious of

64   Calice of a solitary rugose coral.

these is the CARDINAL FOSSULA which is readily identified by the (usually)short cardinal septum which lies in it (fig. 65b, c). Forms in which the primary septa can be distinguished show bilateral symmetry, but in many corals, especially compound forms, the septa are very numerous in the adult stages and develop a dominantly radial symmetry with alternating longer, MAJOR, and shorter, MINOR septa (fig. 65d).

TABULAE (fig. 65h) are flat, or gently dished, or domed plates which represent the floors of successive calices. They were secreted at the base of the polyp during upward growth to seal off the vacated lower part of the corallum. Tabulae are most clearly seen in a longitudinal section, especially in corals in which the septa are short. In corals with fossulae, the tabulae are folded into trough-shaped hollows which slope down to the walls in the vicinity of the fossulae. Forms with minor septa may have a peripheral zone in which there are small arched plates, DISSEPIMENTS instead of tabulae (fig. 65d, e).

While the axial region may remain a clear space, in many forms there is some sort of axial structure. This may take the form of a vertical rod, the COLUMELLA (fig. 65d, e), or a more complex structure (fig. 68e) consisting of plates, some radial, and others concentric about the axis of the coral.

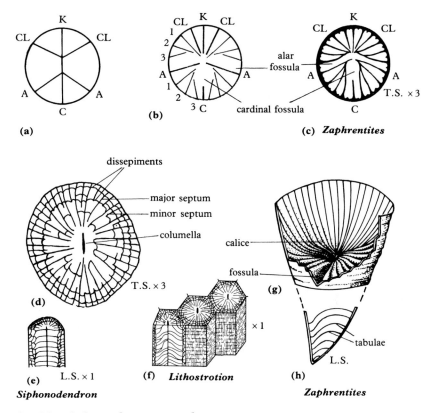

(a)

(b)

(c) *Zaphrentites*

dissepiments

major septum

minor septum

columella

calice

fossula

(g)

(d)   T.S. × 3

(e)   L.S. × 1   (f) *Lithostrotion*   (h)

Siphonodendron   *Zaphrentites*

65   Morphology of rugose corals.

**a, b, transverse sections of a solitary rugose coral, not to scale: a, juvenile stage
with the six primary septa (C, cardinal septum, A, alar septum, K, counter
septum, CL, counter-lateral septum); b, adult stage with later-formed septa
added. c, g, h, *Zaphrentites*: c, transverse section; g, upper part of a corallum,
partly broken to show the deep calice; h, lower part of a corallum sectioned to
show the tabulae. d, e, *Siphonodendron*, one corallite of a phaceloid form; f,
Lithostrotion, cerioid form.**

Increase of corallites

The number of individuals in a compound coral is increased by a vegetative
process, either by splitting (FISSION) of a parent polyp or by the growth of
buds (budding) from the parent polyp. In the latter case the new polyps may
grow out from the side of the parent, or from within the calice, as in fig. 66,
where four corallites have developed in the axial region of the calice, a
process known as axial increase.

Skeletal form in two common rugose corals

Zaphrentites (fig. 65c, g, h).   Solitary; horn-shaped with deep calice and the
cardinal fossula on the concave side; marked bilateral symmetry, the
cardinal and countercardinal septa lying in the plane of symmetry; the other

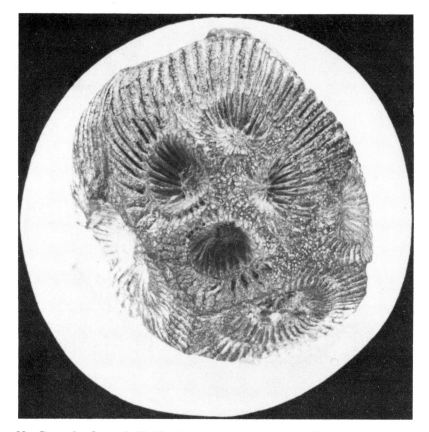

66 Growth of new individuals in a compound coral. Four corallites have developed within the axial region of the topmost calice. *Acervularia*, Wenlock Limestone, Silurian (× 4).

septa slope towards this plane, their inner ends often fusing in the axial region; cardinal fossula a distinct elongated space bounded laterally by major septa, and the short cardinal septum projects into it; usually no minor septa and no dissepiments; tabulae rather irregular and inversely conical in shape.
M. Devonian–U Carboniferous

Lithostrotion (figs. 65f, 67). Compound, massive, ceroid; thin radially arranged major and minor septa; cardinal and countercardinal septa may unite with a thin lath-like columella; wide zone of dissepiments; tabulae somewhat irregular and conical. In the related *Orionastraea* (L Carboniferous) the corallite walls are lacking and the septa of adjacent corallites or, sometimes, the dissepiments, are in contact. *Siphonodendron* (L–U Carboniferous, fig. 65d, e) is similar but fasciculate with cylindrical corallites.
L–U Carboniferous

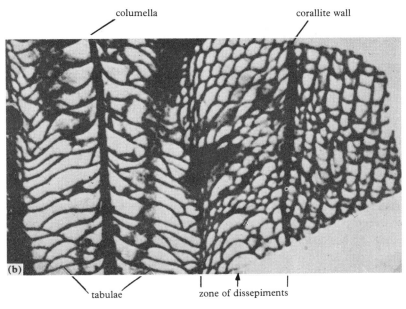

67 *Lithostrotion*, L Carboniferous, cerioid form.
a, a transverse section (× 3.5). b, a longitudinal section (× 5.5).

Mode of life of rugose corals

The clues to the ecology of rugose corals lie in the type of rock in which they occur, the associated fossils and, of course, in analogy with modern corals. Some Rugosa occur mainly in shallow-water limestones and calcareous shales. *Lithostrotion* and *Lonsdaleia* (fig. 68k), for instance, often form extensive sheets which can be traced for many miles in the Lower Carboniferous of the north of England. The enclosing rock is a light-coloured limestone in which brachiopods are next in abundance to corals; and it formed, perhaps, in shallow and relatively warm waters. Some of the small horn-shaped solitary corals like *Zaphrentites* occur in dark limestones and calcareous shales which were formed in deeper, quieter, conditions; they represent an environment from which the compound rugosa are conspicuously absent.

Additional common genera

Solitary Rugosa

Calceola (fig. 68c). 'Slipper-shaped', semicircular in transverse section; calice deep, closed in life by thick lid marked by striae on inner surface; septa form fine ridges on thickened wall.
L–M Devonian

Dokophyllum (fig. 68a).   Conical to top-shaped with deep calice; septa long, becoming discontinuous at higher levels; tabulae flat in axial region and replaced by dissepiments towards periphery; root-like outgrowths from base.
M–U Silurian

Caninia (fig. 68f–h).   Horn-shaped, becoming cylindrical, long; major septa initially long, later withdraw from axis becoming short; cardinal fossula small with short cardinal septum; minor septa very short, may be discontinuous; dissepiments irregular, arranged around periphery; tabulae flattish in axial region and turned down at margins.
L–U Carboniferous

Aulophyllum (figs. 68i, 69b).   Cylindrical with well-defined axial structure consisting of a haphazard arrangement of the radial and concentric plates; seen in transverse section the axial structure has a sharp projection towards the cardinal fossula; numerous major septa extend to axial structure; minor septa half as long; wide zone of numerous small dissepiments; tabulae irregularly domed.
L–U Carboniferous

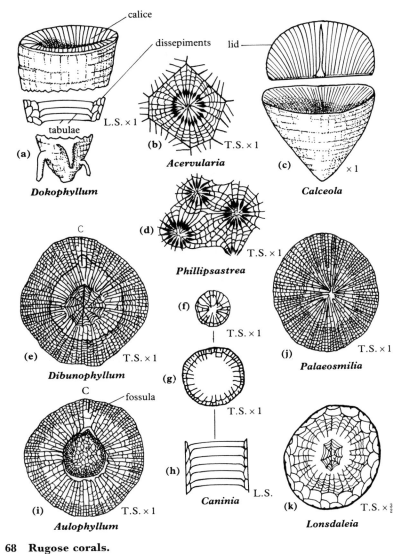

68 Rugose corals.

a, c, and e–j, solitary corals. b, d, k, compound corals (in b and k one individual only is shown).

Dibunophyllum (figs. 68e, 69c). Cylindrical, with axial structure showing spider's-web pattern of concentric and radially arranged plates; prominent median plate aligned towards cardinal fossula; numerous major septa extend to axial structure; minor septa irregular, short; wide zone of closely spaced dissepiments.
L–U Carboniferous

Palaeosmilia (fig. 68j) Cylindrical, long; numerous radial septa extend to axial region where their ends may fuse; minor septa extend half-way to

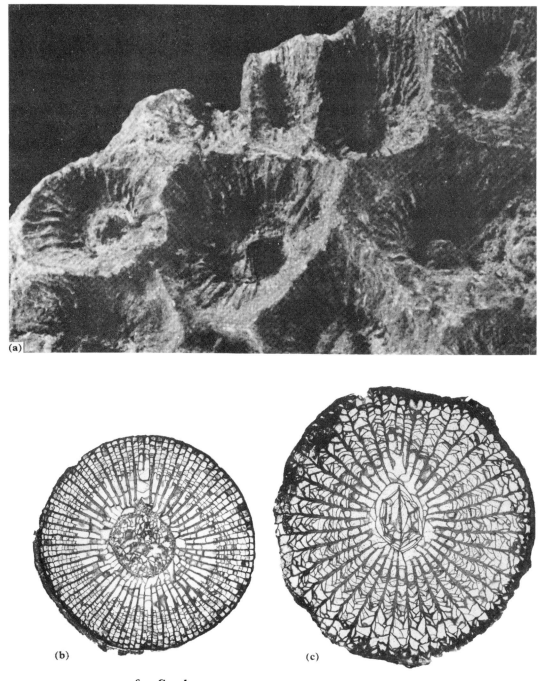

69 Corals.
a, *Actinocyathus*, upper surface (× 4). b, *Aulophyllum*, transverse section (× 3).
c, *Dibunophyllum*, transverse section (× 3). All from L Carboniferous.
(Specimens in Sedgwick Museum.)

centre; cardinal fossula narrow, not easy to distinguish; wide zone of abundant dissepiments; tabulae small, incomplete. *Palastraea* is similar but compound, cerioid.

L–U Carboniferous

Compound Rugosa

Acervularia (fig. 68b). Massive, cerioid; corallite wall hexagonal in transverse section, well-defined inner circular wall formed by thickening of septa; longer septa extend to centre where their ends may fuse; peripheral zone of 'bubbly' dissepiments; incomplete tabulae; small domed plates in axial region.

M–U Silurian

Phillipsastrea (fig. 68d). Similar to *Acervularia* but astraeoid; inner wall encloses smaller area.

M–U Devonian

Lonsdaleia (fig. 68k). Fasciculate, phaceloid; axial structure similar to that of *Dibunophyllum* (p. 107) with median plate; major and minor septa confined to space between axial structure and peripheral zone of large dissepiments, tabulae flat or gently declined; *Actinocyathus* (L Carboniferous, fig. 69a) is similar but cerioid.

L–U Carboniferous

Geological history

M Ordovician–U Permian

The Rugosa appeared in mid-Ordovician times and by the Silurian were common and widespread. They were, however, subordinate to the tabulate corals (p. 110) and stromatoporoids (p. 229) in the clear-water calcareous facies, e.g. the Wenlock Limestone (M Silurian). Here forms like *Dokophyllum* (fig. 68a) and *Acervularia* (fig. 68b) are common examples. There was a further expansion in numbers and diversity in the Devonian when new families appeared including both solitary (e.g. *Calceola*, fig. 68c) and compound (e.g. *Phillipsastrea*, fig. 68d) forms. In the later Devonian numbers plummetted and many families became extinct. More disappeared at the close of the Devonian, few surviving into the Carboniferous. There was, however, a dramatic recovery in the lower Carboniferous with the arrival of new families, some showing new features, e.g. complex axial structures, not seen in earlier forms. During this time rugose corals were at their peak of abundance. Compound forms such as *Lithostrotion*, *Lonsdaleia* and their relatives occur extensively at some levels in the later part of the lower Carboniferous.

Rugose corals are not found after the lower Carboniferous in Britain, suitable marine rocks being absent. Elsewhere, for instance in Europe, Asia, North America and Australia, they persisted, though with reduced diversity, until the end of the Permian.

Tabulata

The Tabulata are extinct compound corals in which the corallum is built up of slender corallites partitioned transversely by many tabulae.

Morphology

The corallum is calcareous. It is usually small, perhaps a few centimetres in diameter, but it can be as much as 2 m across. It may be little more than an

70   Tabulate corals.
(a, and b, reproduced from an exhibit in the Sedgwick Museum.)

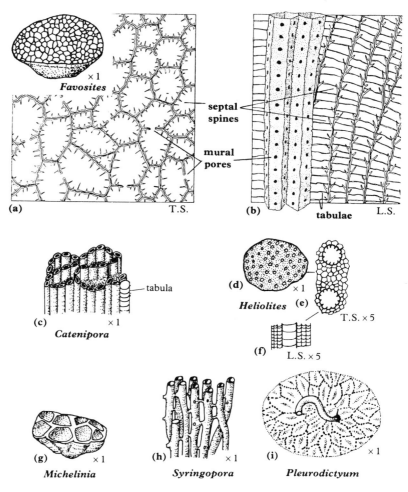

encrustation, but more usually is fasciculate (with individual corallites branching freely) or massive and of irregular bun-shape. Fasciculate forms may branch irregularly and may be united at intervals by connecting tubes (fig. 70h). Massive forms may have round or oval holes, MURAL PORES, in the walls between adjacent corallites. In some massive forms the corallites are separated by a zone of common skeletal tissue, COENENCHYME, made up of narrow tubes divided by close-set transverse plates (fig. 70, f). Coenenchyme looks vesicular in longitudinal section.

Individual corallites are rarely more than a few millimetres across and may be round, oval or polygonal in transverse section. SEPTA, if present, often number 12. They are alike, usually being short spines projecting from the walls. TABULAE are the most characteristic feature of these corals. They are numerous and typically horizontal (fig. 70b), but they may be domed or funnel-shaped.

Skeletal form of a common tabulate coral

Favosites (fig. 70a, b).   Massive, irregularly hemispherical form, typically only a few centimetres across; cerioid with long polygonal corallites measuring about 2 mm across; adjacent corallites communicate through mural pores arranged in rows in the walls; septa are usually present as vertical rows of spines; tabulae are numerous and evenly spaced.
U Ordovician–M Devonian

Exceptional fossils of *Favosites*, with calcified polyps *in situ* in their calices, have been described from Silurian rocks in Quebec. They show about 12 tentacles in retracted state. The polyps are valuable and unexpected confirmation of tabulates as corals, albeit with a morphology distinct from other corals.

Additional common genera

Michelinia (fig. 70g).   Corallum small, sometimes with root-like processes at base; corallites relatively large, thick-walled; tabulae numerous, vesicular. Distinguished from *Favosites* by larger corallites.
L. Devonian–U Permian

Pleurodictyum (fig. 70i).   Similar to *Michelinia*, with large corallites (about 5 mm across) but walls thick with many pores; tabulae may be absent. A curved worm tube often present in centre: the association was probably commensal, i.e. the worm probably benefited (e.g. by gaining shelter) without incommoding the coral.
U Silurian–M Devonian

Vaughania.   Like *Michelinia* but smaller, discoid, with shallow corallites; walls thickened; no tabulae.
L Carboniferous

Alveolites.   Like *Favosites* but encrusting with inclined corallites opening obliquely at surface, rounded polygonal in transverse section.
M Silurian–U Devonian

Heliolites (fig. 70d–f).   Massive, with cylindrical corallites separated by coenenchyme of narrow polygonal tubes; with or without 12 short septa; tabulae flat, regular.
M Ordovician–M Devonian

Catenipora (fig. 70c).   Phaceloid; corallites oval in transverse section, arranged in single series which join at intervals to enclose irregular spaces; spinose septa in 12 vertical rows; tabulae regular.
M Ordovician–U Silurian

Halysites (fig. 71b).   Similar to *Catenipora* but has coenenchyme of a single tube separating adjacent corallites, each tube with closely spaced transverse plates.
M Ordovician–Silurian

Syringopora (figs. 70h, 71).   Phaceloid; corallites cylindrical, thick-walled, united at intervals by transverse tubes; tabulae numerous, more or less funnel-shaped.
U Ordovician–L Carboniferous

Geological history

L Ordovician–U Permian
Tabulate corals appeared in the early Ordovician and became varied and widespread in the later part of that period. They occur most commonly in shallow-water limestones and calcareous shales and they played a significant role in reef formation in the Silurian and Devonian. They outnumbered rugose corals in reefs of that time. Heliolitids and favositids were the dominant forms. Halysitids were also common and widespread but did not survive the Silurian. Heliolitids and many other tabulates disappeared when numbers dropped abruptly in the late Devonian. There was a modest recovery in numbers in the lower Carboniferous, the most important examples being forms like *Syringopora* and *Michelinia*. Later tabulates are not found in Britain. Elsewhere, a small number of genera, including *Michelinia*, occur in the late Carboniferous and Permian, finally disappearing at the close of the Permian period.

71   Tabulate corals.
a, *Syringopora*, L Carboniferous (× 1.4), corallum is silicified. b, *Halysites*,
Silurian (× 7); vertical section showing coenenchyme of single tubule between
corallites. (From an exhibit in the Sedgwick Museum.)

Scleractinia

The Scleractinia are solitary or colonial corals with an aragonite skeleton in
which septa form cycles of six or multiples of six. They first appeared in the
mid-Trias and are important reef-formers in tropical seas today.

Morphology

The coral skeleton, the CORALLUM, is secreted by the outer wall of the body and is formed by the fusion of minute fibres of aragonite. Each polyp sits in a cup-shaped structure, the CALICE, from which the septa project between the pairs of mesenteries which partition the enteron.

Soft body

A solitary polyp is basically like a sea-anemone, sac-shaped with an oval mouth surrounded by one or more rings of tentacles (fig. 72). The tentacles are armed with nematocysts which are partly for defence, and partly are used to paralyse small animals. These are then grasped as prey by the tentacles and thrust into the mouth. Food passes from the mouth through the gullet to be digested in the enteron, and waste is ejected through the mouth. The mesenteries radiate in pairs from the wall of the enteron. There are six pairs of primary mesenteries and a varying number formed later. In colonies adjacent polyps are connected by a sheet of common soft tissue, COENOSARC, which covers intercorallite skeletal tissue.

Skeleton

The terms used to describe the form of the corallum have been defined in the section on rugose corals (p. 100). The corallum of a solitary coral is commonly conical or cylindrical in shape. The corallum of colonial corals varies greatly in shape. It may be encrusting, massive, branching or plate-like (fig. 74). Individual corallites may be similar in shape to those of rugose

72 Coral sectioned to show relationship between soft parts and skeleton.
a, vertical section with part of base cut to show septa. b, transverse section with two cycles of six septa (1, first cycle of six septa each one lying between two mesenteries; 2, second cycle of six septa).

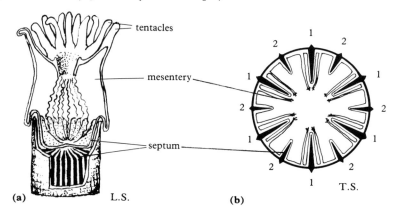

corals, or may be laterally elongated and winding (MEANDROID); or be separated by skeletal tissue, COENOSTEUM. The first-formed part of the corallum is a basal plate by which it is attached to the sea-floor. On this the polyp secretes the series of vertical and transverse plates from which the corallum is built up. The polyp is in contact with the upper surface only of its corallum, the calice, and this is in effect a mould of the base of the polyp. From time to time, as the polyp outgrows its calice, it moves upwards and secretes a new one.

Typically, the scleractinian structure is simpler than that of rugose corals. Septa are the dominant skeletal elements. There are six primary septa (fig. 72b). These are the longest and may mark the corallite into sextants. Later septa are inserted in cycles, each comprising a multiple of six septa. There are usually no more than four or five such cycles. A rod or plate-shaped columella may be present, but more elaborate axial structures are uncommon. Dissepiments occur between the septa. They are formed during growth by the polyp to cut off the lower part of the corallum from which it has withdrawn.

The scleractinian and rugose corals, though similar in some aspects of their morphology, are distinct groups. They show one major difference: the rugose septa, after the first six protosepta were formed, were inserted at *four* sites only, and typically show *bilateral* symmetry in transverse section; whereas the scleractinian septa are *radially* arranged, the first-formed dividing the corallite into 60° sectors, and subsequent septa being added in regular cycles of *six* and then multiples of six (fig. 72b). There is no known record of either group in the lower Trias, the rugosans having disappeared in the upper Permian and the earliest scleractinians appearing in the mid-Trias.

Skeletal form and mode of life of two scleractinian corals

Goniopora (figs. 73c, 76). Colonial, massive or sometimes branching; corallites polygonal in transverse section; septa arranged in three cycles, upper edges serrated; longest septa unite with small columella.
M Cretaceous–Recent

Goniopora is a reef-building (HERMATYPIC) coral. The related *Porites* (Eocene–Recent) is one of the more important colonial forms which today build reefs, often of great extent, in the tropical and subtropical zones between latitudes 35 °N and 32 °S.

Reef-living corals will not tolerate a wide range in salinity, temperature or depth. They require a salinity of about 36‰, a temperature between about 18 and 29 °C, and a depth of less than about 70 m. They grow most luxuriantly in well-lit, sediment-free, and well-oxygenated water between 25 and 29 °C. These stringent requirements are due, in part, to the presence in the polyp walls of minute photosynthetic algae, ZOOXANTHELLAE,

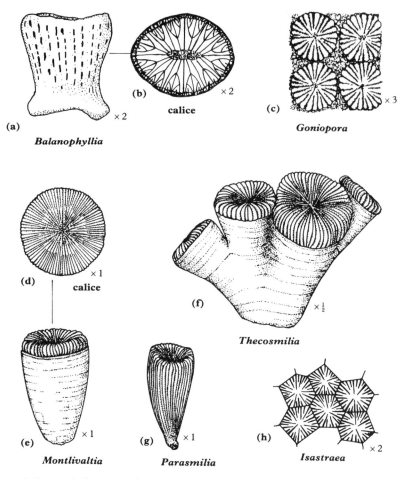

73 Scleractinian corals.
a, b, d, e, g, solitary corals. c, f, h, compound corals.

belonging to a species of dinoflagellate (p. 244). Such corals are said to be zooxanthellate. The algae have a symbiotic relationship with the polyps. (Symbiosis is an association of mutual benefit between unlike organisms.) The algae use waste products as nutrients, including CO_2, the removal of which may facilitate aragonite secretion in the coral skeleton. The coral gains a supply of photosynthetic oxygen plus soluble metabolic products formed by the algae, and may also digest some of the latter directly.

Although there is no direct evidence of symbiotic algae in fossil reef-building corals, they are likely to have been present in the close ancestors of modern forms of this kind.

The shallow-water (photic zone) environment of fossil reef-building corals is directly demonstrated by their association with encrusting calcareous algae, often forming a robust reef-apron behind which the coral framework was built. A great variety of other organisms are associated with the reefs,

74   A coral reef.
An underwater photograph of the Key Largo Coral Reef Preserve, Florida Keys, USA.

which provided shelter and an abundant food supply in well-oxygenated, near-surface waters. They include varied algae, cyanobacteria, foraminifera, sponges, gastropods, bivalves, polychaetes and other 'worms', echinoids, arthropods and an assortment of fish (fig. 74).

Balanophyllia (fig. 73a, b). Solitary, cylindrical coral attached to rocks by broadened base; septa crowded, inner edges partly fused; longest septa join a columella of rather spongy texture.
Eocene–Recent
Balanophyllia is a 'cup'-coral found today in rock pools and ranging down to about 1000 m. It resembles a brilliant scarlet sea-anemone with yellow tentacles, until it contracts within its calice and reveals the skeleton. It, and colonial forms like *Dendrophyllia* (Eocene–Recent), are 'cold-water' (AHERMATYPIC) corals which lack zooxanthellae. They do not form reefs, and typically live in deeper, relatively cold waters, e.g. off the coasts of

Norway and Scotland. They are best developed near the edge of the continental shelf and over the slope in depths of 175–800 m (though they may range deeper) and in temperatures between about 4 and 10 °C. Here they may form coral banks but these lack the rich diversity of invertebrate fauna associated with the hermatypic corals.

Additional common genera

Montlivaltia (fig. 73d, e). Solitary; corallum cup-shaped to cylindrical; many septa project above shallow calice; no columella; many arched dissepiments. Probably zooxanthellate.
L Jurassic–Cretaceous

Thecosmilia (fig. 73f). Like *Montlivaltia* but forms small colonies. Dendroid to phaceloid.
M. Jurassic–Cretaceous

Isastraea (figs. 73h, 75). Colonial; massive, cerioid corallum; corallites hexagonal to pentagonal in transverse section; about four cycles of septa; thin dissepiments. Probably zooxanthellate.
M. Jurassic–Cretaceous

Parasmilia (fig. 73g). Solitary cup-coral fixed at base; elongate becoming cylindrical in mature stage; surface of calice ridged by many septa arranged in four cycles; prominent spongy columella; non-zooxanthellate; common in the Chalk (U Cretaceous).
L Cretaceous–Recent

Geological history

M Trias–Recent
By late Trias times, scleractinians were widespread and varied. They continued to expand in the Jurassic and Cretaceous, reaching peak numbers in the late Cretaceous. In Britain they are represented by only a few genera in the earlier Jurassic, usually of solitary forms like *Montlivaltia* (fig. 73e). The upper Jurassic saw the first real development of coral reefs, especially in the Tethys area and extending to about 54 °N, present day latitude. This is about 20° further north than today's reef belt. Reef-like structures with *Isastraea* (fig. 75) occur, for instance, in the Corallian of southern Britain. There was little reef formation in the early Cretaceous, but in mid and (more so) in late Cretaceous, reefs were again extensive. In Britain, however, corals were not an important component of the fauna, mainly being represented by solitary cold-water cup-corals like *Parasmilia* (fig. 73g). Numbers fell sharply at the end of the Cretaceous and did not begin to revive until the Eocene.

75 Jurassic reef-living coral, *Isastraea*, Corallian, Jurassic, Upware, Cambridgeshire (× 4).
This is an external mould of the upper surface, and the calices appear in inverted relief.

There was then a renewed expansion of coral reefs, reaching its peak in the Miocene in substantially the areas where the reefs are best developed today. These are the tropical and subtropical regions around oceanic islands and along the east coasts of larger land masses, e.g. the barrier reefs of Australia.

Apart from small specimens of *Goniopora* (fig. 76), a reef-building form occurring in the Eocene (Bracklesham Beds only), only cold-water cup-corals are found in later deposits in Britain, e.g. *Balanophyllia* (fig. 73a) in the Pleistocene Red Crag.

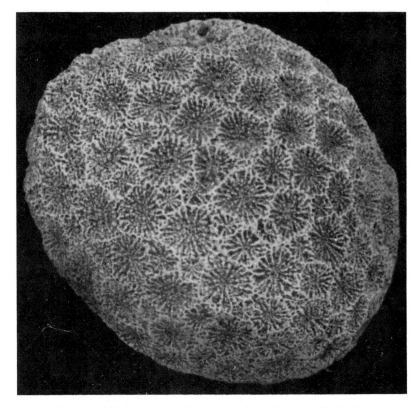

76   Tertiary reef-living coral, *Goniopora*, Bracklesham Beds, Eocene (× 2.6).

Minor cnidarians

Subclass Octocorallia

?Precambrian, L. Jurassic–Recent

These are marine, sessile, colonial forms with polyps projecting from shared soft tissue. Polyps have eight feathery tentacles and eight mesenteries. They may secrete a horny skeleton, sometimes with calcareous spicules, or be calcified.

Octocorallians include sea-fans, e.g. *Gorgonia*, and sea-feathers, which are important members of recent coral reefs; and sea-pens, the Pennatulacea. In the latter a primary polyp produces a long stem from which numerous secondary polyps are budded, often forming lateral leaf-like structures. It also forms a down-growing stalk which fixes the colony in soft sediments. Octocorallians are rare as fossils. Possible sea-pens have been identified from the late Precambrian, e.g. *Charniodiscus* (fig. 77).

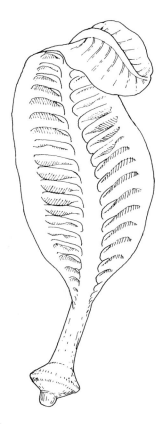

(a)   (b)

77 Possible sea-pen fossils.
a, *Charnia*, Precambrian (Charnian), Charnwood Forest, Leicestershire (× 0.7).
b, *Charniodiscus*, a reconstruction by M.F. Glaessner.

Class Hydrozoa

U Precambrian–Recent
Mainly marine, sessile or free-swimming; usually colonial with tiny polyps
sharing common tissue. Typically their life history involves an alternation of
generations between the polyp phase and free-swimming medusae
(jellyfish). The skeleton, if any, is horny (chitinous) or occasionally
calcareous.

Marine hydrozoans range down into deep water, about 8000 m. Some,
with a massive or encrusting calcareous skeleton, like *Millepora*
(Tertiary–Recent) are important reef-builders.

Class Scyphozoa

U Precambrian–Recent

These are cnidarians in which the free-swimming medusoid (jellyfish) stage is dominant and the fixed polyp stage is inconspicuous or absent. The umbrella-like medusae show four-rayed symmetry and the enteron is divided by four partitions into digestive pockets. They are predators which capture their prey by long tentacles equipped with stinging cells. They occur mainly in shallow seas in cool and temperate waters.

Jellyfish lack a skeleton and their fossils occur as impressions, e.g. various forms from the upper Precambrian and from the Jurassic Solnhofen Limestone, Germany

Technical terms

ASTRAEOID describes a massive corallum in which there are no walls separating the individual corallites; the septa are thus in contact (fig. 68d).

AXIAL REGION the central area of the corallite through which the axis of radial symmetry passes.

AXIAL STRUCTURE any vertical structure in the axial region of a corallite, e.g. columella (fig. 65d). More complicated structures are shown in fig. 68e.

CALICE the cup-shaped depression in the upper surface of a corallite in which the polyp sat in life (fig. 64).

CERIOID massive corallum in which the walls of adjacent corallites are united (fig. 67).

COENENCHYME a zone of common skeletal tissue which may separate individual corallites in a colonial form (fig. 70e, f).

COLUMELLA an axial structure in the form of a vertical rod or plate (fig. 65d).

CORALLITE the exoskeleton secreted by an individual coral which may be solitary or one of a colony.

CORALLUM the entire skeleton of a solitary or of a colonial coral.

DENDROID describes a fasciculate corallum in which the corallites branch irregularly in a tree-like pattern (fig. 63e).

DISSEPIMENTS small convex plates lying between the septa in the periphery of a corallite (figs. 63a, 65d).

FASCICULATE describes the corallum of a colonial coral in which the corallites are cylindrical and not in contact.

FOSSULA (pl. fossulae) a larger than usual space between the septa in rugose corals; it is usually associated with one of the primary septa, in particular the cardinal septum (fig. 65c).

MASSIVE the corallum of a colonial coral in which the corallites are in close contact.

MESENTERY one of several radial infoldings of the body wall which forms a vertical partition in the coelenteron (fig. 63a).

MURAL PORE small hole in the wall separating adjacent corallites (fig. 70b).

PHACELOID a fasciculate corallum in which the corallites are more or less parallel in growth (fig. 63f).

SEPTA (sing. septum) radial plates formed on the floor of the calice within the mesenteries and extending vertically through the corallite. The longer septa are called MAJOR septa, and the shorter ones, usually alternating with the major, are MINOR septa (fig. 65d).

TABULA (pl. tabulae) a flat or convex transverse plate which extends (wholly or in part) across the corallite (fig. 63a).

6

Brachiopoda

The brachiopods are sessile marine animals which secrete an external shell consisting of two dissimilar but equilateral valves. The plane of symmetry bisects the valves in an anterior–posterior direction. The group is of minor importance today, but fossil brachiopods are abundant and varied in Palaeozoic and Mesozoic rocks.

The shell may be calcareous or chitinous, and is secreted by the mantle which consists of two extensions of the body wall, one from the dorsal side and one from the ventral side of the body. The two valves are thus DORSAL and VENTRAL in position relative to the body. The brachiopod is attached to the sea-floor at its posterior end, usually by a stalk, the PEDICLE (fig. 78d). The body lies in the posterior part of the shell, and from it the mantle lobes extend forwards to enclose a space, the MANTLE CAVITY, much of which is occupied by the LOPHOPHORE, a feeding device which collects suspended particles. There is a well-developed coelom. The nervous and circulatory systems are not highly organised. The sexes are typically separate; gametes are shed into the sea, or in some cases eggs may be brooded in the mantle cavity. The larvae are free-swimming for a short time before they settle down on to the sea-floor and metamorphose.

Brachiopods are distributed widely throughout the world in cool and temperate waters; they are common in eastern waters, e.g. around Japan, South Australia and New Zealand, and a rich fauna lives in the North Atlantic, including coastal areas of west Scotland. They are diverse at depths to about 500 m and a few range to greater depths down to about 6000 m. They are typically stenohaline (p. 18) but *Lingula*, for instance, tolerates brackish water for a time.

Brachiopod shells range in size from a few millimetres to as much as 30 cm in width in the case of a Carboniferous form called *Gigantoproductus giganteus*. Most measure between about 2 and 7 cm.

Morphology

Brachiopods are divided into two classes: the Articulata, with a hinge

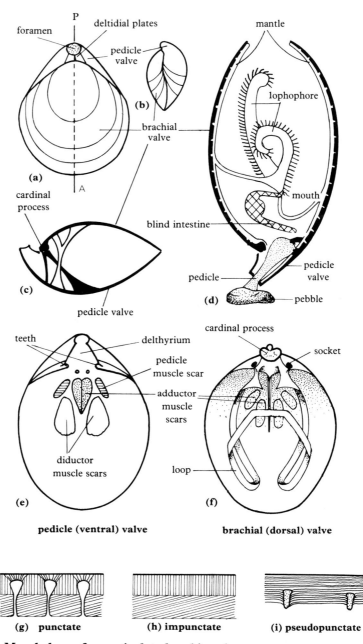

78 Morphology of an articulate brachiopod.

a, dorsal view (A, anterior; P, posterior; line A–P, plane of bilateral symmetry).
b, side view. c, section to show the disposition of the adductor muscles
(unshaded) and the diductor muscles (black). d, simplified section to show the
general relationship of the soft body to the shell. e, interior of the pedicle valve.
f, interior of the brachial valve. g–i, simplified sections showing shell structure.

structure; and the Inarticulata with no hinge, the valves being united by muscles. Their general morphology is similar.

The brachiopod shell encloses the body except for the pedicle (fig. 78d). The valve on the ventral side of the body is known as the PEDICLE VALVE, since the pedicle commonly emerges through it. The valve on the dorsal side, the BRACHIAL VALVE, takes its name from the BRACHIA, the arm-like projections of the lophophore which it carries. Commonly the pedicle valve is the larger, projecting at its posterior end beyond the brachial valve. The pedicle emerges from the shell at its POSTERIOR margin, and the opposite margin is ANTERIOR. The valves open slightly along the anterior margin during feeding (fig. 79f) but remain in contact along the posterior margin by means of a HINGE of teeth and sockets in the Articulata; in the Inarticulata they are held together by a system of muscles only.

Soft body

The mantle lines the shell and in some groups, soft tissue extends into the shell wall by minute tubules (used in food storage and oxygen absorption). Small sensory bristles (setae) extend from the mantle edges. The body is small and the anterior two-thirds of the mantle cavity is taken up by the lophophore (figs. 78d, 80). This is fleshy and may be a lobed disc, or two coiled or folded arms called BRACHIA each of which has a groove leading to the mouth and is fringed with ciliated filaments. The cilia, by beating, maintain currents of water along three paths: a median outgoing flow, and an intake flow on either side. They also filter out minute organisms, and organic particles, from the incurrent water and these are passed, entrapped in mucus, along the lophophore grooves to the mouth and thence to the digestive tract. The intestine opens via an anus in inarticulates, but ends blindly in living articulates. In the latter, waste in the form of pellets is disposed of by reversing the current direction and a snapping action of the valves.

Most brachiopods are attached by a PEDICLE (fig. 78d) which is fixed to the pedicle valve by ADJUSTOR muscles; it is a fleshy stalk in inarticulates but is horny in articulates. Its distal end adheres to a rock or shell, or may diverge into rootlets to secure a hold in soft sediment. These forms are able to swivel and reorient the shell as current directions change. In other forms the pedicle is reduced to a tether; or may be absent and the shell may lie unattached, or may become cemented to a firm surface (fig. 81d). Some extinct forms appear to have been anchored in soft sediment by spines (fig. 82i).

The opening and closing of the valves is controlled by a system of muscles which are attached to the inner surface of the valves towards the posterior end where they may leave MUSCLE SCARS (fig. 78e). The muscle system is simplest in the articulate brachiopods (fig. 78c), consisting commonly of a pair of ADDUCTOR muscles which run across the shell cavity from the interior of the pedicle valve to the interior of the brachial valve; and of two pairs of DIDUCTOR muscles which run obliquely from the pedicle valve to a projection, the CARDINAL PROCESS, from the hinge line of the brachial valve. Both sets of muscles work by contracting. The hinge line acts as a fulcrum and the cardinal process as a lever, so that as the diductor muscles contract they pull down the cardinal process and the valves open. As the diductor muscles relax, the adductor muscles contract and pull the valves together. Inarticulate brachiopods have a quite different and more complex system of muscles which leave only indistinct scars in the shell. As well as those which close the shell, some work obliquely to control lateral movements of the valves.

Shell

In most inarticulate brachiopods the shell has a horny appearance and is composed of alternate layers of chitin and calcium phosphate, but in a few forms it is calcareous. In the articulate brachiopods the shell is calcareous. A thin chitinous layer (periostracum) overlies the shelly part which is made of two layers of calcite crystals: a thin outer, primary sheet, and an inner, secondary layer which thickens with growth, and from which all internal structures are formed. In some articulate groups the shell is PUNCTATE (fig. 78g), being perforated by fine tubules (with soft tissue extensions); while in other groups the shell lacks these tubules and is IMPUNCTATE (fig. 78h). In one group, strophomenids, the shell is PSEUDOPUNCTATE. Here there are no tubules, but the fine structure of the secondary shell shows tiny conical deflections in which are rod-like units directed inwards and appearing on the inside as tubercles (fig. 78i). In weathered shells the deflections may be picked out as minute pits which simulate punctae hence pseudo-punctate.

The shape of the shell may show some correlation with the arrangement of the lophophore and the feeding currents, and markings on the inner surface may provide information about the disposition of muscles, the lophophore, and canals in the mantle in a number of extinct forms. In inarticulate brachiopods the shell is approximately oval or circular in outline with gently convex valves (fig. 81a, b). In articulate forms the shell may be ovate (fig. 78a), tapering slightly at the posterior end with a short curved hinge line, or it may be semicircular in outline with a straight wide hinge line (fig. 81f). The pedicle valve is typically larger than the brachial valve. Both valves may be convex, or one may be convex and the other flat or concave. The shell may be folded along its mid-line so that a ridge or FOLD is formed at the anterior margin of one valve (fig. 84b) with a corresponding depression or SULCUS in the other. The fold and sulcus serve to keep separate the incoming and outgoing currents of water.

The brachiopod shell grows by increments which typically are greater along the anterior and lateral margins and which form concentric growth lines on the outer surface. Thus the initial shell remains at or near the posterior margin and may form the tip of a pointed BEAK. The curved convex area around the beak, which is generally a more prominent feature in the pedicle valve, is the UMBO, and in many forms a curved or flat INTERAREA is interposed between it and the hinge line (fig. 83a). The shell surface may be smooth or bear an ornament (concentric or radial lines; or ribs, tubercles or spines) which is useful in distinguishing species.

In inarticulate brachiopods the PEDICLE emerges through a gape between the valves, or by a groove or slit in the pedicle valve. In the articulate brachiopods the pedicle opening, the DELTHYRIUM, is a triangular gap in the posterior margin of the pedicle valve (fig. 78e). Commonly it is constricted by a pair of DELTIDIAL PLATES (fig. 78a), or by a single plate, the DELTIDIUM, leaving a circular hole, the FORAMEN, for the passage of the pedicle.

The hinge apparatus consists of two TEETH in the pedicle valve (fig. 78e) which fit into two SOCKETS in the brachial valve (fig. 78f). The teeth are short projections from the hinge line, one on each side of the delthyrium, and they may be supported in some genera by DENTAL PLATES projecting from the floor of the pedicle valve. The sockets lie one on each side of a small projection, the CARDINAL PROCESS, to which the diductor (or opening) muscles are attached (fig. 78f).

In most articulate brachiopods there are distinct SCARS left on the floor of the valves by muscles. The degree to which they are defined and their relative positions, however, may vary in different genera. In the pedicle valve, the pedicle and diductor muscle scars are grouped round two close-set adductor muscle scars (fig. 78e). In the brachial valve, four adductor muscle scars are grouped on the floor of the valve, and the diductor muscle scars are on the cardinal process (fig. 78f). In a small number of brachiopods the muscles are attached to a trough-shaped structure, the SPONDYLIUM, in the pedicle valve (fig. 81l). This consists of two enlarged dental plates which converge and may unite on the floor of the valve to form a V-shaped trough. It may be supported by a SEPTUM. The muscle scars in the inarticulate brachiopods are more complicated and often indistinct.

In some groups of articulate brachiopods the lophophore is supported by a calcareous framework, the BRACHIDIUM. This is a feature of systematic importance. It arises from the hinge line of the brachial valve on either side of the cardinal process and takes several forms: (i) CRURA (sing. crus), two curved plates or rod-like processes; (ii) SPIRALIA (fig. 83b, d), two calcareous ribbons coiled in helical spirals; (iii) LOOPS, two calcareous ribbons which unite to form a loop which may be short (fig. 79g) or longer and reflexed (fig. 78f).

Shell form and mode of life of two typical brachiopods

Lingula (fig. 79a–d). Inarticulate; shell chitinophosphatic; biconvex; outline more or less oval to spatulate, tapering towards posterior end; valves almost equal in size; pedicle emerges between valves, the pedicle valve being slightly grooved to accommodate it; ornament of faint ribbing; regular growth lines; length up to about 5 cm. Individual species are short-ranged. U Ordovician–Recent

Inarticulate brachiopods occur predominantly in tropical–subtropical waters. *Lingula*, the best-known example, is an unusual brachiopod in two respects: (i) it is tolerant of ranges of salinity (euryhaline, p. 18); and (ii) it is infaunal. It is found today mainly in Japanese coastal waters, burrowing in

79 Morphology and mode of life of brachiopods.
a–d, an inarticulate brachiopod, *Lingula*: a, in feeding position at mouth of its burrow; b, slit-like opening of the burrow (the arrows indicate the incurrent and excurrent flow of water); c, a lingulid in processes of burrowing; d, fossil *Lingula*. e–h, an articulate brachiopod, *Terebratulina*: e, brachial valve; f, in feeding position with valves open; g, interior of brachial valve; h, attached to stone and other shells.

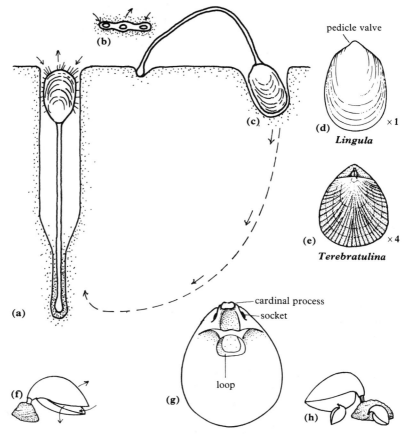

sand or muddy deposits at depths of less than 40 m and often in mudflats which are exposed at low tide. It can survive for a short time in tidal water made brackish by river flood-waters. It lives, anchored by a long worm-like pedicle, in a burrow which may be 30 cm deep (fig. 79a). It feeds, anterior end uppermost, at the slit-like opening of its burrow, bristles on the mantle margin forming three funnels, one for the outflowing and two for the

80   Brachiopod lophophore and brachidia

a, *Terebratulina retusa*: pedicle valve removed to show lophophore extended while feeding. b–d, brachial valves with calcareous loops for support of the lophophores. (Recent shells from the Firth of Lorne; Photographs, courtesy of Dr G.B. Curry.)

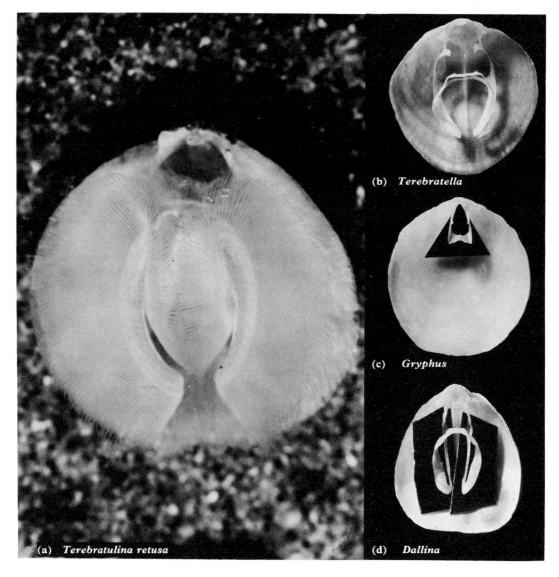

(b)   *Terebratella*

(c)   *Gryphus*

(d)   *Dallina*

(a)   *Terebratulina retusa*

inflowing (feeding) currents. At low tide, or when disturbed, it withdraws down its burrow by contracting the pedicle. A related form makes its burrow as follows. Initially it burrows, anterior end first, downwards into the sediment by scissors-like, rotary and sliding movements between the valves. At the appropriate depth it makes a U-turn, returning vertically upwards until its anterior end just reaches the sediment surface (fig. 79c). Fossils of *Lingula* are common, often without other brachiopods, in shaley or poorly sorted deposits. It belongs to the Lingulida (L Cambrian–Recent), a relatively unspecialised order which has persisted for an unusually long time during which it has shown little change.

Terebratulina (fig. 79e–h).   Articulate; shell calcareous, punctate; biconvex; more or less oval in outline, tapering towards the beak; anterior margin gently folded; hinge line short, gently curved; foramen large; below it, small deltidial plates; ornament of fine radial ribs; brachidium a short ring-like loop, developed from crura.
Jurassic–Recent
Articulate brachiopods are typically found in temperate and cold waters. *Terebratulina*, for instance, is found in shallow seas around the North Atlantic coastline and is common in sea lochs on the west coast of Scotland (fig. 80). It may occur down to about 1400 m but is commonest between 100 and 200 m. It lives gregariously, often attached by a short pedicle to shells of the horse mussel, *Modiolus modiolus*, or in association with sponges and hydroids. Empty shells are uncommon, however, since the organic matrix decays rapidly and the shell disintegrates. This illustrates the danger of assuming that a 'death assemblage' of shells represents the original make-up of a community. *Terebratulina* belongs to the Terebratulida, the last order of brachiopods to appear.

Examples illustrating shell form in the main brachiopod orders

Inarticulata

There are five inarticulate orders of which only two are mentioned here.

Lingulida

L Cambrian–Recent

Lingulella (fig. 81a).   Shell chitinophosphatic; gently biconvex; elongate oval or rather squat quadrate, tapering towards umbones; striated triangular area on posterior margin of pedicle valve, divided by pedicle groove.
L Cambrian–M Ordovician

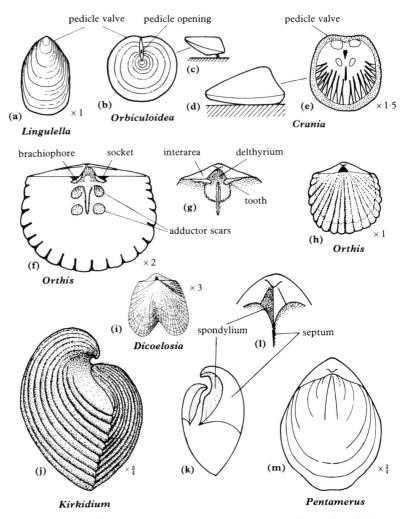

81 **Examples of inarticulate brachiopods, pentamerids and orthids.**
a–e, inarticulates: a, b, pedicle valves; c, d, side view of shells in life position; e,
interior of pedicle valve. f–i, orthids: f, interior of brachial valve; g, interior of
pedicle valve; h, brachial valve; i, brachial valve, j–m, pentamerids: j, side view;
k, almost median longitudinal section showing part of the spondylium
supported by a septum; l, spondylium, m, brachial valve showing the trace of
two plates through the shell.

Acrotretida
L Cambrian–Recent

Orbiculoidea (fig. 81b).   Shell chitinophosphatic; subcircular in outline,
1–2 cm across; brachial valve widely conical; pedicle valve flatter with
furrow from apex to posterior margin, marking the trace of pedicle opening
as it closed with growth; well-marked concentric growth lines.
U Ordovician–Permian

Crania (fig. 81d, e). Shell calcareous; subcircular in outline, up to about 1 cm across; no pedicle; cemented by flat pedicle valve to hard surface, e.g. other shells; brachial valve gently conical.
Cretaceous–Recent

Articulata

All articulates have calcite shells. Most are referred to seven orders; the affinities of a few are not resolved.

Orthida

L Cambrian–Permian
Orthis (fig. 81f–h). Shell small, about 2 cm; almost semicircular with straight hinge line shorter than maximum width of shell; brachial valve almost flat; pedicle valve convex with narrow interarea and open delthyrium; a similar opening in hinge line of brachial valve; ornament of sharp radial ribs; inside, pedicle valve with teeth, brachial valve with sockets, brachiophores (p. 141) and small cardinal process.
Ordovician
Orthids, the first articulates to appear, represent the ancestral stock for later groups. The shell, in most, is impunctate. They are numerous and varied in the Ordovician and Silurian. Examples include *Dicoelosia* (U Ordovician–L Devonian, fig. 81i), *Dalmanella* (M Ordovician–L Silurian), and *Schizophoria* (U Silurian–Permian).

Strophomenida

L Ordovician–Trias

Leptaena (fig. 82a–d). Shell semicircular in outline with long straight hinge line forming maximum width; posterior part of both valves almost flat; towards the anterior the valves make an almost right-angle bend, so that pedicle valve is convex and brachial valve concave; ornament of radial ribs with concentric corrugations on earlier flat part of shell; interarea on each valve; deltidium in pedicle valve, and a similar plate in brachial valve; foramen very small; pedicle non-functional in adult.
M Ordovician–U Devonian

Productus (fig. 82e, f). Shell semicircular in outline; hinge line straight, relatively short; pedicle valve strongly convex with rounded umbo; brachial valve flat initially but anterior part is bent almost at right angles and lies parallel and very close to pedicle valve; this part of shell, known as a 'trail', is often broken away in the fossil giving a misleading appearance that the shell is complete; ornament of close-set radial ribbing; concentric corrugations on

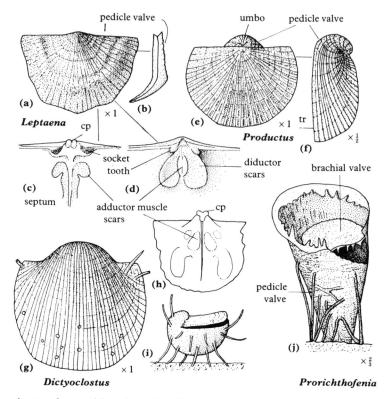

82 A strophomenid and productids.

a–d, strophomenid: a, pedicle valve; b, lateral view; c, inside of the posterior end of the brachial valve and d, of the pedicle valve (cp, cardinal process). e–j, productids: e, brachial valve; f, side view of a specimen with the trail (tr) preserved; g, pedicle valve; h, inside of the posterior end of the brachial valve; i, side view of a spinose shell to show its presumed life attitude, supported by its spines off the sea-bed; j, an aberrant productid cemented to the sea-bed and anchored by spines (the pedicle valve is broken to show the position of the lid-like brachial valve within the pedicle valve).

posterior part of shell; small stubby knobs on pedicle valve mark the sites of long hollow spines which anchored the shell on a soft substrate in life; the spines are occasionally preserved (fig. 82i); no delthyrium; pedicle atrophied; no teeth; strongly marked muscle scars.

Carboniferous

Strophomenids have a pseudopunctate shell. They form the largest group of brachiopods and show great variety. Earlier members like *Leptaena* and *Strophomena* (M–U Ordovician) are very common in shallow-water rocks from the Ordovician to the Devonian. Generally the shell has a thin, gently curved profile, most with the brachial valve concave and the pedicle valve convex (fig. 82b). Frequently the pedicle atrophied so that, when adult, the shell rested on the pedicle valve. In forms like *Chonetes* (fig. 85d), spines developed along the hinge margin of the pedicle valve anchored the shell, posterior end down, in mud.

Productus, with its profusion of anchoring spines on the very convex pedicle valve, is the typical strophomenid of the Carboniferous, being abundant and varied on muddy and calcareous substrates. *Dictyoclostus* (fig. 82g) is a related form. Such forms gave rise to the extreme aberration shown by *Richthofenia* (M–U Permian). This has a cone-shaped pedicle valve, cemented by the apex to the sea-floor and buttressed by spinose rooting processes. The brachial valve is flat. In the related *Prorichthofenia* (fig. 82j) spines grow around the aperture, and the flat brachial valve is recessed within the pedicle valve. These shells, like the bivalve rudists (p. 45), mimicked in shape the corallum of a solitary rugose coral. They lived, often in closely clustered communities, in a reef-like environment.

Pentamerida

M Cambrian–U Devonian

Pentamerus (fig. 81k–m). Shell large, up to 8–9 cm; suboval in outline; strongly biconvex; prominent pedicle umbo incurved over short, curved hinge line; no interarea; delthyrium present but pedicle non-functional in adult shell; surface smooth; impunctate; inside pedicle valve spondylium supported off floor of valve by long median septum; brachial valve with two close-set subparallel plates (fig. 81k).

L Silurian

The pentamerids, an offshoot from early orthids, are a relatively small group with impunctate shell, two diverging plates in the brachial valve and, notably, a spondylium. The latter was the outcome of the great depth of the valves. Without it the muscles would have been unduly long and thus less efficient. By developing a raised platform for their attachment the muscles remained short. The internal plates are often seen in broken fossils since they form a plane of weakness. Many pentamerids are smooth-shelled, though *Kirkidium* (U Silurian, fig. 81j) has prominent ribbing. Pentamerids were common in Silurian and early Devonian times occurring on muddy, calcareous or variable substrates, often in great numbers. The adult shells lay umbones down on a soft substrate. They have been used in identifying depth-related ecological zones in the lower Silurian of Wales and elsewhere.

Rhynchonellida

L Ordovician–Recent

Tetrarhynchia (fig. 84a–c). Shell small, about 1–1.5 cm width; outline subtetrahedral; strongly biconvex; brachial valve with strong anterior fold; pedicle valve with corresponding deep sulcus; beak small, incurved above short curved hinge line; foramen under beak, restricted by small deltidial plates; ornament of sharp radial ribs develops towards anterior margin and produces a sharp zigzag line of contact between the valves (probably helps restrict size of particles entering shell); inside, strongly developed teeth supported by dental plates; no cardinal process; simple crura.

L Jurassic

Rhynchonellids, derived from early pentamerid stock, are a long-lasting successful group still widespread today. The shell in most is impunctate. Their distinctive internal feature, crura, form the support for the lophophore at its posterior end. The group was most diverse in the Jurassic and takes its name from *Rhynchonella* (U Jurassic–L Cretaceous) which is often used in the broad sense. Externally, the basic simple design has varied little in the course of time. Many genera are distinguished only by internal details (e.g. of muscle scars, septum or crura) which are revealed by serial sections of the shell. Some common genera which are distinctive include *Sphaerirhynchia* (L Silurian–L Devonian, fig. 84f), *Pugnax* (U Devonian–U Carboniferous, fig. 84e), the spinose *Acanthothyris* (M Jurassic), and *Cyclothyris* (Cretaceous, fig. 84d). Rhynchonellids are generally found on varied shallow-water substrates (limey muds, silts and sands). They are attached by a small pedicle to other organisms or to shell fragments, sometimes occurring in clusters.

Atrypida

M Ordovician–U Trias

Atrypa (fig. 83b, c). Shell small, about 2 cm; outline oval; short curved hinge line; biconvex; brachial valve very inflated; weak sulcus in pedicle valve and fold in brachial valve; ornament of many fine ribs crossed by well-marked concentric growth lines; inside, short cone-like spiralia with apices directed towards the deep brachial valve; strong teeth and sockets; pedicle non-functional in adult shells.

L Silurian–U Devonian

Atrypids, the first group of spire-bearers, diverged from rhynchonellid stock in the Ordovician. They are distinguished by their curved hinge line. The spiralia, extensions from crura, are variously disposed: they may be directed laterally as in the smooth-shelled *Dayia* (U Silurian–L Devonian) and *Composita* (U Devonian–Permian, fig. 83e); or towards the pedicle or the brachial valve, or towards the median plane.

Spiriferida

L Silurian–L Jurassic

Spirifer (fig. 83a–d). Shell up to about 12 cm; outline subtriangular; straight hinge line marks maximum width; biconvex; pedicle valve with strong median sulcus and corresponding fold in brachial valve; prominent umbo on pedicle valve, incurved above well-defined interarea; delthyrium

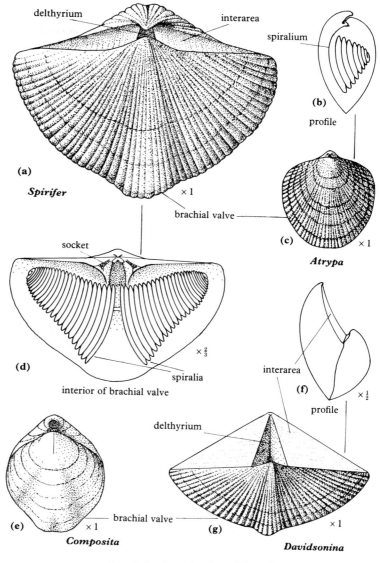

delthyrium

interarea

spiralium

(b)

profile

(a)

Spirifer × 1

brachial valve

(c) × 1

Atrypa

socket

× $\frac{2}{3}$

(d)

interarea

spiralia

(f) × $\frac{1}{2}$

interior of brachial valve

profile

delthyrium

brachial valve

(e) × 1

(g) × 1

Composita

Davidsonina

83 Spire-bearing brachiopods.

with small deltidium; ornament of many radial ribs; inside, spiralia with axis
parallel to hinge line and apices directed laterally; teeth with stout dental
plates.

Carboniferous

Spiriferids are possibly an offshoot from orthid stock. They are a distinctive
group with straight hinge line, biconvex shell with folded anterior margin,
and spiralia arising from crura. This shell design probably arose as an
efficient way of channelling the through-currents of water. The shell is
impunctate. Variations on the basic plan include the development of a very

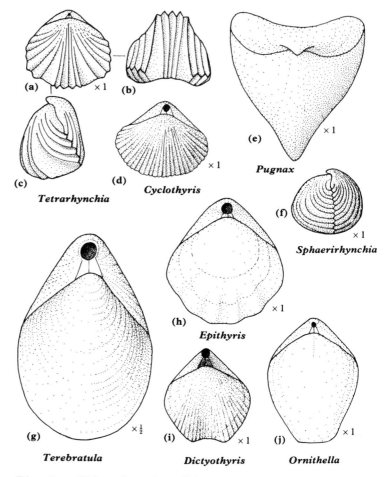

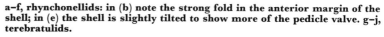

84 Rhynchonellids and terebratulids.
a–f, rhynchonellids: in (b) note the strong fold in the anterior margin of the
shell; in (e) the shell is slightly tilted to show more of the pedicle valve. g–j,
terebratulids.

high interarea on the pedicle valve as in *Davidsonina* (L Carboniferous,
fig. 83f, g) and *Syringothyris* (U Devonian–L Carboniferous). In the latter
the delthyrium is filled in by a plate bearing an interior tube for the pedicle.
The pedicle remained functional in the spiriferids; they lived, hinge down,
on varied substrates, limey muds, silts and sands.

Terebratulida
U Silurian–Recent

Terebratula (fig. 84g).   Shell large, 5–10 cm; outline elongate oval, tapering
to posterior end; pedicle umbo projects above short curved hinge line;
pierced by large foramen with deltidial plates below it; biconvex; anterior
margin gently folded; surface smooth; fine growth lines; inside, securely

interlocking teeth and sockets; brachial valve with cardinal process and short loop; shell punctate.

Miocene–Pliocene

Terebratulids, the last order to appear, are thought to be derived from atrypids. For the most part they are smooth-shelled though in some there is an ornament of ribbing. They are characterised by the short curved hinge line and the more or less complex brachidium in the form of a loop which extends as a calcareous ribbon from the crura. The loop may be short, as in *Terebratula*, *Dielasma* (Carboniferous), *Epithyris* (M Jurassic, fig. 84h), *Dictyothyris* (M–U Jurassic, fig. 84i) and *Terebratulina* (Jurassic–Recent, fig. 79h), and may be associated with a median septum in the brachial valve. In other forms the loop is long (fig. 78f), e.g. *Ornithella* (M Jurassic, fig. 84j) and *Obovothyris* (M Jurassic).

Geological history

Basal Cambrian–Recent

Brachiopods appeared in the earliest Cambrian. Their numbers increased greatly during the lower Palaeozoic to a maximum in the Silurian and Devonian, and then declined gradually during the upper Palaeozoic. Only a few stocks survived into the Mesozoic during which two groups were common, especially in the Jurassic. They were less important in the Tertiary and today are a minor group.

In the early Cambrian, brachiopods were already established and diverse. In the initial radiation both classes, the articulates (orthids only) and the inarticulates were represented. The latter diversified rapidly, all five orders occurring in the lower Cambrian. Their dominance was short-lived and only the lingulids and acrotretids survived beyond the mid-Ordovician. These occur sporadically in later rocks, retaining to the present day their simple unmodified shell morphology.

The orthids were joined in the mid-Cambrian by simple pentamerids. Further radiation in the Ordovician brought greater variety with strophomenids, rhynchonellids and the first of the spire-bearers, the atrypids. These were joined in the early Silurian by the spiriferids, and numbers and variety rose to a maximum in the Silurian and early Devonian.

Brachiopods are of stratigraphic importance in the shelly facies of the lower Palaeozoic. Orthids and strophomenids, with lesser numbers of pentamerids are the characteristic Ordovician forms. In the Silurian they were joined by increasing numbers of spire-bearers, and by the end of that time the first terebratulids had developed.

In the Devonian, spire-bearers, strophomenids and, to a lesser extent, rhynchonellids were of most importance. In the later Devonian, numbers and variety declined as the pentamerids disappeared and orthids were reduced in number.

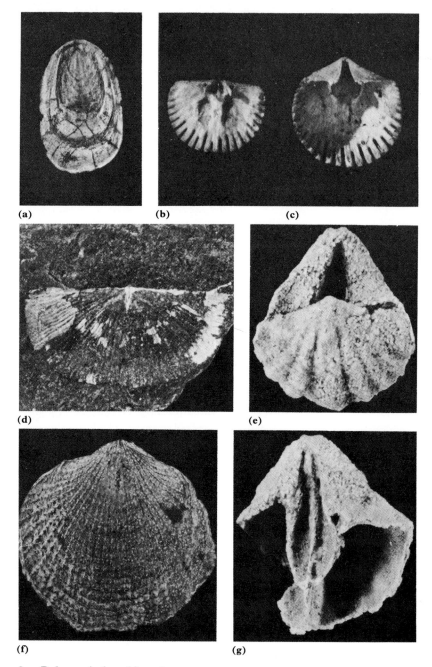

(a)   (b)   (c)

(d)   (e)

(f)   (g)

85   Palaeozoic brachiopods.

a, *Lingula squamiformis*, L Carboniferous (× 2). b, c, and orthid, *Hesperorthis*,
M Ordovician (× 2). d, a strophomenid, *Chonetes*, U Silurian (× 2). e, g, a
spiriferid with spondylium, *Cyrtina carbonaria*, L Carboniferous (× 4). f, *Atrypa
reticularis*, U Silurian (× 2).

In the lower Carboniferous, strophomenids like *Productus* (fig. 82e) and *Chonetes* (fig. 85d) became the dominant brachiopods. Rhynchonellids like *Pugnax* (fig. 84e) were also common. The productids at this time were highly diverse, the many genera being distinguished by their surface ornament and shell form. They are stratigraphically useful in the limestone facies.

'Productids' and, to a lesser extent, spire-bearers continued in importance in the Permian. In England they occur in the upper part of the Magnesian Limestone. Elsewhere, *Richthofenia* and related forms, offshoots from productid stock, adapted to a reef-like environment by developing a coral-like morphology. These aberrant forms occur widely in a band eastwards from Sicily through India to China, and also in North America (Texas).

Several stocks survived the Permian extinctions. These include spire-bearers which lingered on into the lower Jurassic. Most important, however, are the rhynchonellids and terebratulids. The rhynchonellids, with their stable and successful shell design, had from their inception been common, especially in the Devonian. In the Mesozoic, they were numerous and varied in the Jurassic but numbers declined in the Cretaceous. The terebratulids were, in comparison, relatively insignificant until the Mesozoic, when their numbers increased reaching a maximum in the Jurassic and then declining slowly in the Cretaceous. Both groups are of minor importance in the Cainozoic.

Technical terms

ADDUCTOR MUSCLES Paired muscles which, on contracting, close the valves. They leave a pair of muscle scars in the pedicle valve (fig. 78e), and two pairs in the brachial valve (fig. 78f).

ANTERIOR MARGIN the margin of the shell along which the valves open.

BEAK pointed tip of a valve at its posterior end.

BRACHIAL VALVE the dorsal valve to which the brachidium or the lophophore is attached; it is usually the smaller valve (fig. 78f).

BRACHIDIUM a calcareous support for the lophophore in some group of articulate brachiopods (figs. 78f, 79g, 83b, d).

BRACHIOPHORE a diverging plate below each socket in orthids (fig. 81f).

CARDINAL PROCESS a knobbly projection from the hinge line of the brachial valve to which the diductor muscles are attached (fig. 78c, f).

CRURA (sing. crus) a brachidium in the form of two short calcareous prongs.

DELTHYRIUM a triangular gap along the hinge line of the pedicle valve, through which the pedicle emerges (fig. 78e).

DELTIDIAL PLATES two plates which partly or completely fill in the delthyrium (fig. 78a).

DELTIDIUM a single plate which fills in the delthyrium except for the pedicle opening.

DENTAL PLATES plates which project from the floor of the pedicle valve to support the teeth.

DIDUCTOR MUSCLES Paired muscles which, by contracting, open the valves. They run from the floor of the pedicle valve to the cardinal process in the brachial valve (fig. 78c).

FORAMEN a circular opening in, or near, the umbo of the pedicle valve through which the pedicle emerges (fig. 78a).

HINGE LINE line of articulation between the valves along the posterior margin of the shell.

LOPHOPHORE the feeding mechanism consisting of a pair of grooved lobes or arms (BRACHIA) with a fringe of ciliated filaments (fig. 78d, 80).

MANTLE two extensions of the body wall which enclose the viscera and secrete the valves (fig. 78d).

MUSCLE SCARS the impressions on the inside of the valves which mark the area of attachment of the adductor, diductor and pedicle (adjustor) muscles (fig. 78e, f).

PEDICLE the stalk by which the brachiopod is attached to the sea-floor (fig. 78d).

PEDICLE VALVE the ventral valve to which the pedicle is attached (fig. 78d, e).

SEPTUM a narrow plate rising from the floor of a valve.

SOCKET one of two pits in the hinge line of the brachial valve into which the teeth of the pedicle valve fit (fig. 78f).

SPIRALIA spirally coiled calcareous ribbons forming brachidium in some brachiopods (fig. 83d).

SPONDYLIUM a V-shaped arrangement of plates in the pedicle valve, to which muscles were attached (fig. 81k, l).

SULCUS a downfold of the anterior region of one valve opposite a complementary upfold, FOLD, in the other valve.

TOOTH one of two projections along the hinge line of the pedicle valve which fit into sockets in the hinge line of the brachial valve (fig. 78e).

UMBO (pl. umbones) the rounded portion of either valve around the beak (fig. 82e).

7

Arthropoda

Arthropods are bilaterally symmetrical, segmented animals with a chitinous external skeleton and paired jointed limbs. They comprise more than three-quarters of known living species of animals. Today they are enormously varied, including, for instance, the familiar sea-living crabs and lobsters, and the land-living flies, spiders and centipedes. Diversity, too, is the keynote of their past history. Their fossils include an abundance of the extinct trilobites with a well-mineralised skeleton, and many others, some with the most delicate organic structures preserved under unusual circumstances.

The arthropod body consists of a number of articulating segments, each bearing a pair of limbs which may have widely different functions. The exoskeleton, of chitin and proteins, may be strengthened by calcium carbonate or, in some cases, calcium phosphate. It is flexible over the joints between the segments to allow movement, but over the segments themselves it is rigid. Consequently growth cannot occur except during periodic moulting (ECDYSIS) when the exoskeleton comes away and a new one is formed. The nervous system is highly developed, with a brain from which ventral nerve cords arise and give off branches to each segment. Blood is circulated in the body cavity by a heart, and it provides a hydrostatic support in forms with a flexible chitinous skeleton. Eyes, either simple or compound, occur in most forms. Respiration in aquatic arthropods usually takes place by means of gills but may occur through the surface of the body; in land-dwelling forms it is through tubes (tracheae) opening to the outside and ramifying within the body. The sexes are separate in most forms. The arthropod body organisation has proved highly flexible, and members of the group have adapted to all known environments and modes of life including flight and parasitism. They tolerate extremes of temperature: the wingless jumping springtails can live in freezing arctic conditions, and a variety can withstand heat or the desiccation of deserts.

The fossil record of the skeletalised arthropods dates from early in the lower Cambrian. While most Cambrian fossil arthropods are trilobites, it is clear that they were only a part of the picture. The rich fauna of the Burgess Shale (M Cambrian), preserved under unusual conditions, includes, in addition to trilobites, a great range of lightly skeletalised arthropods which were not normally fossilised. It gives us a glimpse of the true diversity of body-plan in Cambrian arthropods, including many which are otherwise unknown.

Traditionally, arthropods have been treated as a single phylum. However, the evolutionary relationship between the major groups is not clear and many authorities believe they represent more than one phylum.

Most fossil arthropods belong to one of the following groups of which only the trilobites, with a major fossil record, are treated in any detail:

Trilobita: trilobites (L Cambrian–Permian)
Crustacea: Branchiopoda (water fleas) (L Devonian–Recent); Ostracoda (Cambrian–Recent, see microfossils, p. 266); Cirrepedia (barnacles, copepods) (L Devonian–Recent); Malacostraca (crabs, prawns) (Cambrian–Recent)
Chelicerata: Merostomata (xiphosurids (*Limulus*); eurypterids) (Ordovician–Recent); Arachnida (spiders, scorpions) (Silurian–Recent)
Uniramia;   Myriapoda (centipedes, millipedes) (Silurian–Recent); Hexapoda (insects, springtails, fleas) (Devonian–Recent)

Trilobita

The extinct trilobites are assigned to the arthropods by virtue of their segmented bodies with chitinous exoskeletons and paired, jointed limbs. They are distinguished by the body being divided longitudinally into three parts, an axial and two lateral regions; by the grouping transversely of the segments in three regions, the CEPHALON, THORAX and PYGIDIUM; and by the form of their two-branched limbs.

The trilobites did not survive the Palaeozoic, and understanding of their body parts and possible behaviour is based largely on the study of modern arthropods like the crustaceans. Trilobites appear to have been entirely marine and mainly benthic, living for the most part in shallow waters. They were, on average, small creatures, measuring about 5–8 cm in length and with an overall range in size from about 5 mm to 70 cm.

Morphology

The exoskeleton enclosed and supported the soft body, providing a firm surface of attachment for the muscles. The part usually preserved is the exoskeleton which covered the dorsal and a part of the ventral side of the body. It consists of a two-layered cuticle of chitin (p. 12) hardened by impregnation with calcium carbonate. Most of the ventral region appears to have been covered by a soft membrane, and the limbs (which are only exceptionally preserved) by chitin. There is ample evidence that trilobites

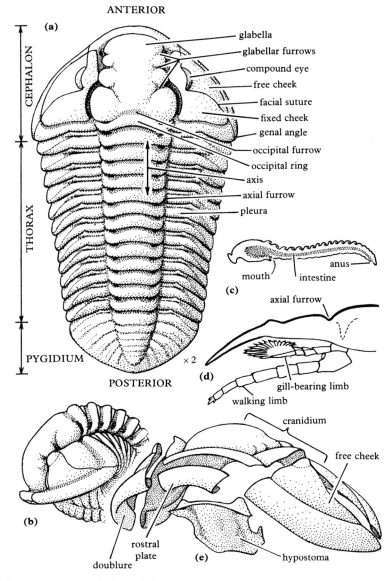

ANTERIOR

CEPHALON

THORAX

PYGIDIUM

POSTERIOR

glabella
glabellar furrows
compound eye
free cheek
facial suture
fixed cheek
genal angle
occipital furrow
occipital ring
axis
axial furrow
pleura

×2

anus
mouth   intestine

axial furrow

gill-bearing limb
walking limb

cranidium

free cheek

rostral
plate
doublure

hypostoma

(a) (b) (c) (d) (e)

86   Morphology of the trilobites.

a–c, e, *Calymene*: a, dorsal view of the exoskeleton; b, lateral view of an enrolled
specimen; c, longitudinal section showing the supposed form of the digestive
tract; e, oblique frontal view of the cephalon 'exploded' along the suture lines to
show the relationship of the various regions. d, a transverse section of part of a
thoracic segment showing a reconstruction of the limbs in *Ceraurus* (anterior
view).

moulted periodically, so that each one in its lifetime may have discarded
many exoskeletons each of which was a potential fossil.

The dorsal exoskeleton is typically a flattened or gently convex shield with
the edges turned under on the ventral side to form a rim, the DOUBLURE

(fig. 86e). It is grooved longitudinally by two AXIAL FURROWS which separate an arched central region, the AXIS, from two side regions, the PLEURAE (fig. 86a). The anterior and posterior parts consist of fused segments forming the head-shield, or CEPHALON, and the tail-shield, or PYGIDIUM, respectively. Between them is the THORAX containing freely articulating segments allowing, in many cases, the trilobite to roll-up, or ENROLL, like a woodlouse (pill-bug) with the head- and tail-shields pressed together to protect the soft ventral side of the body (fig. 86b).

The CEPHALON is typically semicircular in outline. The axial region is called the GLABELLA and is separated by well-defined axial furrows from the side regions, the CHEEKS (fig. 86a). The glabella varies widely in size and shape. It is generally strongly convex and, in many trilobites, shows its originally segmented character in the form of partial or complete transverse divisions, the GLABELLAR FURROWS. The last of these is the OCCIPITAL FURROW which defines the posterior segment, the OCCIPITAL RING (fig. 86a).

The cheeks may be continuous in front of the glabella, or may be separated by it. In the majority of trilobites each cheek is crossed by a suture, the FACIAL SUTURE (fig. 86a) along which the cephalon splits during moulting. The facial suture defines two regions: one, the FIXED CHEEK, remained attached to the glabella, forming with it the CRANIDIUM; while the other is the FREE CHEEK which separated from the cranidium on moulting (fig. 86e). In trilobites which possess eyes these lie on the margin of the free cheek against the facial suture, and must accordingly have been freed at an early stage in moulting (fig. 86a, e).

The facial suture follows one of several different courses. Starting from the anterior margin it passes on the inner side of the eye, and then cuts either the posterior margin, OPISTHOPARIAN condition (fig. 90a); or the lateral margin, PROPARIAN condition (fig. 90g); or, in a very few forms, the angle (genal angle) between the posterior and lateral margins of the cephalon, GONATOPARIAN condition (fig. 86a). The facial suture is not conspicuous in a number of trilobites because it runs along the margin of the cephalon (fig. 90b).

Most trilobites had eyes, though in a number of forms these were rudimentary or absent (blind, fig. 90b). They are located at the inner margin of the free cheeks, abutting against a raised portion of the fixed cheek, the PALPEBRAL LOBE. In early forms, there may be a thin ridge (EYE RIDGE) between the eye and the glabella (fig. 93b). The eyes are more or less kidney-shaped and are compound. The many lenses each consist of a single calcite crystal with its C-axis normal to the lens surface. This orientation would reduce problems of double refraction in calcite. In most trilobites, the eye consists of a very large number (sometimes several thousand, fig. 90c) of tiny, polygonal lenses, closely packed and in contact (holochroal), and covered by a clear cornea formed by a thin extension of

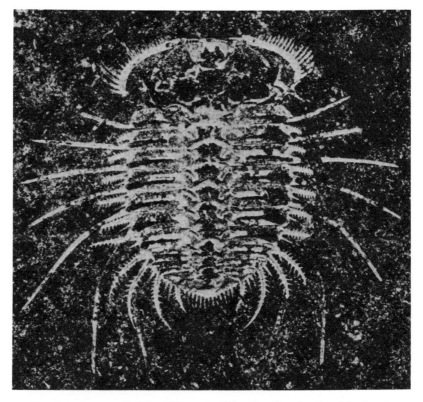

87 **A spinose trilobite,** *Miraspis,* **Silurian, Czechoslovakia (× 4).**

cuticle. This design would have provided mosaic vision as in recent arthropods (in which the eyes are chitinous).

In *Phacops* (fig. 93h) and related forms, however, the eyes are modified with fewer, larger lenses. Each one is thick, biconvex, has its own cornea, and is separated from its neighbour by a chitinous membrane (i.e. schizochroal). The lens design is unique among animals, and may have provided improved focussing for clearer vision.

The genal angles (fig. 86a) are commonly drawn out into GENAL SPINES (fig. 90a), and some genera carry additional spines on the margins of the cephalon or on the glabella (fig. 87).

On the ventral side of the cephalon there may be two or three small plates. One of these, the HYPOSTOMA (fig. 86e), lies in front of the mouth region; and a second lies between the hypostoma and the anterior margin. The hypostoma, a shield-shaped structure, was joined to the cephalon at the hypostomal suture and separated from it during moulting.

The THORAX contains free segments varying in number from 2 to over 40. In related genera the number is often constant. The segments are generally similar except in size, becoming narrower towards the pygidium so that the outline of the thorax tapers posteriorly. Each segment is divided by the axial furrows into a strongly arched AXIAL RING flanked on each side by flatter PLEURAE (fig. 86a, d). The pleurae may have rounded or pointed ends. In specimens with the inner surface exposed clear of rock matrix, paired processes (apodemes) to which the limb muscles were possibly attached, may

88 A trilobite with some of the appendages (preserved as a thin, silvery film) extending from under the dorsal exoskeleton. *Olenoides serratus*, **Burgess Shale, M Cambrian, British Columbia (× 1.75). (Fossil in the United States National Museum.)**

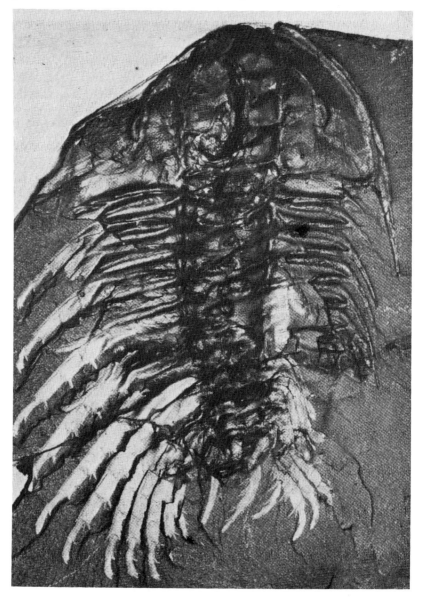

be seen near the axial furrows. (The latter appear as ridges on the inner surface.)

The PYGIDIUM is a semicircular or triangular shield which covered a number of fused segments (fig. 86a). These are usually indicated by transverse furrows on both the axial and pleural regions, which give some indication of the number of segments involved, this varying from about 2 to about 30. In some genera, however, the furrows may be partly or wholly effaced, in which case the pygidium may be smooth. The size of the pygidium relative to the cephalon varies. It may be smaller (MICROPYGOUS); about the same size (ISOPYGOUS); or larger than the cephalon (MACROPYGOUS).

Trilobite LIMBS are known from a small number of species, mostly preserved in fine-grained black shale. They include species of *Olenoides* (M. Cambrian, Fig. 88), *Triarthrus* (Ordovician) and *Ceraurus* (Ordovician, Figs. 86d, 89); and *Phacops* (Devonian). The basic structure is similar in all. There are two types of limbs: (i) one pair of many-jointed whip-like ANTENNAE lying one on each side of the hypostoma (fig. 89); and (ii) a series of two-branched limbs (BIRAMOUS), one pair to each segment and all similar except in length; those on the pygidium being smaller. The two-branched limbs consist of (i) a lower stout branch of about seven joints with spinose inner margins; and (ii) an upper shaft bearing a comb-like fringe of fine slats or filaments lying close under the ventral surface and directed backwards. These paired branches arise from a basal segment, the COXA, which is attached on the ventral surface of each segment. Compared with the limbs of other aquatic arthropods, the trilobite limbs are simple. In the crustaceans, for instance, the 'head' appendages include antennae, or 'feelers', and nipping pincers which hold and cut food; while other appendages may be modified for walking, swimming, breathing, filtering food or creating currents of water. By analogy, in the trilobites the lower jointed limbs are walking legs while the upper fringed appendages are likely to have been primarily respiratory. Specialised feeding limbs are lacking in trilobites. However, from the basal segment (coxa) of each limb a spinose process projects inwards towards the mid-line. These processes (gnath-obases) probably functioned in the manipulation of food, passing it forwards to the mouth, macerating and chewing it in the process.

The development (ontogeny) of some trilobites has been established by the comparison of specimens of the same species in different stages of growth. The earliest stage, about 1 mm long, consists of an almost circular shield with no free segments, but with the axis partly defined. In some genera the eyes are in a marginal position so that the larva may at this stage have been free-swimming like larvae of many modern arthropods. Later stages show the separation of the pygidium from the cephalon, followed by the addition of free thoracic segments increasing in number at each stage until the adult complement of segments is reached. Beyond this stage, there is only an increase in size with each moult.

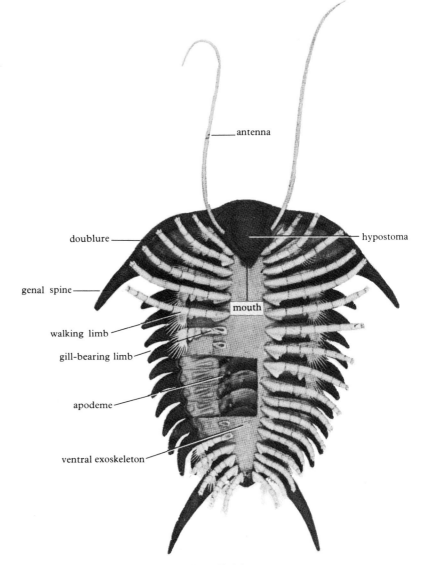

89   Model of the ventral side of a trilobite.
(Based on *Ceraurus* by Mr F. Munro and Dr J.K. Ingham.)

Skeletal form and possible life-style of some trilobites

Calymene (fig. 86a–c, e).   Cephalon semicircular with rounded genal angles; glabella strongly convex, defined by deep axial furrows uniting in front; glabellar furrows incomplete, separating knob-like lobes on each side; eyes small; facial suture cuts genal angle; thorax with 13 segments, gently tapering towards rear; pygidium small with six axial rings and strongly marked pleural furrows.

L Silurian–M Devonian

Finely preserved specimens of *Calymene*, often in enrolled state (fig. 86b), occur commonly in Silurian limestones (e.g. Wenlock) along with benthic animals like corals, brachiopods and bryozoa, and also calcareous algae. The association suggests that the limestone was laid down in clear, relatively shallow warm water. *Calymene* itself was probably benthic. The evidence suggesting this includes the dorso-ventrally flattened shape of the body, with the mouth on the under (ventral) side, and the eyes set high on the cheeks so that vision was restricted to a field overhead.

It must be assumed that its legs were of the normal trilobite walking/breathing type, and that it could crawl around or scrabble in the substrate for food. Its pace, and that of trilobites in general, was probably rather slow. It might also have been able to swim, using its legs, but again only slowly. Trilobites lack the streamlined torpedo shape of fast swimmers.

The possible feeding habits of trilobites are not immediately obvious. The mouth lies on the ventral surface behind the hypostoma and opens towards the rear, so that food must have been moved forwards. Such an arrangement, plus their lack of special feeding limbs, and their probable lack of speed, makes it unlikely that they were active predators or able to cope with hard-shelled prey. Their food is likely to have consisted of soft-bodied sessile animals, seaweed, or perhaps organic-rich sediment. Such food could have been obtained by using the spines of the walking legs to scrape soft tissue, or excavate worms from the substrate, or simply to stir up the sediment. Food was probably passed forwards to the mouth along the mid-ventral line between the legs, the spinose basal segments functioning as gnathobases ('jaw' processes).

Bumastus (fig. 92).   Body strongly convex with rounded-oblong outline; surface smooth with trilobation almost effaced; glabella and axis very broad; facial suture opisthoparian; eyes large; ten thoracic segments; isopygous.
M Silurian

The most notable feature of *Bumastus* is its smooth well-rounded skeleton. This is characteristic of a number of trilobites, e.g. *Illaenus* (fig. 90e) and *Trimerus* (fig. 93g), though not usually so extremely developed. The common occurrence of such an effaced skeleton suggests that it was an adaptation to some particular life-style. Burrowing has been suggested. Unlike many infaunal creatures, in which eyes are reduced or rudimentary, *Bumastus* has well-developed eyes. It is, therefore, unlikely to have been wholly covered but to have lain in sediment so that the eyes were protruding. Thus placed, its field of vision would cover a semicircular arc on each side providing awareness of movement on or near the substrate around it. Like most trilobites, *Bumastus* could enroll, in this state becoming almost ball-like in shape. This would be advantageous if *Bumastus* lived in high-energy areas, e.g. near reefs, or in shallow water, and liable to be washed out of sediment

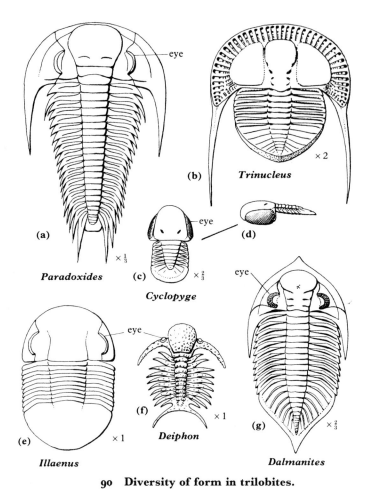

90 Diversity of form in trilobites.

by waves. When enrolled, its globular shape would offer minimal resistance to wave action, and its soft underbody would be entirely protected. *Bumastus* is often found in calcareous rocks, sometimes in crinoidal limestones.

Agnostus (fig. 93c).   Tiny isopygous form, about 5–10 mm length; two thoracic segments; no eyes; no facial suture. Valuable index fossil. Related genera, mostly blind (L Cambrian–U Ordovician).
U Cambrian

Agnostus has a distinctive body form, unusual among trilobites. It is one of a long-ranged group of similar forms, most of them blind, often geographically widespread, which may be valuable as index fossils on an almost global scale.

Rare examples of the hypostoma have been found. It is a fragile structure of unique design, unlike the shield-like hypostoma usual in other trilobites, and appears to have been attached by tissue and, presumably, was movable.

Its modified nature implies a specialised mode of feeding, the interpretation of which depends on its mode of life, pelagic or benthic.

There are differing views on how *Agnostus* lived. It usually occurs in black, fine-grained oceanic deposits, but related forms may occur in varied rock types; limestones, sandstones, silts, chert, and frequently in black organic-rich shales, indicating anoxic conditions. Occurrence in such varied lithologies has been claimed as indicating a pelagic life, and the suggestion has been made that these forms could swim by clapping cephalon and pygidium together, the thorax of only two segments being essentially a hinge between the two parts. This suggestion is refuted on the grounds that no comparable adaptation is known among living arthropods. The nearest analogy might be swimming bivalves, like the scallop, which swim in jerky manner, clapping the valves open and shut (p. 40). An alternative possibility is that they were epifaunal, living among floating algae. If pelagic, the unique hypostoma might have been modified for plankton feeding. However, most forms were blind, which suggests life, perhaps benthic, in dark conditions, in which case the hypostoma may have been used to suck up detritus.

Cyclopyge (fig. 90c, d). Large cephalon with wide glabella; greatly enlarged convex eyes occupy most of the narrow free cheek area and extend over part of the ventral surface; five or six thoracic segments with axis, initially wide, tapering towards the rear; pygidial axis very short.
Ordovician
Cyclopyge is one of various small trilobites in which the eyes are enormously enlarged. Because of this, they have long been thought of as pelagic forms. A number of these trilobites are geographically widespread, occur in an assortment of different rock types, and are associated with different assemblages of fossils. *Cyclopyge* (and related forms), however, is of more restricted geographical distribution and occurs in dark, fine-grained shales. Support for a pelagic mode of life for both of these groups comes from the occurrence of similarly, unusually large eyes in several living pelagic crustacean amphipods. Some of these small animals live at depths of 100–200 m; others at greater depths where the light intensity is much reduced, and large light-gathering eyes are an asset. The amphipods are scavengers and predators which swim only slowly. By analogy, those large-eyed trilobites occurring in a range of deposits are likely to have inhabited shallow depths, ranging from near-shore to open oceanic waters; and *Cyclopyge*, restricted to fine-grained rocks, may have lived in deeper waters in oceanic regions. All these trilobites were probably able to swim, albeit slowly (they show no streamlining), which implies that their appendages were appropriately modified. Appendages, however, have not yet been found in these forms.

Additional common genera

Olenellus (fig. 93a). Wide semicircular cephalon with short genal spines; glabella extends to anterior border and has distinct lateral furrows; large crescentic eyes joined to front of glabella; no dorsal suture, but one on underside of margin of cephalon; thorax with 14 segments tapers to the rear, and axis ends in long sharp spine; pleurae spinose, those on third segment being larger and longer; several posterior segments without pleurae end with small plate.
L Cambrian

Callavia is similar to *Olenellus* but its spines are shorter, and the long axial spine is absent.
L Cambrian

Paradoxides (fig. 90a). Body elongate with large semicircular cephalon; genal spines long; glabella, expanded towards anterior margin, has well-defined furrows; facial suture opisthoparian; eyes large, crescentic; thorax with about 16–21 segments, tapers backwards; long sharp pleural spines; pygidium a small plate with several segmental furrows on axis; large form reaching almost 50 cm.
M Cambrian
Long pleural spines are a feature of many Cambrian trilobites and might be thought of as protective. However, similar spines are not common in later trilobites and it is possible that they had some adaptive function which later became unnecessary.

Shumardia (fig. 93d). Tiny (under 5 mm) blind form without facial sutures; thorax with five to seven segments of which the fourth bears long pleural spines curving back beyond pygidium.
Topmost Cambrian–Ordovician

Olenus (fig. 93b). Small (about 3 cm) with almost rectangular cephalon wider than rest of body; genal spines sharp; eyes connected by eye ridges to front of glabella; opisthoparian facial suture; 10–15 thoracic segments with spinose pleurae; pygidium small.
U Cambrian

Asaphus (fig. 93i). Isopygous; cephalon with rounded genal angles; glabella widens towards anterior margin; occipital furrow distinct; opisthoparian facial suture; eyes close to glabella; free cheeks wide; eight thoracic segments, pleurae rounded; pygidial axis tapers with well-defined segmental furrows, but pleural regions smooth.

L–M Ordovician

Related forms range throughout the Ordovician, e.g. *Ogygiocaris* (fig. 93e).

Illaenus (fig. 90e).   Isopygous, strongly convex, smooth exoskeleton, oval in outline; trilobation not strongly marked; eyes large; opisthoparian suture divergent anteriorly; ten thoracic segments; often found in limestones. Ordovician

Trinucleus (fig. 90b).   Cephalon semicircular, large, wider than rest of body; genal spines long, projecting well beyond pygidium; margin of cephalon forms a wide two-layered brim (fringe) pierced by many radially disposed pits; glabella highly convex with prominent swollen anterior lobe and, posteriorly, three pairs of short lateral furrows; cheeks between fringe and glabella convex, smooth; no eyes; facial suture follows margin; thorax with six segments; pygidium widely triangular; small, up to 3 cm. L – M Ordovician

Trinucleus usually occurs in mudstones. It represents a group of especial importance as index fossils in the Ordovician; related genera include *Onnia*, *Cryptolithus* and *Tretaspis* (fig. 91).

Phillipsia (figs. 93j, 94).   Small, isopygous, outline oval; glabella parallel sided, cut by short furrows; large crescentic eyes set close to glabella;

91   A trinucleid. A lateral view of *Tretaspis sortita*, Starfish Beds, Ashgill Series, U Ordovician, Girvan, Scotland (× 6).

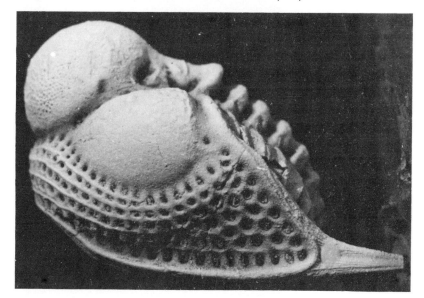

opisthoparian facial suture; short genal spines; nine thoracic segments; many axial rings on pygidium.

L Carboniferous

Phillipsia belongs to the Proetidae (L Ordovician–U Permian) a family which survived longer than any comparable group of trilobites and showed comparatively little skeletal modification during their history.

Phacops (fig. 93h).   Glabella inflated, markedly expanded towards anterior margin and covered with tubercles; genal angles rounded; eyes large, kidney-shaped with lenses distinct; proparian facial suture; 11 thoracic segments (a characteristic of many related genera); pygidium small with axial rings more numerous than pleurae; often found enrolled.

Devonian

Dalmanites (fig. 90g).   Rather flattened exoskeleton, isopygous; cephalon extended backwards into quite long genal spines; glabella widens towards anterior, cut by three pairs of short furrows; eyes large, kidney-shaped, close to glabella and posterior margin; proparian facial suture; 11 segments in

92   Trilobite with exoskeleton from which the axial furrows have been almost completely effaced. *Bumastus barriensis*, Wenlock Limestone, U Silurian, Benthall Edge, Shropshire (× 1.7).

thorax, pleurae with spinose tips; pygidium drawn out to a point at posterior end.

L Silurian–M Devonian

Deiphon (fig. 90f).   Spiky exoskeleton; cephalon with cheeks reduced to long curved spines; eyes set on anterior margins close to glabella; glabella greatly inflated, balloon-like; nine thoracic segments with spinose pleurae curved laterally; pygidium with two pairs of spines, the more posterior long, splayed laterally; small, about 2.5 cm.

Silurian

Encrinurus (fig. 93f).   Surface of cephalon covered by coarse tubercles; genal spines extend obliquely backwards; glabella much inflated, and expanded in front; thorax with 11–12 segments; pygidium triangular with many axial rings and fewer pleural ribs.

L Ordovician–L Devonian

93   Trilobites.

In a, c, d, the facial suture does not appear on the dorsal surface. In the remaining forms the facial suture is either opisthoparian as in b, e, i, j; proparian as in f, h; or gonatoparian as in g.

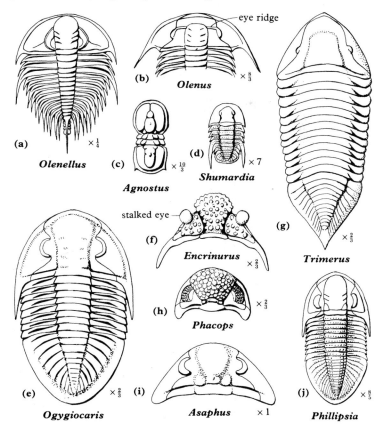

(b) **Olenus** ×⅔

eye ridge

(a) ×¼

Olenellus   (c) ×¹⁰⁄₃   (d) ×7

Agnostus   **Shumardia**

stalked eye

(f) (g) ×⅔

Encrinurus ×⅔   **Trimerus**

(h) ×⅔

Phacops

(e) ×⅔   (i)   (j) ×⅔

Ogygiocaris   **Asaphus** ×1   **Phillipsia**

Trimerus (fig. 93g). Isopygous with cephalon and pygidium subtriangular; exoskeleton smooth and trilobation indistinct; glabella not clearly defined and rather short; facial suture cuts genal angle; thorax with 11–13 segments; axis wide.

L Silurian–M Devonian

Geological history

L Cambrian–U Permian

From the time of their first appearance early in the Cambrian, trilobites were the dominant skeletal animals of the period. They increased in diversity to a late Cambrian peak when there was a considerable turnover, old forms dying out and new ones appearing in the early Ordovician. Many more extinctions occurred in the late Ordovician followed by generally declining diversity which reached a low at the close of the Devonian. Only a single order ranged through the Carboniferous to total extinction in the late Permian.

Trilobites are both complex and diverse when they appear in the lower Cambrian, which implies a previous history, as non-skeletal arthropods, leaving no fossil evidence. *Olenellus* is one of the earliest fossils and shows features characteristic of many trilobites of this time: a large cephalon with large eyes; furrowed glabella; many spinose thoracic segments, and a tiny pygidium. Another lower Cambrian group, the agnostids, are quite different, being tiny, isopygous and blind. The contrast underlines the diversity of this early radiation which continued into the upper Cambrian and ended with the multiple extinctions of that time. Some Cambrian lines, however, persisted including the agnostids and relatives of *Olenus*. To these were added new trilobites such as the trinucleids, and forms related to *Asaphus*, which are especially characteristic of the Ordovician; and others, like the illaenids which ranged on into the Silurian. Most Ordovician trilobites had opisthoparian facial sutures. In the Silurian there was a marked increase in importance of proparian forms, *Dalmanites* and *Encrinurus*, for instance. Forms with more or less effaced trilobation, like *Illaenus* and *Bumastus*, were common at times. By the Devonian, diversity had reduced considerably. Many families disappeared abruptly in the middle of the upper Devonian. Phacopids, however, the best-known Devonian forms, persisted until the end of this period. Only one order, Proetida (L Ordovician – U Permian), continued thereafter, little changed from its earliest forms.

In Britain, various small proetids, like *Phillipsia*, are locally common in shales and limestones of the lower Carboniferous. The last known occurs in a marine band in the Coal Measures.

Many trilobites have a short time-range and are valuable as index fossils defining biozones. They occur in a variety of different facies but are

94 A Carboniferous trilobite, *Phillipsia gemmulifera*, L Carboniferous (× 4).

particularly valuable in the so-called 'shelly' facies of the lower Palaeozoic: shallow-water rocks, including limestones, with a benthic fauna of trilobites, corals, brachiopods, echinoderms and others. On a global scale some trilobites were cosmopolitan, but many were provincial in distribution during certain parts of the Palaeozoic. In Britain, for example, the lower Cambrian of north-west Scotland is characterised by *Olenellus*, whereas in the Welsh Borders its relative, *Callavia*, is found. In Cambrian times these two areas, lying on different plates, were a considerable distance apart, with faunas which, in part, evolved in slightly different ways because of their isolation from each other. Provincialism was well marked until the upper Ordovician when faunas became generally cosmopolitan.

Technical terms

ANTENNAE many-jointed whip-like appendages. Trilobites have one pair attached one on each side of the hypostoma (fig. 89).

APODEME a paired process on the inner surface of the dorsal exoskeleton, near the axial furrow, to which the limb muscles may have been attached (fig. 89).

AXIS central region of the trilobite defined on each side by a longitudinal furrow (fig. 86a).

CEPHALON the anterior part of the dorsal exoskeleton which covered the head region (fig. 86a).

CHEEK the area of the cephalon on each side of the glabella. The cheeks may be separated by the facial suture into the fixed and free cheeks (fig. 86a).

CRANIDIUM the central region of the cephalon comprising the glabella and fixed cheeks (fig. 86a, e).

DOUBLURE the margin of the dorsal exoskeleton which is reflexed onto the ventral side (fig. 86e).

ECDYSIS moulting of the exoskeleton in arthropods.

EYE RIDGE a fine ridge between the eye and the anterior part of the glabella in some trilobites, e.g. *Olenus* (fig. 93b).

FACIAL SUTURE a fine suture which separates the free cheek from the cranidium on each side of the head (fig. 86a); it runs from the posterior or lateral margins of the cephalon across the cheeks, and along or under the anterior margin.

GENAL ANGLE the angle between the posterior and lateral margins of the cephalon (fig. 86a). It may be rounded or produced into a genal spine (fig. 90a).

GLABELLA the arched axial region of the cephalon. It is separated from the cheeks by the axial furrows and may be cut by transverse glabellar furrows (fig. 86a).

HYPOSTOMA a small plate lying in front of the mouth region on the ventral side of the cephalon (fig. 86e).

LIMBS a pair of appendages attached to each segment of the body on the ventral side. They comprise (i) one pair of single-branched antennae, and (ii) many pairs of similar two-branched limbs, each consisting of an outer gill-bearing branch, and an inner walking leg made up of about seven joints (fig. 86d).

OCCIPITAL FURROW a transverse groove across the glabella which separates the posterior segment, the OCCIPITAL RING, from the rest of the glabella (fig. 86a).

OPISTHOPARIAN a facial suture which cuts the posterior margin of the cephalon (fig. 90a).

PALPEBRAL LOBE a raised area on the fixed cheek against which the eye abuts.

PLEURAE the lateral parts of a thoracic segment (fig. 86a).

PLEURAL FURROWS furrows, usually oblique, on the pleurae.

PROPARIAN a facial suture which cuts the lateral margin of the cephalon (fig. 90g).

PYGIDIUM the tail-shield of the exoskeleton covering the fused posterior segments of the body (fig. 86a).

THORAX the separate, freely articulating segments interposed between the cephalon and the pygidium (fig. 86a). The number of segments varies in different genera.

Non-trilobite arthropods with a fossil record

While trilobites, with their well-mineralised skeletons, are the dominant arthropod fossils of the Lower Palaeozoic, they are only a fraction of the real diversity of arthropods which existed then. A glimpse of this variety is seen in the fauna of the Burgess Shale (M Cambrian, British Columbia). Out of 44 species of arthropods occurring there, only one-third are trilobites, the remainder being lightly skeletalised forms not usually preserved. That sample is not an isolated one. Similar fossils to those of the Burgess Shale have recently been found in lower Cambrian rocks in China. These discoveries are reinforced by others in later rocks, including a wealth of finely preserved

95 Diverse arthropods.
a–c, crustaceans: a, branchiopod (Trias); b, cirripede (Eocene–Recent), c, 'prawn', Solnhofen Limestone (U Jurassic). d, xiphosurid, Solnhofen Limestone (U Jurassic). e, primitive palaeodictyopterid insect (U Carboniferous).

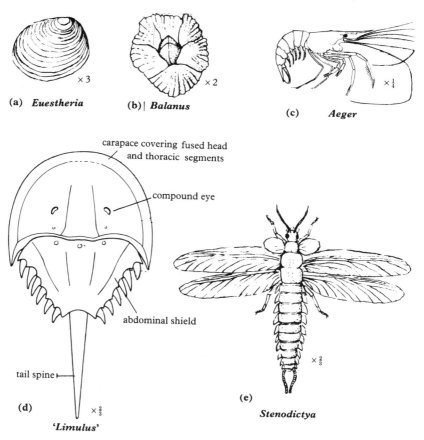

(a) *Euestheria*   (b)| *Balanus*   (c)   *Aeger*

carapace covering fused head and thoracic segments

compound eye

abdominal shield

tail spine

(d)

'*Limulus*'

(e)

Stenodictya

fossils in upper Cambrian dark limestones in Sweden (fig. 96) and in the lower Devonian Hunsrück Slate. Here, too, non-trilobite arthropods predominate.

The non-trilobite forms are of considerable variety. Many of them cannot be assigned to known groups of arthropods, and apparently had a short time-range. In body-plan, however, there are enough features in common to suggest an origin in soft-bodied worm-like stock at a much earlier time.

The forms mentioned below are a sample of those which occur sporadically during the Phanerozoic. They represent some of the main groups of arthropods still living, a few which are extinct, and include aquatic and terrestrial animals.

Crustacea

Cambrian–Recent

Crustaceans are a large and very varied group of aquatic arthropods (marine and fresh-water) and a small number are terrestrial. They are typically gill-breathing with a segmented body, the segments bearing two-branched limbs. The body is divided in to three parts: a head (with compound eyes), thorax and abdomen. The head may bear a shield (CARAPACE), and in some forms it is joined to the thorax. The limbs are variously modified. Those on the head total five pairs: two pairs of antennae (sensory), and three pairs involved in feeding. Other limbs, varying in number, are used in locomotion, may bear gills for breathing, and may be used for filtering detritus.

Examples

Water fleas (Branchiopoda, L Devonian–Recent, fig. 95a) are tiny primitive crustaceans with a one- or two-valved carapace, or rarely, no carapace. They are widespread in temporary pools, often brackish, e.g. in playa lakes in South Africa. *Euestheria* (L Devonian–U Cretaceous), which may occur in abundance in Triassic rocks, for example, is closely similar to a modern form. They are detritus feeders.

Barnacles (Cirripedia), U Silurian–Recent, fig. 95b) are aberrant, sessile crustaceans, familiar objects adhering to rocks on the shore. They are filter-feeders, many of which secrete a wigwam of calcareous plates; these may occur as discrete fossils.

Crabs and prawns (Malacostraca, Cambrian–Recent) are crustaceans of a type (decapods) which appeared in the Trias. Many possess nipping pincers, structures not seen in earlier, more primitive forms. These enabled crabs and

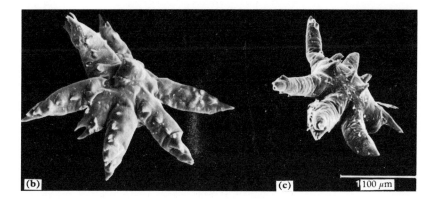

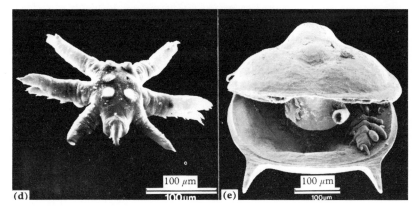

96   Varied, uncrushed arthropods from black nodular limestone 'orsten', U Cambrian, Sweden.

a, *Martinssonia*; this shows crustacean-like features but cannot be assigned to a known arthropod group. b–d, a larval form showing marked similarity to a recent crustacean nauplius larva; b, ventral view; c, d, dorsal views seen from side (c) and back (d), note biramous limbs in (d) e, *Agnostus*, partly enrolled but showing an appendage unlike any known in trilobites. (See also fig. 167, ostracods.) These fossils (*none more than 2 mm in length*) have been preserved by a phosphatic coating which may be extra thick on spines and hairs, making these appear stubby. The limbs are hollow and, where broken, look like hollow tubes. (SEMs courtesy. Professor K.J. Müller.)

97 *Aeger tipularis*, **Solnhofen Limestone, U Jurassic, Bavaria** (× 0.8).

lobsters to become formidable predators in the Mesozoic, preying on, for instance, gastropods. Crabs are crawling, benthic forms, which are common in a few restricted horizons, e.g. nodules in the Gault Clay (Cretaceous) and London Clay (Eocene). Prawns, typically swimmers, tend to be more varied in feeding habit, including burrowing to process mud for its organic detritus. A primitive example from the Solnhofen Limestone (U Jurassic) is shown in figs. 95c and 97.

Chelicerata

Ordovician–Recent
Chelicerates include two major groups, the merostomes (essentially aquatic) and arachnids (mostly land-living). The body comprises an anterior head/thorax shield with six paired appendages, and a posterior abdomen which may end in a tail spine (TELSON). They differ from other arthropods

in having no antennae, the first appendages being pre-oral pincer-like feeding limbs (CHELICERAE). Behind the mouth lie further appendages: one pair for walking, feeding or sensory functions, followed by four pairs of walking legs. Appendages on the abdomen are reduced, or modified as gills.

Merostomes

(Ordovician–Recent)
These comprise xiphosurids and eurypterids.

Xiphosurids (Silurian–Recent) are represented today by the primitive marine kingcrabs such as *Limulus* (Tertiary–Recent) with a semicircular trilobed head/thorax shield, an abdomen of more or less fused segments, and a long tail spine. A Jurassic form from the Solnhofen Limestone (U Jurassic, fig. 95d) is not greatly dissimilar from *Limulus*. More primitive non-marine xiphosurids shown in fig. 98 are widespread in the Coal Measures of the upper Carboniferous; traces of their burrows, made when seeking food, are preserved.

Eurypterids (Ordovician–Permian) were archaic, mainly aquatic and relatively streamlined forms with a long, rather narrow body tapering to the rear. The head/thorax shield was relatively small and the segments of the abdomen articulated freely. The telson was a long spine in some but was spatulate in others. The last pair of walking legs were commonly paddle-like and modified for swimming as in *Pterygotus* (Silurian–Devonian). A related form is shown in fig. 99. Eurypterids were initially marine, their fossils occurring in near-shore sediments. Later they invaded brackish and fresh waters. They were diverse in Siluro–Devonian times. There is evidence that some may have become partly terrestrial. Emergence on land depended on

98 Xiphosurids.
a, *Euproops rotundata* (× 0.8). b, *Belinurus baldwini* (× 1.6). Both from U Carboniferous, Rochdale. (Bronze replicas in the University of Nottingham.)

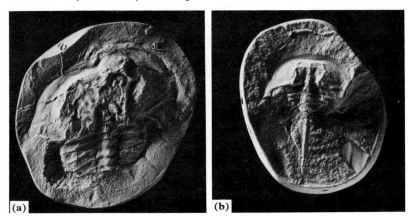

(a)   (b)

99   A eurypterid, *Slimonia*, Silurian, Lesmahagow
(A model, by Mr F. Monro and Dr W.D.I. Rolfe, is included.)

their having adequate protection against desiccation of their gills. They were predators and scavengers usually less than 30 cm long, but some, reaching over 2 m, were giants among arthropods.

Arachnids

Silurian–Recent
These include scorpions, spiders and mites, essentially land-living, air-breathing and predatory animals, with four pairs of walking legs.

Scorpions have a segmented abdomen and a modified stinging telson with a poison gland. The early forms were aquatic; later they became terrestrial. *Palaeophonus* (Silurian, fig. 100) is the earliest known. They are usually small but one form is estimated to have been nearly 1 m long.

Spiders and mites lack a telson and the spider body has a 'waist', clearly defining the anterior and posterior regions; the abdomen is not segmented. They are typically predators feeding on other arthropods and, in later times, parasitic on higher animals (e.g. ticks); some mites eat plants. Remains of mites, spider-like forms and possibly spiders are recorded from the Devonian. Mites, for instance, occur in the Rhynie Chert (L Devonian). As land animals with soft bodies their chances of preservation were poor. They are best known from the Carboniferous, when the humid conditions were

100 A scorpion, *Palaeophonus caledonicus*, Silurian, Lesmahagow, Scotland (× 4)
(The specimen is in Kilmarnock Museum.) (Courtesy Hunterian Museum.)

congenial to them; and from Tertiary amber. *Eophrynus* (U Carboniferous, fig. 101) represents an archaic spider-like Devonian–Carboniferous group. The main deployment of spiders had occurred by the Carboniferous, since when there has been little major change.

Uniramia

Silurian–Recent

Uniramians (insects, centipedes, springtails) are a major group of essentially land-living, air-breathing arthropods with one-branched limbs and with tracheae. They are sufficiently distinct from other arthropods to raise the possibility that they represent a separate line of descent from annelid stock. Their establishment on land was an event which influenced the way in which plants and other animals became established and evolved. A host of tiny forms like springtails, together with fungi, take part in the breakdown of plant debris, and return nutrients to the soil; while insects (p. 169) are inextricably involved in the evolution of plants and animals. Uniramians comprise Myriapods and Hexapoda.

Myriapods

Silurian–Recent

These are the familiar centipedes (Devonian–Recent, predators) and millipedes (Silurian–Recent, plant-eaters). They are long and slender with many trunk segments, each with one or two pair(s) of walking legs. They live in and on plant litter, rotting wood and soil, and are the first known land uniramians. Fossils are rare (fig. 102). A giant fossil, *Arthropleura* (U Carboniferous), nearly 2 m in length, represents an archaic myriapod-

101 An arachnid, *Eophrynus*, a spider-like member of an extinct order, dorsal view. Replica of a fossil preserved in an ironstone nodule, Coal Measures, U Carboniferous (× 3).

like group (Devonian–Carboniferous). The presence of tracheids in its gut indicates its plant diet.

Hexapods

Devonian–Recent

Hexapods, including various wingless forms and the winged insects, are the most varied arthropods, and have the most extensive record. They differ from other uniramians in having the body divided into three distinct sections: head, thorax and abdomen. The thorax contains three segments only, each with one pair of walking legs (hence hexapods). They have no legs

102   A millipede (Diplopoda), U Carboniferous, Lancashire (× 3).
Millipedes have two pairs of legs to each segment, hence Diplopoda. The
specimen is exposed from the left side and is loosely coiled. It shows
overlapping dorsal plates with spine bases, and ventral plates each with
breathing pores and articulating legs. (Courtesy J.E. Almond.)

on the abdomen. Small wingless forms, like springtails (Collembola) and
bristletails, were the earliest, and occur at several horizons in the Devonian:
e.g. *Rhyniella*, a tiny springtail, 1.5 mm in length, much like living forms, in
the Rhynie Chert (L Devonian). Living forms are plant-eaters common in
damp vegetable litter.

Winged insects (Carboniferous–Recent), despite a poor potential for
preservation, have a very extensive fossil record. They occur at relatively few
horizons, mainly in lacustrine deposits, but the number of fossils present may
be enormous. The finest specimens are found in amber where preservation
may be exquisite. The arrival of the flying insects marks stage two in
hexapod evolution. They first occur in the lower Carboniferous in Germany
(primitive forms of dragonflies and cockroaches) and in the early upper
Carboniferous where they are already markedly diverse. The most primitive
forms were those with wings which could not be folded back over the body
(fig. 95e), and in which the larvae were aquatic. They included mayflies (U

103 **A dragon fly, *Protolindenia wittei*, Solnhofen Limestone, Bavaria**
(× **1**).
(Courtesy Professor F.M. Carpenter.)

Carboniferous–Recent) and archaic dragonfly-like forms, some with a wing
span of over 70 cm. Others, with a different wing articulation, were able to
fold their wings over the body, and the young broadly resembled the adults
except in size, e.g. cockroaches (Carboniferous–Recent). The ability to fold
the wings was an asset which enabled them to creep into crevices, a retreat
from predators.

Further diversification followed, and in the Permian, new groups
appeared with a larval stage as worm-like grubs which, after a non-feeding
pupal stage during which the body was redesigned, emerged in adult form.
The Permian record is good. The fauna consists of: (i) survivors from the
Carboniferous many of which became extinct at the close of the Permian;
and (ii) new groups, many of which are still represented today, e.g. true
dragonflies (fig. 103, carnivores), bugs (plant-suckers) and beetles with
varied feeding habits; lacewings and caddis flies also occur.

In the course of the Mesozoic, when fewer fossils occur, other forms are
first noted. These include flies and wasps (Trias); earwigs (Jurassic); social
insects like termites, ants (fig. 104) and bees (Cretaceous); and also moths
and butterflies (Cretaceous, fig. 105). By Cretaceous times the insect fauna
was essentially of modern aspect.

The Tertiary record is extensive. It is contained in several richly
fossiliferous deposits: e.g. the Geiseltal Lignite (Eocene), volcanic ash in
Colorado (Miocene), and amber from the Baltic (Eocene–Oligocene).
Many of the families and genera found in these beds are still extant.

The insect fossil record is a great success story. At an early stage the insects

104 A worker ant in amber, *Sphecomyrma freyi*, U Cretaceous, New
Jersey, USA
(Courtesy Professor F.M. Carpenter.)

105 A butterfly showing colour markings. *Prodryas persephone*,
Oligocene, Colorado.
(Courtesy Professor F.M. Carpenter.)

had developed basic adaptations for terrestrial life: efficient air-breathing tracheae; a waterproof cuticle which prevented desiccation; efficient walking legs and, later, wings, providing mobility. Their initial emergence from water was in the wake of their basic plant food. On land they found and exploited new niches. As plants evolved and diversified, so did insects. The lush plant growth and congenial humid climate of the Carboniferous was ideal for their diversification. Their abundance in the later Carboniferous must have been significant in connection with the development and diversification of contemporary reptiles which found them nutritious food-stuff (as indeed, in later times, did the flying pterosaurs, birds and bats). Insect development, in association with plants, continued in the Mesozoic. It accompanied and contributed to the great radiation of the flowering plants (angiosperms, p. 365) during the Cretaceous, in the shape of important pollinating forms like flies, bees, beetles and butterflies. This association continued in the Cainozoic together with further developments connected with the radiation of mammals: ever opportunistic, the first flea appeared during this time (fossil fleas are found in Oligocene amber).

8

Echinodermata

Echinoderms are exclusively marine animals with a calcareous skeleton and distinctive five-rayed symmetry. They include the varied and familiar sea-urchins, starfish, brittle stars, sea-cucumbers and crinoids, all with a long fossil record. There are also many extinct groups.

Echinoderms have a number of unique features. Their skeleton (TEST), which is *internal*, consists of a variable number of small plates, each of which is a single calcite crystal. They possess a hydraulic system of tubes, the WATER VASCULAR SYSTEM, through which water is circulated in the body. Their FIVE-RAYED symmetry is typically radial, as seen, for instance, in the starfish, but they may also show bilateral symmetry. Internal organs include a gut, with mouth and anus, but there is no head, heart or special excretory system. The nervous system is a network of nerve cords following the plan of symmetry of the body, and sense organs are poorly developed. The various organs are enclosed in the body cavity, or coelom, and an extension from this forms the water vascular system (fig. 106j). This consists of internal tubes through which fluid, mainly sea water, is circulated in the body. The system is closed except for a porous plate, the MADREPORITE, which leads via an axial tube (stone canal) to a central ring vessel around the oesophagus. From this arise five radially disposed branches, the RADIAL WATER VESSELS, which in turn give off side branches to supply soft tentacles, the TUBE FEET. The tube feet extend through pores in the skeleton to the surface where they are located in five bands, the AMBULACRA, overlying the radial water vessels. They are actuated by variation of hydraulic pressure and are used, for instance, in locomotion and respiration.

In echinoderms, the sexes are separate and fertilisation of the eggs takes place in the sea. The larvae swim freely among the plankton for a while, becoming widely dispersed in the process. Most mature echinoderms are benthic, though a small number are pelagic, and range from shallow to abyssal depths in conditions of normal salinity.

Living echinoderms are a small sample of the diversity which marked their early history, and which is indicated by the recognition of over 20 classes (a new class has recently been found in New Zealand waters). They fall naturally into two groups (subphyla): Eleutherozoa, vagrant forms which move in search of food; and Pelmatozoa, essentially sessile suspension feeders. Those classes which have a significant fossil record, and are dealt with here, are:

(i) Eleutherozoa: Echinoidea (sea-urchins)
                       Asteroidea (starfish)
                       Ophiuroidea (brittle stars)
(ii) Pelmatozoa: Crinoidea (crinoids)
                       Cystoidea (a varied superclass including blastoids)

Class Echinoidea

The echinoid skeleton (TEST) is a rigid structure which may be hemispherical, disc-shaped or heart-shaped, and consists of many interlocking plates which are covered by skin. It is, therefore, an ENDOSKELETON. The outer surface is covered by spines which serve as a protection, and which can also be used for walking. There are no arms.

Echinoids are gregarious benthic forms, vagrant to some extent or burrowing in sediment. They live for the most part in shallow coastal waters. Common living examples include the so-called edible sea-urchin, *Echinus*, and the heart urchin, *Echinocardium*.

Morphology

The test of a typical echinoid, such as *Echinus*, is hemispherical in shape. Most of it consists of many interlocking plates, arranged in ten sets of paired columns (i.e. 20 columns in all) which radiate from the apex of the upper (ABORAL) surface to the mouth in the centre of the lower (ORAL) surface (fig. 106). Five sets of columns carry tube feet and are known as AMBULACRA. They alternate with columns in which there are no tube feet and which are known as INTERAMBULACRA. The ten columns together make up the CORONA. Situated at the apex of the test is the APICAL SYSTEM consisting of about ten small plates which are connected with specialised functions, and one of which is the MADREPORITE. The central part of the apical system consists of a membrane, the PERIPROCT, which surrounds the anus. In the centre of the oral surface of the test is a similar membrane, the PERISTOME, which surrounds the mouth. Most of the surface of the test is covered by spines and tiny pincer-like organs (PEDICELLARIAE) which are attached to tubercles of various sizes.

Soft body

The test is filled with fluid in which the rather insubstantial soft organs are suspended (fig. 106j), the gut being looped around inside the body wall as it

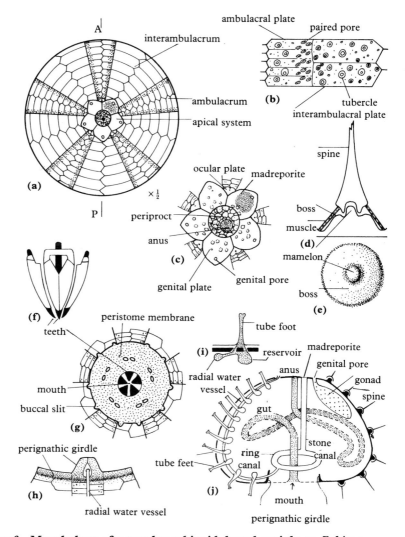

106 Morphology of a regular echinoid, based mainly on *Echinus*.
a, aboral view (A, anterior; P, posterior; line A–P, plane of symmetry). b,
ambulacral and interambulacral plates, enlarged. c, view of apical system,
enlarged. d, section through a tubercle and spine. e, plan of a tubercle. f, side
view of the jaws showing a part of the complex of plates which holds the five
teeth (shaded black). g, oral view showing the peristome membrane (stippled)
surrounding the mouth. h, a part of the perignathic girdle seen from inside the
test. i, a section through a tube foot to show its relationship with the paired
pores of an ambulacral plate (thick black line). j, a simplified section through
an echinoid test to show the arrangement of the internal organs (on the left the
section cuts an ambulacrum and on the right it cuts an interambulacrum); jaws
omitted.

ascends from mouth to anus. The radial water vessels of the water vascular
system underlie the ambulacral areas, and the tube feet penetrate through
paired pores in the ambulacral plates to the exterior. These tube feet,
typically, end in suckered discs, with which the echinoid can cling to a firm

surface. In use, the tube foot is extended by fluid forced into it from a reservoir (ampulla) (fig. 106i) and exerts a suction grip on an object which it touches, by the withdrawal of fluid. In this way the tube feet may be used either for clinging or in locomotion. They are also used for respiration (often modified as leaf-like structures); for feeding; in the construction of burrows; and as chemo-sensors. There are five gonads and their products are discharged through pores in five plates (GENITAL PLATES) of the apical system (fig. 106c).

The outside of the test is covered by skin, and the spines, which are attached by ball and socket joints to the tubercles on the outer surface of the plates, are worked by muscles. The plates and spines consist of a 3-D meshwork of calcite in which the micropores are permeated with soft tissue. Each of the plates and spines is a single crystal, and when on death the soft body decays, the minute pores are usually filled by precipitation of further calcite during the process of fossilisation. Accordingly, the individual plates of a fossil, when broken, show the characteristic calcite cleavage.

Orientation

When viewed from the upper (aboral) surface, the outline of the type of echinoid test just described is a circle, with the apical system at its centre, and the paired columns of plates forming radii (fig. 106a). The ambulacra are said to be RADIAL in position, since each overlies a radial water vessel of the water vascular system (p. 173); and the interambulacra are said to be INTERRADIAL. The only asymmetric feature in the radial symmetry of this test is the position of the madreporite, and this is used to define the ANTERIOR-POSTERIOR orientation. The test is conventionally aligned as shown in fig. 106a with the madreporite on the right, towards the anterior end. The mouth lies on the under (oral) surface.

Test

The pattern of symmetry shown by the test is a convenient basis for separating the echinoids into two groups: REGULAR forms, in which the coronal plates show *radial* symmetry (fig. 106a); and IRREGULAR forms, in which the five rays are arranged with *bilateral symmetry* (fig. 112a, c). In regular forms the anus (surrounded by the periproct) lies *within* the apical system (fig. 106a); and the peristome, with the mouth, is at the centre of the oral surface. The peristome is large and jaws are present. In irregular forms the anus lies *outside* the apical system in the *posterior* interambulacrum (fig. 118a); the peristome is small and may lie either in the centre of the oral surface and have jaws (fig. 116l), or towards the anterior margin and lack jaws (fig. 118g). The plates of the apical system and of the corona may be preserved intact in a fossil echinoid. The position of the periproct and the peristome, however, are usually indicated by a space.

Apical system. In regular echinoids (fig. 106c), the apical system contains ten plates, arranged in one or two rings around the periproct. Five of these, the OCULAR PLATES, are situated radially; and alternating with them are five inter-radial GENITAL PLATES. The ocular plates each bear a pore through which passes the sensory end of the radial water vessel. The genital plates are the larger and each has a pore through which eggs or spermatozoa are discharged. One genital plate (the right anterior) is also the MADREPORITE and is finely perforated. In irregular echinoids (fig. 116i, j), the apical system does not enclose the periproct and it is small and compact; it may contain less than five genital plates, but always has five oculars.

Corona. In Mesozoic and Cainozoic echinoids, with rare exceptions, the ambulacra and interambulacra each consist of paired columns of plates. In Palaeozoic echinoids, however, each may contain more than two columns.

The AMBULACRAL PLATES are small and each is pierced by one pair of pores, a PORE PAIR (fig. 116d), except in those regular echinoids which have COMPOUND ambulacral plates (fig. 106b). The latter consist of two or more plates fused together and have two or more pairs of pores. The pores are round and close-set, except in the irregular echinoids which possess specialised respiratory tube feet; in these, one or both of each pair may be elongated (fig. 112a). Interambulacral plates (fig. 106b) are large and have

107 A modern regular sea-urchin with long spines, up to about 75 mm, *Astropyga magnifica*, underwater photograph at 26 m off Florida Keys, USA.

no pores. Their surface is covered by many tubercles and granules to which, in life, movable spines are attached by muscles. The spines are rarely preserved *in situ* in fossil echinoids, but may occur as discrete fossils separated from the test.

A TUBERCLE consists of a shallow mound, the BOSS, topped by a knob, the MAMELON, and surrounded by a shallow groove, the AREOLE (fig. 106d, e). Tubercles occur on both ambulacral and interambulacral plates, but they are more numerous and larger on the latter. In regular echinoids they vary in size from large primary tubercles to small granules; but they are rather small and close-set in irregular echinoids.

A SPINE (fig. 106d) consists of a shaft with a socket at its proximal end which fits over the mamelon of a tubercle. It is operated by two rings of muscles, an inner 'catch' or holding muscle attached to the boss; and an outer 'quick' or rapidly reacting muscle attached to the areole. Spines are appropriate in size to the tubercles with which they articulate. Thus, in regular echinoids, spines are of several sizes; whereas in irregular forms they tend to be more uniform in size, and are close-set like a bristly fur. Spines are in part a protective device and in part are used for walking or burrowing. Thus they vary widely in shape: they may be needle-like (fig. 107), rod-

108 A fossil echinoid with spines preserved *in situ*: *Cidaris clavigera*, Upper Chalk, U Cretaceous (× 1.6).

shaped, club-shaped (fig. 108) or spatulate. PEDICELLARIAE are tiny organs attached to GRANULES; they consist of a thin stem bearing a flexibly attached, three-piece, pincer-like head. They are used for defence (they may carry poison) and for cleaning the test of detritus and settling larvae. They are rarely preserved.

The lower edge of the corona, to which the peristome membrane is attached, is referred to as the PERISTOME MARGIN. This may be an entire margin, i.e. a simple circular opening; but in regular echinoids with jaws it is notched by BUCCAL SLITS for passage of bushy outgrowths of soft tissue (fig. 106g). These allow movement of fluid when the jaws are in active use. In both regular and irregular echinoids which have jaws, the peristome margin projects inside the test as a series of processes, the PERIGNATHIC GIRDLE, which may form more or less complete arches over the radial water vessels (figs. 106h, 109).

The jaws (Aristotle's lantern, figs. 106f, 110) lie within the mouth, and

109 Peristome margin with perignathic girdle and buccal slits, viewed from inside the test of a regular echinoid: *Hemicidaris*, **Coral Rag, Corallian, U Jurassic, Wiltshire (× 3).**

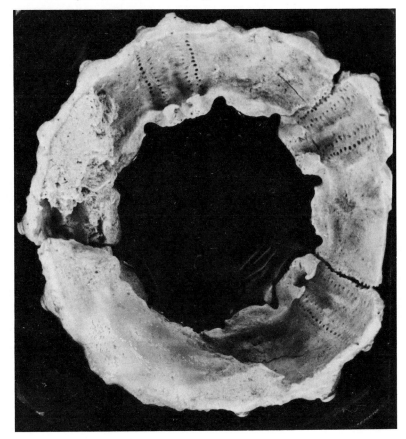

110   Echinoid jaws with teeth. An oral view of 'Cidaris', Coral Rag, U Jurassic (× 1.5).

consist of a complex of calcareous pieces, usually 40, which are disposed in five similar groups each supporting a sharp chisel-like tooth. They are held in place and worked by an intricate system of muscles and ligaments which are inserted between the units and the perignathic girdle. The teeth can be moved out and in, together and apart, to rasp and chop food. They are strengthened on the inner face by a keel.

Modifications of the test occurring in irregular echinoids

Position of the anus.   In irregular echinoids the anus may remain on the aboral side, either flush with the surface or in a groove (fig. 118a); or it may lie on the margin; or on the oral side of the test (fig. 118g).

Ambulacra and pore pairs.   In most irregular echinoids the form of the pore pairs and of the ambulacra shows some modification, especially on the aboral side adjacent to the apical system where the tube feet are specialised for respiration. Some forms may show only a slight elongation of the *outer* pore of each pore pair. In others, the two rows of pore pairs in each ambulacrum diverge and then converge, so that together the ambulacra resemble a five-rayed flower; hence they are described as PETALS (or PETALOID) (fig. 112a).

Oral surface. In those irregular echinoids in which the mouth lies towards the anterior end, jaws and perignathic girdle are lacking. The posterior interambulacrum is extended towards the mouth as a broad ridge, the PLASTRON, and at its forward end may project as a lip, the LABRUM, below the mouth (fig. 119e). The plastron is flanked by similarly extended posterior ambulacra. FASCIOLES are relatively smooth bands on the test of certain irregular echinoids like *Echinocardium* (fig. 112c). The bands are covered by minute granules to which, in the living echinoid, very fine ciliated spines are attached. Fascioles are named according to their location, e.g. a subanal fasciole lies below the anus.

Three echinoids: form of test and mode of life

Echinus (fig. 106). Regular; test hemispherical, flattened on oral side; outline circular as seen from apex; apical system with plates in two rings; genital plates in contact with periproct; ambulacra about half width of interambulacra; plates compound, each with three pairs of pores arranged in three vertical rows; primary tubercles in vertical series with smaller tubercles and granules interspersed; peristome wide with buccal slits; perignathic girdle; jaws with keeled teeth.
Pliocene–Recent
Echinus esculentis (fig. 111) is a common British echinoid which lives mainly on a hard sea-floor in the turbulent sublittoral zone down to about 50 m; a deeper-water species ranges down to about 1000 m. The tube feet, which extend beyond the spines, and their suckered discs can exert a powerful grip on rocks so that the urchin is not readily detached by waves. It can move in any direction and climb steep rock surfaces using its oral tube feet; but on sand walks using its oral spines. It is partly carnivorous, eating, for example, sponges and bryozoa, but also browses on seaweed.

Clypeaster (fig. 112a, b). Irregular; test flattened; outline oval with bilateral symmetry; apical system a star-shaped plate formed by fusion of genital plates including madreporite; ocular plates minute; ambulacra petaloid; periproct on oral side near posterior margin; five radial, mid-ambulacral food grooves converge on centrally placed small recessed mouth; jaws present; internal supporting pillars.
U Eocene–Recent
Clypeaster lives in shallow tropical water about 0.5–3 m deep (fig. 113). It is semi-infaunal and moves only in a forwards direction, using the spines on

111 *Echinus esculentis* **clinging to rock; some of the tube feet can be seen extending beyond the spines; diameter about 90 mm.**
The small echinoid at the bottom right, holding a shell over its apical system, is *Psammechinus miliaris*. Photographed in shallow water, at low spring tide, on the west coast of Argyll.

its oral surface. A dense cover of tiny aboral spines prevents sediment clogging the respiratory tube feet in the petals. Outside the petals, numerous tiny tube feet pick organic detritus out of the sediment, transferring it to the food grooves along which it is passed by ciliary currents to the mouth. The jaws do not protrude and the teeth are used with a horizontal crushing action.

Clypeaster represents a group which appeared in the later Palaeocene and became diverse and widespread by the Oligocene. On several occasions they have given rise to highly specialised forms with very flat, disc-shaped tests (sand-dollars) living in shallow water and sieving fine detritus from the surface layer of sediment in which they are just covered. In the most advanced forms the test margin is notched, and the notches may later develop into enclosed perforations (lunules).

Echinocardium (fig. 112c–e). Irregular; test heart-shaped; anterior ambulacrum (with simple pores) in deep groove leading to mouth; other ambulacra subpetaloid; periproct on posterior margin; fascioles within

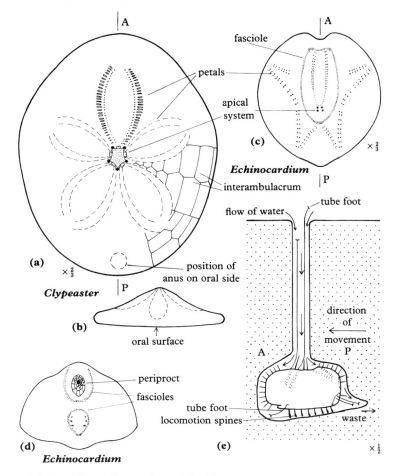

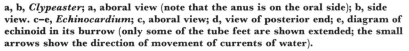

112 Mode of life of irregular echinoids.
a, b, *Clypeaster*; a, aboral view (note that the anus is on the oral side); b, side view. c–e, *Echinocardium*; c, aboral view; d, view of posterior end; e, diagram of echinoid in its burrow (only some of the tube feet are shown extended; the small arrows show the direction of movement of currents of water).

petals and also subanal; plastron with quite coarse tubercles; mouth anterior, small, opening forwards; dense cover of aboral and lateral spines, short, curved and directed backwards; spines on plastron longer and paddle-shaped.

Oligocene–Recent

Echinocardium cordatus lives, infaunally, in sandy substrates around the coast of Britain, is commonest between low-tide mark and 50 m, but may range down to 200 m. It burrows forwards in the sand at a depth of about 15 cm and keeps contact with the sea water through a slender funnel in the overlying sand (fig. 112e). Both spines and tube feet are used in burrowing, and the sand is stabilised by mucus secreted from tiny specialised spines on the fasciole within the petals, and by funnel-building tube feet. It moves

113 *Clypeaster* **in its natural environment.** *Clypeaster rosaceus*
**almost completely covered with sand and shells, including the empty
test of an echinoid.**
Photographed underwater at 7 m depth, off the Florida Keys, USA.

forwards by pushing sand aside and back with a rowing action of the stouter
spines on the oral surface. The funnel is constructed afresh periodically as
immediate food supplies are used up. It is formed by specialised tube feet
concentrated in the upper part of the anterior ambulacrum. These are
highly extensible and have mucus-secreting fringed tips. Within its burrow
Echinocardium maintains a through-flow of fresh oxygenated water over the
test, collects organic particles for food, and disposes of waste. Water is drawn
in by currents produced by the ciliated spines of the fasciole, and is passed
over the respiratory tube feet of the petals down the anterior ambulacrum.
Clogging of these tube feet is avoided by a mucus coating on the dense
canopy of spines which overlies the petals. The current passing down the
anterior ambulacrum conveys food to the mouth, including particles in
suspension and further detritus raked in from the surface by the long tube
feet. Additional food is picked out of the sand by sticky tube feet near the
mouth. Water currents pass on towards the back where waste is flushed
along a sanitary pipe, aided by further currents induced by cilia on the
subanal fasciole. The pipe is constructed by subanal tube feet.

Summary

Echinoids may be assigned broadly to one of three groups, each following a
different mode of life: (i) vagrant regular forms, with a more or less globular

test, which use tube feet and spines for locomotion; (ii) irregular forms, with a more or less flattened test, which live covered, partly or completely, by a thin layer of sediment (fig. 114); and (iii) irregular forms with a heart-shaped test, which live in burrows in soft sediment. All forms are gregarious.

114 Sand-dollars in motion.

Two specimens of *Encope michelini* in the process of disappearing under a thin layer of sediment as they move forwards; the time interval between the two exposures is about five minutes. Photographed underwater at 4 m depth, off the Florida Keys, USA.

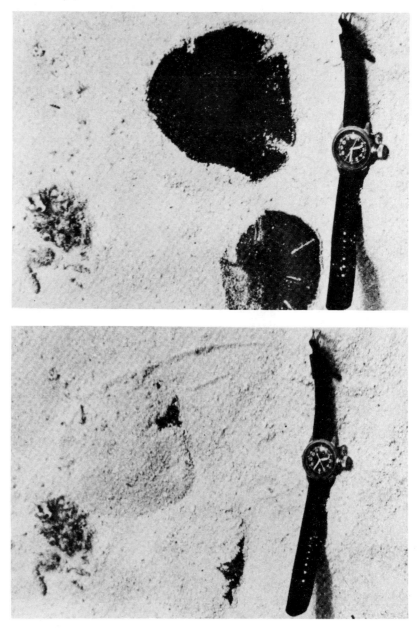

Additional common genera

Palaeozoic echinoids

Palaeozoic echinoids were regular forms in which the corona consisted typically of more than 20 columns of plates (fig. 115). The number of columns varied between 2 and 20 in each ambulacrum, and 1 and 14 in each interambulacrum. The test appears to have been flexible in many genera, with the plates arranged in overlapping series (imbricate), rather like fish scales. In some lower Carboniferous forms the test was a mosaic of thick plates. The ambulacral plates were simple, each with one pore pair. Jaws have been found in many genera. They are rather broad, low-angled, and the teeth are grooved. There is no perignathic girdle. Compared with later forms they were probably less strong and active. Some forms, with many pore pairs around the mouth, may have fed on detritus passed in by tube feet. Fossils are rarely of whole tests but isolated plates may be common. They are readily distinguished as echinoid by the presence of pore pairs or tubercles.

Archaeocidaris (fig. 116a).   Ambulacra with two columns of simple plates; interambulacra with four columns of plates, and large primary and several

115   A Carboniferous echinoid, *Melonechinus*, L Carboniferous, Missouri, USA. (× 0.8). Lateral view of a flattened test.

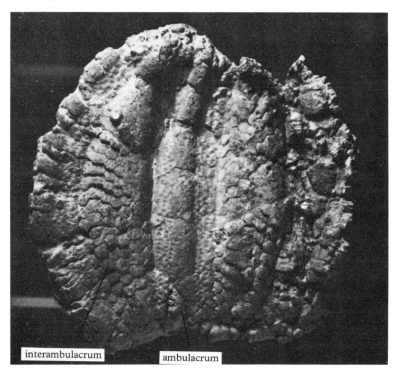

interambulacrum   ambulacrum

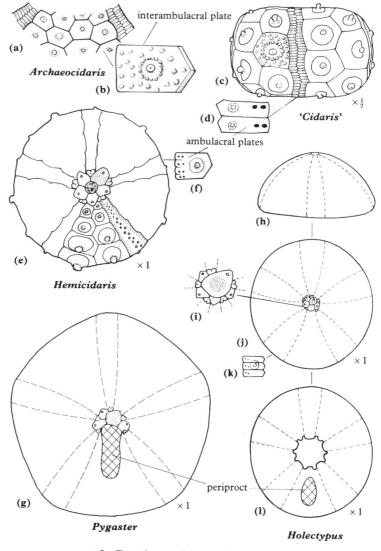

interambulacral plate

(a)

Archaeocidaris

(b)

(c)

(d)

'*Cidaris*'

×½

ambulacral plates

(f)

(h)

(e)   ×1

Hemicidaris

(i)

(j)

(k)   ×1

periproct

(g)   ×1

(l)   ×1

Pygaster

Holectypus

116   Regular and irregular echinoids.

smaller tubercles on each plate; test flexible; jaws present; is a primitive member of the long-ranged order Cidaroidea (U Silurian–Recent).
L Carboniferous

In the later Palaeozoic two events in echinoid history are noteworthy: a marked decline in diversity; and the appearance of *Miocidaris* (Permian–Trias), a cidaroid, the first known echinoid to have 20 columns of plates in the test, and a perignathic girdle. *Miocidaris* is the only echinoid known to have survived the Permian and may represent the stock from which the Mesozoic echinoids developed. The contrast between the Palaeozoic and later forms is used to define two subclasses: Perischoechinoidea (Ordovician–Recent) containing the Palaeozoic forms and all cidaroids; and Euechinoidea (U Trias–Recent), forms with a rigid test of 20 columns.

Post-Palaeozoic echinoids

Post-Palaeozoic echinoids have, typically, a rigid test containing 20 columns of plates in the corona. They fall into two groups: the cidaroids, representing the perischoechinoid stock; and the 'new-look' euechinoids. The cidaroids are regular forms with simple ambulacral plates and without buccal slits. They are long-ranged and show little change in their main characters in the course of their history. The euechinoids contain the vast majority of the post-Palaeozoic echinoids. These show great diversity, including regular forms with compound ambulacral plates and buccal slits; and a wide range of irregular forms.

Perischoechinoid

Cidaris s.l. (figs. 116c, 108). Regular; hemispherical; ambulacra narrow, sinuous, each plate with one pore pair; interambulacral plates each with a large primary perforate tubercle encircled by a ring of granules; primary spines long, stout, spinose or ridged; apical system may have oculars insert between genital plates; peristome wide, with jaws and perignathic girdle. U Trias–Recent

Cidaris s.s. is found today in warm waters; its oral spines are stilt-like, used for walking.

Regular euechinoids

Hemicidaris (figs. 116e, 117). Hemispherical; ambulacra narrow, slightly sinuous with aboral plates simple but, towards the oral surface, becoming compound, each with two to four fused plates; interambulacral plates each with large perforate primary tubercle of cidaroid type; spines long, slender; apical system with two rings of plates; wide peristome with perignathic girdle and buccal slits (fig. 109); jaws. L Jurassic–U Cretaceous

Irregular euechinoids

Pygaster (fig. 116g). Depressed with pentagonal outline and bilateral symmetry; ambulacra straight, narrow with simple plates; interambulacra with numerous small regularly arranged tubercles; periproct outside, but in contact with apical system; four genital plates; mouth central, with jaws. M Jurassic–U Cretaceous

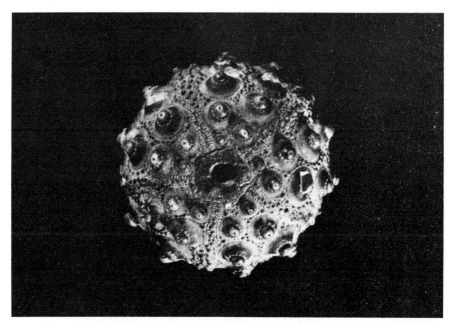

117 *Hemicidaris*, **Coral Rag, U Jurassic, Wiltshire (× 3). Aboral view.**

118 Irregular echinoids.

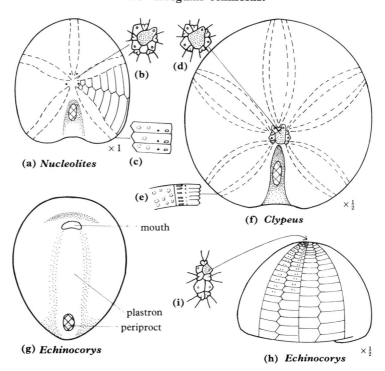

Holectypus (fig. 116i–l).   Hemispherical; outline circular; ambulacra narrow, straight, with compound plates on oral surface; interambulacra with
small scattered tubercles, larger on oral surface; periproct on oral surface;
apical system with five genital plates (one imperforate); mouth central, with
jaws; buccal slits.
L Jurassic–U Cretaceous

Nucleolites (fig.118a–c).   Depressed; heart-shaped outline; ambulacra subpetaloid with outer pore slit-like; interambulacral tubercles small; periproct
in groove immediately behind compact apical system; posterior genital plate
without pore; peristome anterior.
M Jurassic–U Cretaceous

119   Changes in the test in *Micraster*.
**a, aboral view of a late form. b, apical system. c, d, a comparison of the
positions of the apical system and the mouth, and also the depth of the anterior
groove, in a late *Micraster* (c) and an early *Micraster* (d). e, oral surface of a
late form. f, g, pore pairs and cross-section of an ambulacrum of an early form
(f) and a late form (g). h, side view of a late form.**

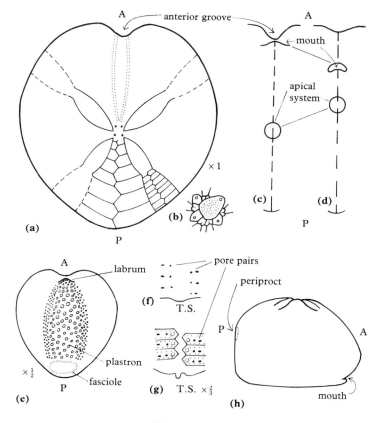

Micraster

Clypeus (fig. 118d–f). Similar to *Nucleolites* but test more depressed; ambulacra more petaloid.
M–U Jurassic

Echinocorys (fig. 118g–i). High, with strongly convex aboral surface; outline oval; ambulacra not petaloid, the anterior three being well separated from posterior two by elongated apical system; four genital plates; tubercles very small; periproct on oral surface; peristome anterior.
U Cretaceous

Holaster differs from *Echinocorys* mainly in its heart-shaped outline with the anterior ambulacrum in a groove leading to the mouth; the other ambulacra subpetaloid.
L Cretaceous–Eocene

120 *Micraster*, aboral view of a late form (× 2).

Micraster (figs. 119–122).   Heart-shaped; anterior ambulacrum in deep groove leading down to the mouth; the other ambulacra subpetaloid; plastron developed on posterior interambulacrum (oral surface); apical system compact with four genital plates; periproct on truncated posterior margin; subanal fasciole; peristome near anterior margin with prominent labrum.

U Cretaceous–Palaeocene

Micraster occurs throughout much of the upper Cretaceous Chalk which represents a time span of several tens of millions years. The lithology of the Chalk varies in detail but much of it was deposited as an exceptionally fine-grained calcite ooze formed from the remains of coccoliths (p. 248). This ooze must have contained a proportion of organic matter which would provide food for detritus-feeding animals. Most infaunal echinoids live in

121   *Micraster*, petals of a late form (× 4).

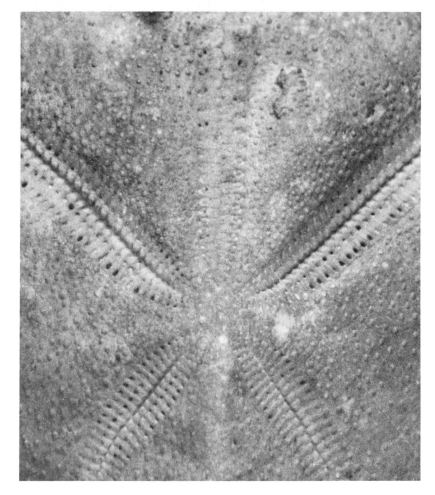

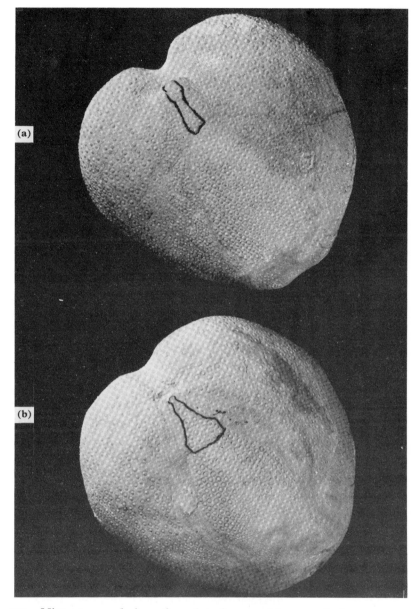

122 *Micraster*, **oral view of a, a late form and b, an early form**
(× **1.5**).

Compare the depth of the notch formed in the outline by the anterior groove;
the development of the labrum; and the nature of the tubercles on and around
the plastron. The ink lines are drawn round the plate forming the labrum.

coarser-grained sediments such as sands or silty sands, but *Micraster* adapted
successfully to a burrowing life in this fine-grained medium.

Some of the changes which occurred in one lineage of *Micraster* have been
interpreted as being connected with improvements in locomotion, food-

collecting and respiration. They may be summarised as follows. (i) Increased size and robustness of test. (ii) Increased height accompanied by formation of a keel (ridge) on the aboral posterior interambulacrum and a backwards shift of the apical system. (iii) Emphasis of the heart-shaped outline as the anterior ambulacral groove became deeper and longer; the mouth, with enlarged labrum, moved forwards. (iv) Lengthening of the petaloid ambulacra, and development of granules on the smooth areas of the plates between the pore pairs. (v) Increased development of tubercles on the oral surface around the plastron. (vi) Widening of the subanal fasciole.

These changes have been interpreted with reference to the morphology of living echinoids such as *Echinocardium* (p. 182) as probably indicating: (a) Improved efficiency of movement within a somewhat sticky fine-grained sediment (i, ii and v). (b) Improved feeding; with emphasis on food, in suspension or collected from the substrate surface, being passed in ciliary currents down the anterior ambulacrum (iii). (c) Improved respiration and circulation of water over the test by increasing the number of respiratory tube feet; and also by the development of small current-creating ciliated spines on the granules (iv). (d) Improved sanitation (vi).

Geological history

U Ordovician–Recent

Echinoids appeared in the later Ordovician but diversity remained low during the remainder of the Palaeozoic, with a minor peak in the lower Carboniferous followed by decline and near-extinction at the end of the Permian. A major adaptive radiation in the upper Trias led to greater diversity, including many irregular forms, in the remainder of the Mesozoic and during the Cainozoic. Today they are still a flourishing group.

The status of *Bothriocidaris* (M Ordovician) which, uniquely, has only 15 columns of plates and was once regarded as the earliest echinoid, is now uncertain. It may represent a dead-end experiment in echinoid-like life-style. By the upper Ordovician, however, echinoids were certainly launched. Three genera are known. They show a primitive stage of ambulacral evolution in having the radial water vessels enclosed within the ambulacral plates instead of lying internally.

Echinoids are inconspicuous in Palaeozoic faunas. Fossils tend to be sporadic, occurring in rocks deposited in quiet offshore habitats, and their remains are seldom well preserved. This is not surprising. They were epifaunal and the plates of their flexible tests readily fell apart to be scattered on death and decay. The greatest number of genera occur in the lower Carboniferous, usually in limestones where *Archaeocidaris*, with long spines, is notable as a primitive cidaroid. Diversity began to decline in the upper Carboniferous, a trend continued in the Permian at the end of which only *Miocidaris* escaped extinction. It remained the sole known genus until the upper Trias, and it set the pattern of the test with 20 columns of plates followed by subsequent echinoids. Beyond this time the cidaroids, having evolved a rigid test, changed little. The first euechinoids appeared at the end of the Trias (Rhaetian). They were regular and marked the start of a major adaptive radiation during which echinoids developed new life-styles both epifaunal and infaunal.

The main changes in the epifaunal regular echinoids concerned the compounding of ambulacral plates, and the evolution of more efficient, stronger jaws. Fossils are, however, less common than those of the infaunal irregulars. The latter appeared in the lower Jurassic and are found in increasing numbers in the later Mesozoic. Their success was the outcome of more efficient food-gathering (leading to eventual loss of the jaws); and improved sanitation (achieved by movement of the periproct away from the apical system). Some, like *Holectypus*, retained approximate radial symmetry. Generally, however, bilateral symmetry was developed and the test became rather flattened, as in *Clypeus*, or heart-shaped as in *Micraster*.

The Mesozoic echinoids may be very abundant in calcareous rocks, for instance in some Jurassic limestones; they are generally rarer in clayey rocks. In the upper Cretaceous, *Micraster* and *Holaster* have proved useful stratigraphically.

Echinoids, especially irregulars like clypeasteroids, are numerous and diverse in warm-water Cainozoic deposits in some regions of the world. In the limited occurrence of these rocks in Britain, however, they are rare.

Technical terms

ABORAL the upper side of the test opposite the mouth (fig. 106a).

AMBITUS the circumference of the test.

AMBULACRA five, radially arranged columns of pore-bearing plates extending between the mouth and the apical system, and overlying the radial water vessels (fig. 106a).

APICAL SYSTEM small plates (oculars and genitals) terminating the ambulacra and interambulacra at the apex of the test (fig. 106c).

ARISTOTLE'S LANTERN see jaws.

BUCCAL SLITS notches in the peristome margin for the passage of soft tissue connected with the operation of the jaws; found in regular echinoids like *Echinus* (fig. 106g).

COMPOUND PLATE an ambulacral plate composed of two or more plates united by a primary tubercle, and bearing two or more pore pairs (fig. 106b).

CORONA the main part of the test, composed of five ambulacra and five interambulacra.

FASCIOLE a narrow band of ciliated spines which articulate with very minute tubercles; it appears on the naked test as a relatively smooth band (fig. 112c).

GENITAL PLATE a plate lying in the apical system at the top of an interambulacrum, and pierced by a pore for the discharge of eggs or spermatozoa (fig. 106c).

INTERAMBULACRA five columns of plates, lacking pores, which lie, inter-radially, between the ambulacra (fig. 106a).

JAWS a complex series of calcareous pieces holding five sharp teeth, possessed by regular, and some irregular, echinoids (figs. 106f, 110).

LABRUM a lip formed by the end plate of the posterior interambulacrum, which projects below the mouth in forms like *Micraster* (fig. 119e).

MADREPORITE a perforated genital plate in the apical system (right anterior side) through which the water vascular system is in contact with sea water (fig. 106c).

OCULAR PLATES five plates lying (in the apical system) one at the top of each ambulacrum (fig. 106c). Each is pierced by the terminal tentacle of the radial water vessel which was once thought to be light sensitive; this explains the term 'ocular'.

ORAL SURFACE the lower side of the test, on which the mouth lies.

PEDICELLARIAE tiny pincer-like grasping organs, borne on granules, used in defence and in cleansing the test.

PERIGNATHIC GIRDLE an extension, inside the test, from the peristome margin, to which the jaw muscles are attached (fig. 106h).

PERIPROCT a plated membrane which surrounds the anus (fig. 106c).

PERISTOME a plated membrane surrounding the mouth (fig. 106g).

PETAL the part of an ambulacrum, shaped like a petal, lying adjacent to the apical system in irregular echinoids and with respiratory tube feet (fig. 112a).

PLASTRON a broad inflated extension of the posterior interambulacrum occurring on the oral side, towards the mouth, in irregular echinoids like *Micraster* (fig. 119e).

PORE PAIR two close-set openings in an ambulacral plate, through which a single tube foot passes (fig. 116d).

SPINE a calcareous rod, articulating with a tubercle, and used in locomotion and defence; the largest are primary spines (fig. 106d).

TEST the entire echinoid skeleton, comprising periproct, apical system, corona and peristome.

TUBE FEET slender extensible tentacles connected with the water vascular system and found in the ambulacra of all echinoderms. They function mainly in locomotion, feeding or respiration (fig. 106i).

TUBERCLES knobs occurring on most plates of the test, to which movable spines are attached (fig. 106e).

WATER VASCULAR SYSTEM the hydraulic system of tubes, through which water is circulated to the tube feet (fig. 106j).

Class Crinoidea

Crinoids are essentially sedentary echinoderms with a CROWN consisting of five simple or branched flexible ARMS arising from a cup-shaped body. They are usually fixed to the sea-floor, at least for a time, by a jointed stem located on the under (ABORAL) side of the body. All parts are supported by an endoskeleton of many calcite plates. Those enclosing the body form the THECA and are arranged with pentamerous symmetry.

Crinoids are gregarious animals which range in distribution from tropical seas to arctic waters. Most living forms are stemless pelagic 'feather-stars' living in clear shallow coastal waters down to about 200 m. But some, the 'sea-lilies', are attached by a stem and live in deeper water mainly down to about 1000 m; a few may range to 6000 m. Fossil crinoids were, typically, attached shallow-water animals.

Morphology

In living crinoids the soft body is contained in the THECA. This consists of two parts, the CALYX (aboral cup) which is covered on the ORAL surface by a plated membrane, the TEGMEN. The arms articulate freely with the calyx. In many genera they branch, and each branch may bear two rows of small branchlets, pinnules. Together they form an open 'funnel' around the mouth. Each bears a deep ciliated food groove (AMBULACRUM) on its oral side, lined with tube feet which trap food (plankton, e.g. diatoms) in mucus. The food is swept by ciliary action down the food grooves to the mouth.

The stem is of varying length. It may be anchored in sediment by 'rootlets' diverging from its base; or it may have prehensile branches (CIRRI) with which the crinoid clings to seaweed, or it may be lacking.

Skeleton

The plates of the crinoid skeleton are disposed in four main regions: the CALYX, TEGMEN, ARMS and STEM (fig. 123). The calyx may be a relatively flexible structure, but in many fossil genera the plates were united to form a rigid cup or box. The individual plates are usually hexagonal or pentagonal in shape, and are arranged symmetrically in circlets.

The CALYX (aboral cup) consists of two circlets of five plates, the BASALS below, and the RADIALS above. In some forms an extra ring of five plates, the INFRABASALS, is intercalated between the basals and the stem (fig. 123b). In certain crinoids the calyx contains additional plates; for instance, there may be INTER-RADIAL plates intercalated between the radials; or in some cases the lowest arm plates, BRACHIALS, may be incorporated in the calyx (fig. 124a).

The TEGMEN (on the oral surface of the calyx) may be a membrane studded with discrete plates, or a rigid structure doming over the mouth and

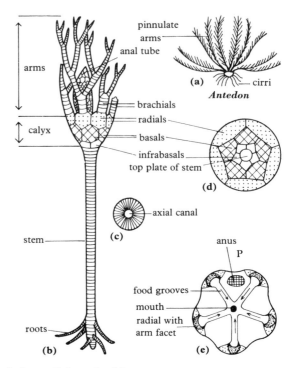

123   Morphology of the crinoids.
a, a free-swimming crinoid. b, an attached crinoid showing the disposition of
the main parts of the body (the arms are incomplete). c, the articular surface of
a stem plate (columnal). d, aboral view of the dorsal cup. e, oral view of the
calyx to show the food grooves converging on the mouth.

adjacent food grooves (fig. 124a). The ANAL OPENING lies in the posterior
inter-radius (fig. 123e) between two food grooves and is either on the surface
of the tegmen or at the tip of a tube, the ANAL TUBE (fig. 123b).

The arm plates (BRACHIALS) articulate freely with the radial plates.
They are cylindrical in shape, with a V-shaped incision on the oral side to
accommodate the food groove. The stem plates (COLUMNALS) also
articulate freely. They are discoid plates of circular (fig. 123c) or star-shaped
outline (fig. 124d), and each has a central hole through which an extension
of the soft parts passes.

Example of a living crinoid

Antedon (fig. 123a).   Calyx tiny; ten pinnulate arms; no stem but many cirri
(prehensile appendages) clustered at base of calyx.
Eocene–Recent
Antedon bifida is the only crinoid found in British waters. It lives from the
littoral zone down to about 200 m in regions swept by currents bearing food
particles and oxygen. In early life it has a stem but later this is lost and it

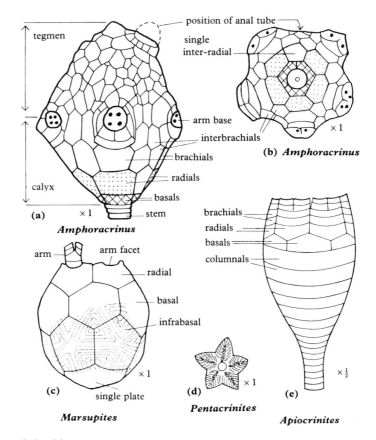

124 Crinoids.
b, aboral view of calyx. c, surface marking drawn on two plates only. d,
articular surface of a columnal.

becomes free-living. It uses the cirri to grip the substrate or crawl about, but
can also 'swim' by slow arm waving.

The arms house extensions of the soft body, including gonads in some of
the pinnules; and some pinnules are sensory. Both arms and pinnules are also
concerned with food collecting, and carry food grooves with sticky, finger-
like tube feet arranged along either side in alternating groups of three. When
feeding, the crinoid bends with the current, its arms and pinnules spread out
fanwise so that the food grooves are facing downstream. The tube feet are
thus exposed to food particles in the water which eddies round them. They
secrete strings of mucus to trap the particles which are passed by ciliary
currents down the food grooves and converge on the five primary food
grooves leading to the mouth.

Most stemmed crinoids live in deep water. There they may be exposed to
moderate currents but others live in quiet conditions and depend on the rain
of plankton from the surface; their arms are spread out to form a passive
collecting bowl.

Some common fossil genera

Palaeozoic crinoids

Most Palaeozoic crinoids were stemmed forms. Most had a rigid calyx, and they show modifications in arm arrangement, including the development of pinnules, which increased the effective area of food-gathering.

Amphoracrinus (fig. 124a, b).   Rigid, ovoid theca; calyx of three basals, five radials and five brachials; a single extra plate (inter-radial) between radials; extra plates (interbrachials) between the brachials; tegmen with polygonal plates forms high domed roof over food grooves; short anal tube; arms (seldom preserved) branch several times; pinnulate.
Carboniferous

Post-Palaeozoic crinoids

Post-Palaeozoic crinoids belong to one subclass (Articulata; see p. 203) and may be stemmed or stemless. They have a simple calyx with a flexible tegmen on which the central mouth and food grooves are exposed.

Pentacrinites (fig. 126).   Calyx small with infrabasals, basals and radials; arms very long, much branched, pinnulate; stem long with cirri at intervals; columnals star-shaped.
Jurassic

Apiocrinites (fig. 124e).   Thick-walled calyx, pear-shaped, tapering aborally to incorporate topmost stem plates; includes basals, radials and some brachial plates; arms (seldom preserved) branch once or twice.
L Jurassic–L Cretaceous

Marsupites (fig. 124c).   Cup-like calyx with large plates which are pentagonal in shape except basals which are hexagonal; base of calyx formed by one plate; above this are five each of infrabasals, basals and radials; plates have a ridged radial surface pattern; radials with notched facets for arms; arms short, branched, rarely preserved.
U Cretaceous

Geological history

M Cambrian, L Ordovician–Recent
Although a single genus is known from the middle Cambrian, crinoids did not appear in numbers until the lower Ordovician. They were in their prime during the Silurian, Devonian and Carboniferous; declined for a time in the

125 *Gissocrinus*, Wenlock Limestone, Silurian (× 2).

Permian and then recovered in the Mesozoic, though not regaining their earlier importance.

Crinoid diversification began in the lower Ordovician but they first became abundant in the Silurian. They remained numerous throughout the Devonian and Carboniferous. They occur most typically in calcareous rocks, such as the Silurian Wenlock Limestone (fig. 125), and lower Carboniferous 'reef' limestones. An entire calyx is rare but separate plates, especially columnals, may be the main constituent of some rocks, appropriately called crinoidal limestones. An entire theca (e.g. of a form like *Amphoracrinus*,

126   *Pentacrinites*, L Lias, L Jurassic (× 2).

127 *Uintacrinus*, **a stemless crinoid, U Cretaceous (× 1).**

fig. 124a) may sometimes be collected from the lower Carboniferous 'reef' limestones of the north of England.

Most crinoids did not survive the late Permian extinctions, but one subclass (Inadunata) continued into the middle Trias. This may represent the stock from which the Mesozoic articulate crinoids were derived in the lower Trias. Earlier Mesozoic forms were mainly stemmed crinoids like *Pentacrinites* (fig. 126); later, in the Cretaceous, stemless forms such as *Marsupites* (fig. 124c) and *Uintacrinus* (fig. 127) became more important. They remain the dominant crinoids today.

Recently, a crinoid resembling the Palaeozoic inadunates has been dredged from below 1500 m in the Indian Ocean.

Minor echinoderms

Asteroidea and Ophiuroidea

L Ordovician–Recent

Asteroids (starfish, fig. 128a) and ophiuroids (brittle stars, fig. 128b) are star-shaped eleutherozoans with a depressed and flexible body, from which five or more arms radiate. The mouth is in the centre of the lower (oral) surface, and ambulacra, with tube feet, radiate from it along the arms. The asteroid arms are not sharply separated from the central body, whereas the ophiuroid arms are clearly differentiated from the central disc. The skeleton consists of discrete internal plates. Living members of each group are usually vagrant, and live on sandy or rocky sea-floors (see fig. 129), but some burrow in sediment.

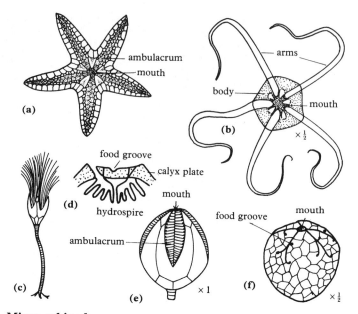

128   Minor echinoderms.
a, asteroid, oral view. b, ophiuroid, oral view. c–e, blastoid: c, restoration of a
complete specimen; d, section across an ambulacrum; e, lateral view of the
calyx. f, cystoid: lateral view of the theca showing the irregular arrangement of
plates.

**129   A starfish, *Oreaster reticulatus*, preying on an echinoid, *Meoma
ventricosa*.**
Photographed underwater at 3 m depth off the Florida Keys, USA.

130   An ophiuroid, *Lapworthura miltoni*, Ludlow, Silurian,
Leintwardine, Hereford (× 1.5).

131   An ophiuroid, *Ophiurella speciosa*, Solnhofen Limestone, U
Jurassic, Bavaria (× 0.7).

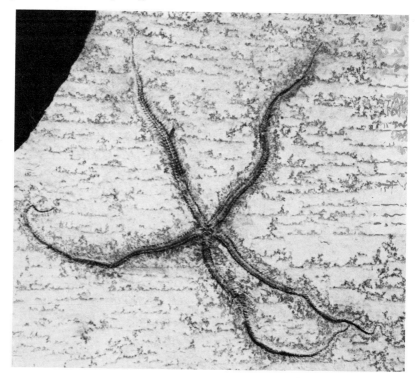

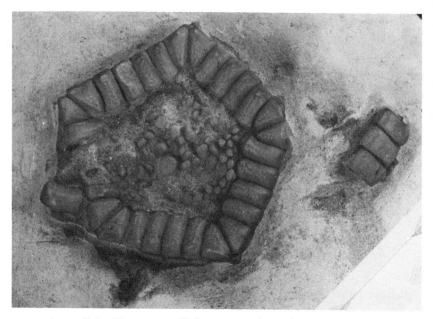

132   A starfish, *Metopaster*, U Cretaceous (× 1.7).
Aboral view of a pentagonal form with very short arms, showing the
arrangement of large plates around the margin, and smaller plates covering the
remainder of the surface.

Asteroids and ophiuroids each appeared in the lower Ordovician. While, in general, they are rare fossils, at some horizons (often referred to as 'starfish' beds) they are relatively numerous, e.g. in the upper Ordovician at Girvan (Scotland) and in the lower Ludlow (U Silurian) in Herefordshire (fig. 130). Mesozoic horizons include the Solnhofen Limestone (ophiuroids, fig. 131) in the upper Jurassic, and the upper Chalk (starfish, fig. 132).

Cystoidea

L Cambrian–U Permian

The cystoids comprise several classes of extinct pelmatozoans (p. 174) with a more or less ovoid or bud-shaped theca from which free, rarely preserved, appendages (BRACHIOLES) arise. The theca was attached to the substrate either directly or by a stem. Most have specialised respiratory structures or pores which are the basis for separating them into classes. The theca consists of many polygonal plates arranged either haphazardly (fig. 128f) or regularly; but in many forms pentameral symmetry is not well developed. The mouth and anus are on the upper (oral) surface. Ambulacra radiate from the mouth over the surface of the theca and, in some forms, may extend along the brachioles. In one class, the Blastoidea (L Ordovician–U Permian, fig. 128e), pentameral symmetry is well developed. The theca was attached either by a stem or directly to the sea-floor. It is bud-shaped and

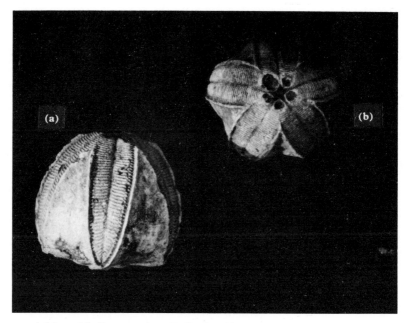

133   A blastoid, *Pentremites*, L Carboniferous (× 2.7).
a, lateral view of calyx. b, oral view.

consists of three circlets, each of five plates. The mouth is in the centre of the
upper surface, surrounded by five pores, one of which is the anal opening
(fig. 133). Five food grooves radiate from the mouth, over the surface of the
calyx and up the delicate arms. Blastoids are distinguished by paired
internal folds of skeletal tissue, the HYDROSPIRES, which underlie the food
grooves and which are presumed to have aided in respiration. Blastoids are
rare fossils; occasionally, however, they may be found in considerable
numbers in shallow-water limestones, often associated with a 'reef' facies as
in the lower Carboniferous.

The Cystoidea appeared in the lower Cambrian and diversity rose to a
maximum in the Ordovician. But most individual classes were short-ranged,
some disappearing at the end of the Ordovician, others in the Devonian. The
blastoids, however, had a longer range, from lower Ordovician to upper
Permian with a diversity peak in the lower Carboniferous.

Technical terms

ANAL TUBE a tube, with the anal opening at its end, which projects from
    the oral surface (fig. 123b).
ARM a jointed appendage bearing an ambulacral groove on its oral side and
    supported by calcareous plates (ossicles), BRACHIALS (fig. 123b).
CALYX (aboral cup) forming the lower part of the theca and consisting of
    circlets of plates including RADIALS above and BASALS below; in some

forms INFRABASALS are intercalated between the stem and the basals
(fig. 123b).

CIRRI small prehensile branches from the stem or base of calyx (fig. 123a).

STEM the flexible stalk, supported by a series of COLUMNAL PLATES, by
which the crinoid is usually fixed to the sea-floor (fig. 123b).

TEGMEN a membrane, with or without plates, which covers the oral side of
the body (figs. 123e, 124a).

THECA the skeleton, comprising the calyx and tegmen but excluding the
stem and free arms.

9

Graptolithina

Graptolites are the remains of extinct colonial marine organisms which are confined to Palaeozoic rocks. They form a class now generally included in the phylum Hemichordata. Living hemichordates are small inconspicuous animals. They include the Pterobranchia which have a fossil record going back to the middle Cambrian and provide a model for inferring the construction of the graptolite skeleton.

The graptolites secreted an organic exoskeleton with a characteristic growth pattern of half-rings joined by zigzag sutures which is otherwise known only in the pterobranchs (p. 221) Each colony originated in a tiny conical cup, the SICULA. From this grew one or more slender branches, called STIPES, which were either free or were linked by short cross-connections, and each of which was made up of a linear series of short overlapping tubes, the THECAE. Each of these housed an individual member (ZOOID) of the colony. The colony, known as the RHABDOSOME, shows bilateral symmetry.

Graptolites comprise eight orders of which only two are important: the Graptoloidea, and the Dendroidea. Graptoloidea have an especial geological importance as zonal fossils in Ordovician and Silurian rocks where they occur, often in great numbers, in dark shales which for the most part lack other types of fossils. They have two characteristics which make them particularly suitable for this purpose: firstly, the various species and genera have comparatively short ranges in time, following one another in relatively rapid succession; secondly, having apparently followed a planktonic mode of life, many are widely distributed throughout the world. Most Dendroidea, by contrast, have a restricted occurrence and their main interest as fossils centres on the clues they provide to relationships between graptolites and pterobranchs. The dendroid graptolites are considered here only briefly.

Order Graptoloidea

In the Graptoloidea the rhabdosome consisted of one or a relatively small number of stipes each comprising a series of similar thecae, and growing from the sicula in a definite pattern.

Preservation. Graptoloids are commonly found in black shales as carbonised impressions, whitish films or pyritised infills resembling tiny fret-saw blades. More rarely, uncrushed specimens are found in limestone or chert (fig. 136). These can be freed by dissolving the matrix with acid (acetic for limestones; hydrofluoric for chert) to which the fossil material is resistant. Often the substance of the skeleton in such specimens is only slightly carbonised and the original structure is preserved. This can be studied in detail using electron microscopy.

Morphology

Nature of skeleton

The skeleton, or P E R I D E R M, is composed of collagen, a fibrous scleroprotein which in the fossil is slightly carbonised, and in transmitted light under the microscope is usually a rich brown or honey colour. The skeleton itself is basically a series of tubes the substance of which consists of two layers, an

134 Morphology of the graptoloids.

a, the proximal end of the stipe of a scandent uniserial rhabdosome (the thecae, of simple type, are numbered 1–5). b, a part of the inner layer of the periderm, showing growth lines arranged in alternating half-rings which meet along a zigzag suture. c, sicula; X is the point of origin of the first theca.

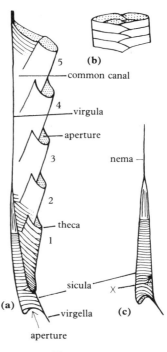

Monograptus

inner fusellar, and an outer cortex. The fusellar layer shows growth bands arranged as alternating half-rings which dovetail along zigzag sutures on the ventral and dorsal sides (or ventral only) (fig. 134b). The cortex is essentially a 'plastering' over the fusellar layer of thin criss-cross bandage-like strips; it probably served to strengthen the skeleton (fig. 142).

Rhabdosome

Stages in the development of the sicula and rhabdosome in a number of different genera are known in some detail. The sicula was secreted by the first

135 Variations in form of the theca and rhabdosome in graptoloids.
The rhabdosome is uniserial in each case except in d, where it is biserial; it is scandent in a–e and in i; reclined in f; horizontal in g, and pendent in h. In d, part of the rhabdosome is broken, exposing the sicula and virgula.

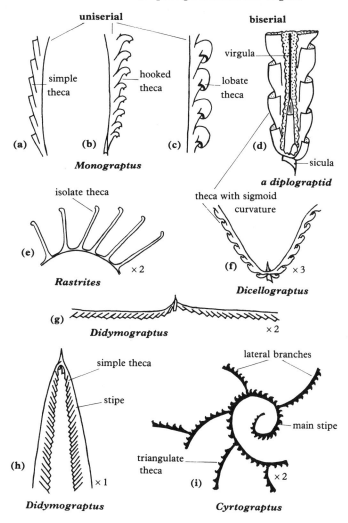

member (a zooid) of the colony, and the thecae were formed by the subsequent zooids using a process of budding which follows a distinct and characteristic pattern in each of several families of graptoloids.

The SICULA (fig. 134c) is a conical structure about 1.5 mm in length. Its apex is extended as a thread-like tube, the NEMA, and its wide end, the

136 Monograptus: an uncrushed specimen extracted from limestone. Lateral view showing the sicula, the first three thecae, and growth lines. *Monograptus chimaera*, Ludlow, Silurian (× 60).

aperture, is open. A spine, the VIRGELLA, projects from one side of the aperture.

The THECAE (fig. 134a) are basically short tubes arranged in an overlapping series along the stipe. Each opens internally in a passage, the common canal, shared with the other thecae; the external opening is the aperture. There is considerable range in shape of thecae, and this is of value in the recognition of species and genera. Thus the thecae may be straight, i.e. simple (fig. 134a), or may show S-shaped or sigmoidal curvature (fig. 135d). In some forms the apertural end is bent over, i.e. hooked (fig. 135b), or may be twisted to one side. The thecae may overlap closely, or may be widely separated, ISOLATE (fig. 135e). The aperture may be circular in shape or constricted. Spines occur on some forms (fig. 136).

The RHABDOSOME may consist of one, two, three, four or many STIPES. The angle at which the stipes diverge from the sicula is fairly constant in a given species, and in most forms with more than one stipe the branching of the stipes is symmetrical (DICHOTOMOUS). The rhabdosome is described as PENDENT (fig. 135h) if the stipes grow downwards from the sicula with the thecae facing inwards and the sicular aperture downwards. It is SCANDENT (fig. 135d) if the stipes grow upwards from the sicula with the thecae facing outwards. Other positions of the stipes relative to the sicula include HORIZONTAL (fig. 135g) and RECLINED (fig. 135f). The stipes may consist of a single row of thecae, UNISERIAL (fig. 135a), or of two, three or four rows of thecae growing back to back, BISERIAL (fig. 135d), TRISERIAL, or QUADRISERIAL (fig. 137h). These forms are also scandent.

The NEMA (fig. 134c) is typically present, although in some graptoloids it may be very short or restricted to early growth stages. In scandent forms the nema is referred to as the VIRGULA (fig. 135d) and it is incorporated in the rhabdosome.

Some common genera

Dichograptus (figs. 137j, 138).   Rhabdosome with eight horizontally disposed stipes, formed by third-order branching close to the sicula; thecae simple, overlap closely.
L. Ordovician

Tetragraptus (fig. 137i, k, l).   Usually four stipes, pendent, horizontal or reclined; thecae simple, overlap closely.
L Ordovician

Phyllograptus (fig. 137h).   Rhabdosome leaf-like, quadriserial, scandent; thecae simple, overlap closely; apertures directed upwards.
L Ordovician

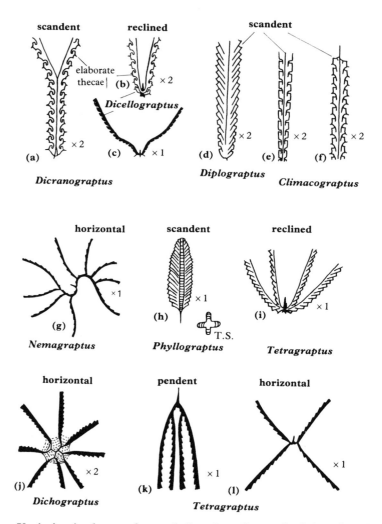

scandent reclined scandent

elaborate
thecae | **(b)** ×2

Dicellograptus

×2

(a) **(c)** ×1 **(d)** **(e)** **(f)**

Dicranograptus *Diplograptus* *Climacograptus*

horizontal scandent reclined

×1

×1

(h) T.S.

(g) **(i)** ×1

Nemagraptus *Phyllograptus* *Tetragraptus*

horizontal pendent horizontal

×2 ×1 ×1

(j) **(k)** **(l)**

Dichograptus *Tetragraptus*

137 **Variation in the number and direction of growth of the stipes in graptoloids.**

The rhabdosome is uniserial in each case except in a, where it is unibiserial; d–f, where it is biserial; and h, where it consists of four stipes. Note also the differences in form of the thecae.

Didymograptus (fig. 135g, h). Two stipes; attitude varied, mainly pendent, horizontal or reclined; thecae simple.
L–M Ordovician

Dicranograptus (fig. 137a). Rhabdosome initially biserial, scandent; stipes diverge distally to become uniserial, reclined; thecae sigmoidally curved with apertures strongly turned inwards.
L–U Ordovician

138 A many-branched graptoloid: *Dichograptus octobrachiatus*, **Ordovician, Victoria, Australia (× 1.5).**

Dicellograptus (fig. 137b, c).   Rhabdosome with two stipes, uniserial, reclined, may be slightly curved or coiled; thecae show pronounced sigmoidal curvatures; apertures may be turned inwards.
L–U Ordovician

Nemagraptus (fig. 137g).   Two main S-shaped slender stipes diverge at about 180° from the sicula; each gives off lateral stipes at intervals; thecae long, low-angled, slender and sigmoidally curved.
M–U Ordovician

Diplograptus (fig. 137d).   Biserial; scandent; nema projects distally; initially thecae have sigmoid curvature, but distally are straight tubes closely overlapping.
Ordovician–L Silurian

Climacograptus (fig. 137e, f).   Like *Diplograptus* but with uniform strongly sigmoidal thecae; apertures narrow; spines in some.
L Ordovician–L Silurian

Monograptus (figs. 134, 135a–c, 136, 139).   Uniserial, scandent; stipe straight or curved; nema in dorsal wall and projects distally; sicula aperture faces down but first theca grows upwards; thecae of diverse shapes in different species, including simple overlapping or isolated tubes, slight or marked sigmoidal curvature, hooked with apertures turned down or curved inwards.
End Ordovician–end L Devonian (Emsian)

Rastrites (fig. 135e).   Curved monograptid with long isolated thecae; thecal tips hooked.
L Silurian

139   a, *Monograptus triangulatus,* **L Silurian** (× 5)**; b,** *Glyptograptus persculptus,* **L Silurian** (× 10)**.**
(Courtesy Dr R.B. Rickards.)

Crytograptus (fig. 135i). Spirally coiled monograptid with lateral branches (cladia) attached; thecae more strongly hooked initially than distally; rhabdosome probably disposed horizontally.
U Silurian

Mode of life

Since graptoloids are extinct and lack close living relatives, any assessment of their life-style is conjectural and based on circumstantial evidence, partly sedimentary, partly biological. The nature of the animal itself is considered later (p. 221).

The graptoloids may occur in clays, calcareous clays, siltstones, sand-stones, conglomerates and, occasionally, in limestones where they are associated with trilobites, brachiopods, cephalopods and other marine organisms. They were therefore marine animals. More commonly, however, they are preserved in great numbers in dark shales which contain few or no other fossils. Such shales were formed in quiet, relatively deep water. Their dark colour is due to organic matter derived from an abundance of planktonic plants and animals living in the photic zone, whose remains sank and accumulated on the sea-bed. Such organic matter is a source of food for many animals and is also susceptible to destruction by oxidation. Its persistence, therefore, is an indication that the source was prolific, that few benthic feeding organisms were present, and that the oxygen concentration at the sea-bottom was low. The shales also contain ferrous sulphide, pyrites, which forms in reducing conditions either within the sediment or, if the bottom waters are completely deficient in oxygen, on the sea-floor itself. In the latter case, the production of H_2S may make the water toxic to most forms of life. This environment, therefore, is one which is unfavourable or even hostile to benthic organisms. On the other hand, it is an environment favourable to the preservation of floating organisms which might sink to the bottom after death, since scavengers would be few or absent. It is most likely, therefore, that the graptoloids were planktonic organisms, living in the well-oxygenated surface waters and likely to be widely distributed by marine currents. Their remains accumulated in quantity, however, only in the dark mud environment. Elsewhere, turbulent conditions and the presence of scavengers would ensure the destruction of the majority of the dead bodies. Graptoloid remains found in near-shore deposits are generally fragmentary.

Observed facts about the wide geographic distribution of the graptoloids, and about the nature of the rhabdosome are in accord with the conclusions reached above. The evidence that they were not benthic creatures lies in the lack of any rooting device by which they might have been anchored to the sea-bed, and also in the bilateral symmetry of the rhabdosome which is not in general a characteristic of sessile animals. Also, there are occasional instances of a number of rhabdosomes being associated in groups, with the nemas

directed towards a centre as if in life the graptoloids hung down from
buoyant material which has not been preserved. It has to be presumed that
graptoloids, like other planktonic animals, were able to adjust their density
in water, e.g. by secreting oil or gas in their tissues, in order to remain at their
preferred level for feeding and illumination.

Geological history

L Ordovician–L Devonian
Graptoloids appeared in the early Ordovician (late Tremadoc) and prolife-
rated rapidly to become a widespread and dominant part of the marine
fauna throughout most of the Ordovician, but diversity dropped sharply
towards the end of that period. In the early Silurian there was an equally
sharp recovery but, in Wenlock times, diversity fell again and remained
generally low, except for a brief revival in the lower Ludlow, until total
extinction at the end of the lower Devonian (Emsian).

Order Dendroidea

Dendroids differ from graptoloids in their much-branched rhabdosome of
uniserial or compound stipes each consisting of a regular alternation of *three*
main types of thecae originating from an internal STOLON. The stipes may
be joined at intervals by transverse bars (DISSEPIMENTS). Most dendroids
were sessile but some were planktonic. Dendroids are usually preserved as
flattened carbonised films, and details of the thecae are often concealed.
Structural details have been revealed in relatively uncrushed material from
limestones and cherts.

   Sessile dendroids were fixed to the substrate by holdfasts or rooting
structures (fig. 141); planktonic forms had a nema. Branching of the
rhabdosome may be irregular or dichotomous. The stolon is a dense,
sclerotised rod-like system within the stipes; at intervals it branches into
three, giving rise to a triad of thecae: an AUTOTHECA, a BITHECA, and a
STOLOTHECA (fig. 140c). The autothecae are simple tubular structures
similar to the graptoloid thecae. The bithecae are similar but smaller. The
stolothecae are essentially tubes which house the stolon as it continues to
generate further thecae and thus extend the stipe.

A common dendroid

Dictyonema (figs. 140a–c, 141).   Rhabdosome more or less conical; net-like;
outline triangular when flattened by compaction; branching dichotomous;
stipes joined at intervals by dissepiments; benthic with discs or root
structures. Planktonic forms with a nema formerly referred to *Dictyonema* are

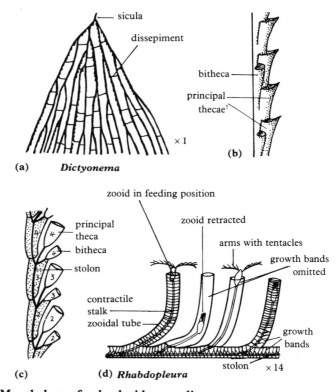

140 Morphology of a dendroid graptolite.
a, part of a rhabdosome. b, part of a stipe showing the arrangement of the
larger and smaller thecae. c, generalised diagram to show the development of
the thecae, in triads, from the internal stolon (stolotheca stippled). d, part of a
living pterobranch.

now transferred by some workers to *Rhabdinopora* (L Ordovician, Family
Anisograptidae).
U Cambrian–U Carboniferous; ? L Permian

Mode of life

The majority of dendroids were sessile animals attached at the sicula end of
the rhabdosome by varied root structures. They occur in small numbers
along with benthic animals (e.g. brachiopods and trilobites) in shallow-
water, inshore sandstones, silty mudstones and limestones. The geographical
distribution of species is very limited but genera may be widespread and very
long-ranging, their essential structures remaining unchanged.

The rhabdosome, composed of collagen, though flexible was kept
relatively rigid by the dissepiments connecting the stipes. It formed a funnel
directed upwards with the apertures of the thecae facing inwards. Currents
of water (natural or induced by the zooids) filtered through the lacy
structure, eddying around while the zooids collected food particles (micro-

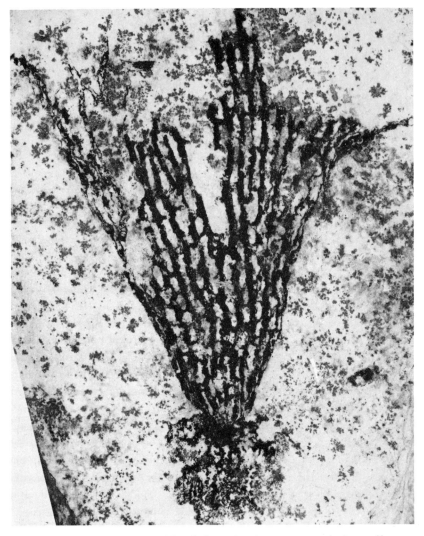

141 *Dictyonema*, **Silurian, North America (approx. × 2.5). A sessile dendroid with basal disc 'holdfast'.**
(Courtesy Dr R.B. Rickards.)

organisms) carried in suspension. The zooids may have been similar to those found in *Rhabdopleura* (p. 221) with a lophophore bearing ciliated tentacles.

Geological history

M Cambrian–U Carboniferous; ? L Permian

Dendroids, initially rare, were well established by the late Cambrian. Sessile forms, like *Dictyonema*, were the first to appear. The only major development in the dendroids came with the appearance in the early Ordovician (Tremadoc) of planktonic forms like *Rhabdinopora*. Later in the Tremadoc an offshoot from planktonic stock gave rise to the major graptolite order of graptoloids. The main line of dendroids persisted without further developments until their extinction which was thought to have occurred in the upper Carboniferous. However, a possible dendroid has recently been reported from the lower Permian in China.

Affinities of the graptolites

Graptolites have been classified in the past with various groups, for instance with the Bryozoa and with the hydroids (Cnidaria). Uncertainty about their affinities is not surprising, for their normal mode of preservation reveals little of their detailed structure. Examination of uncrushed dendroids from chert nodules, first described in 1938, demonstrated that they may be most closely related to the Pterobranchia, a class in the phylum Hemichordata. Living hemichordates are worm-like animals, some colonial, tube-living, and have some features reminiscent of chordates but lack a notochord (p. 286). The pterobranchs are sessile colonial organisms represented today by two genera. One of these, *Rhabdopleura* (fig. 140d) is found, for example, in the North Sea and North Atlantic region. The zooids live in tubes which rise upwards from an irregularly branched, horizontal 'creeping' tube. The substance of the tube is a fibrous organic substance which is almost certainly collagen, the material of the graptolite skeleton. Pterobranchs and graptolites also share a number of other traits. For instance, they both have: (i) a colonial habit; (ii) an organic skeleton of branching tubes; (iii) an incremental growth pattern of half-rings dovetailing along zigzag sutures; and (iv) an internal stolon system (dendroids). The zooids of *Rhabdopleura* (fig. 140d) provide a model which suggests a possible reconstruction of the graptolite animal. They are bilaterally symmetrical with the body joined by a long contractile stalk to the stolon in the creeping stem. They have a lophophore with ciliated tentacles used in collecting micro-organisms from the water as food, and in front of the mouth a fleshy pre-oral lobe.

Rhabdopleura has been observed as it constructs its skeleton. It is secreted by the pre-oral lobe. This lobe bends down to fold over the tube which is built up, a ring or a half-ring at a time, by secretion of a soft jelly-like substance which solidifies quickly. *Rhabdopleura* may not add a cortex as an outer layer. But zooids have been observed to extend from the tube, or even creep out by detaching from the stolon, to repair a damaged tube. Graptolites do have a cortex, (fig. 142), which could have been constructed by zooids capable of extending from the thecae in this way.

Pterobranchs are first known from the mid-Cambrian. Occasional later fossils include *Eorhabdopleura* (lower Ordovician, Tremadoc) and others from the Silurian, Carboniferous, Cretaceous and Eocene. It is thought that they and the graptolites share a common ancestry, perhaps in the late lower or mid-Cambrian.

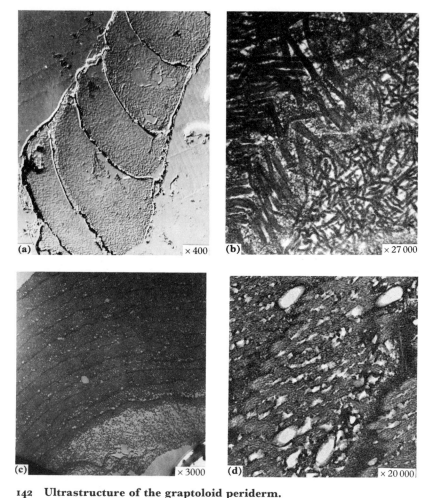

142   Ultrastructure of the graptoloid periderm.
a, *Monograptus dubius*, mid-Silurian, a section showing growth increments in
fusellar layer with fibrils appearing as spongy tissue, b, *Monograptus formosus*,
late Silurian, showing enlarged view of fusellar fibrils. c, *Monograptus
tumescens*, U Silurian, a cross-section of the nema, showing fusellar fabric at
bottom and cortical fabric at top. d, enlarged view of (c) to show collagen
fibrils in cortical layers.

Significant events in the history of the Graptolithina

Graptolites appeared in the mid-Cambrian where three sessile orders, the
dendroids and two minor short-lived groups, were associated. Further
diversification in the lower Ordovician introduced: (i) four minor and short-
lived orders all with encrusting habit; (ii) planktonic dendroids; and (iii)
from the latter, the planktonic graptoloids. Most interest centres on the
graptoloids because their fossils are numerous, diverse, and of great value in
biostratigraphy. The origin of graptoloids from dendroids is undisputed:
there are intermediate forms with characters transitional between the two.

The changes which occurred during their separation include the gradual loss of the small bithecae; the loss of dissepiments between the stipes; and the loss of a sclerotised stolon system. Presumably, a stolon of soft tissue remained.

In detail the evolution of the graptoloids is quite complex. A simplified summary follows. The early Ordovician (Tremadoc) planktonic dendroids (anisograptid fauna) were succeeded later in the Tremadoc by the first graptoloids (dichograptid fauna). In these the stipes were reduced in number (two in *Didymograptus*; four in *Tetragraptus*; usually not more than eight as in *Dichograptus*); the bithecae disappeared; and the uniserial stipes ranged in attitude from pendent to declined to horizontal.

Later Ordovician graptoloids (diplograptid fauna) included mainly reclined two-stiped forms (*Dicellograptus*) and scandent biserial forms (*Diplograptus*, *Climacograptus*). In all these the thecae may show slight or strong sigmoidal curvature.

The reclined two-stiped forms disappeared at the end of the Ordovician but the scandent biserial forms persisted in the Silurian. The first monograptids, uniserial scandent forms with quite simple thecae, appeared in the latest Ordovician. They continued throughout the Silurian with elaboration of thecal shape and aperture. In the lower Devonian diversity fell and the graptoloids became extinct in the Emsian. The dendroids, however, persisted until the upper Carboniferous or, possibly, into the lower Permian.

Technical terms

BISERIAL describes a rhabdosome in which two rows of thecae are united back to back and grow in an upwards direction from the sicula enclosing the virgula.

COMMON CANAL the cavity along the axis of the stipe which contains the stolon, and into which the thecae open.

DENDROID refers to members of the Dendroidea; it also describes an irregular bushy branching habit.

DICHOTOMOUS dividing into two equal branches.

DISTAL END last-formed part of the skeleton.

DORSAL side of the rhabdosome opposite to the apertures of the thecae.

ENCRUSTING describes the habit of some sessile organisms which overlie and adhere to rocks, etc., by the secretion of additional skeletal material.

ISOLATE describes the wide spacing of adjacent thecae along a stipe.

NEMA the thread-like extension of the apex of the sicula.

PENDENT refers to a rhabdosome in which the stipes hang down from the sicula.

PROXIMAL the first-formed part of the rhabdosome.

RHABDOSOME the graptolite colony.

SCANDENT describes stipe(s) growing erect along the virgula.

SICULA skeleton of first member (zooid) of the colony.

SIGMOIDAL S-shaped bending of the thecae.

SIMPLE refers to a theca with straight tubular shape.

STIPE one branch of the rhabdosome made up of a linear series of thecae.

STOLON an internal rod-like structure found in dendroid graptolites, from which the zooids originated.

THECA the tubular structure in which a zooid was housed.

UNISERIAL describes stipes consisting of a single series of thecae.

VENTRAL the side of the rhabdosome bearing the apertures of the thecae.

VIRGELLA spine projecting from the margin of the aperture of the sicula.

VIRGULA name given to the nema when it is incorporated in the rhabdosome of scandent forms.

ZOOID the individual animal in a graptolite colony.

10

Miscellany of minor groups

Porifera

The Porifera (pore-bearers), or sponges, are the simplest of the many-celled animals. They lack defined tissues and are thus excluded from higher grade animals (Metazoa). They are mainly marine, sessile filter-feeders. They may be encrusting and irregular in shape but many have a definite form, e.g. bowl- or vase-shaped with approximate radial symmetry. The skeleton is internal and may be organic (spongin, a protein), or made of opaline silica, or calcite and/or aragonite. Only those with a mineral skeleton are likely to become fossils.

A sponge has no true tissues, being essentially an aggregation of a few types of cells. The simplest form is sac-shaped (fig. 143a). The body wall consists of an inner and outer cell layer sandwiching a jelly-like substance containing 'wandering' cells. The latter secrete the SPICULES (fig. 143g) of which the skeleton is composed. The spicules may remain discrete and thus separate on the death of the animal; or they may unite to form a rigid structure (either calcareous or siliceous) which may be preserved. The body wall is perforated on the outside by small pores (OSTIA) leading by a system of canals to the central cavity. This is lined by special feeding COLLAR CELLS which have thread-like flagella (fig. 143c). These beat to and fro creating a current of water which is drawn in through the pores into the central cavity and passed out by one or more large openings (OSCULUM pl. oscula) on the upper surface. In the process oxygen is absorbed, food particles (micro-organisms) are filtered out by the collar cells, and waste is carried away. In larger forms the area of collar cells is increased by infolding of the body wall to form small U-shaped, or more complexly patterned chambers (fig. 143b).

Examples

Siphonia (fig. 143d, e). The skeleton is tulip-shaped and was attached by a stalk; minute surface pores lead to a system of canals which traverse the thick

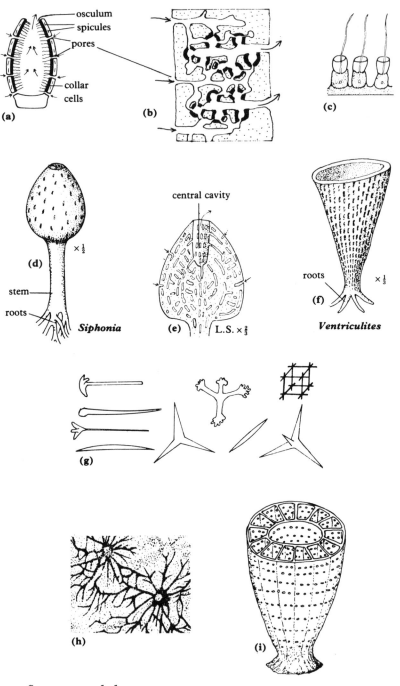

143   Sponge morphology.

a, longitudinal section through a simple type of sponge to illustrate the general structure of the body. b, part of a more complex sponge. c, enlarged view of collar cells. d, e, tulip-shaped sponge; e, is a longitudinal section showing the system of radial and vertical canals (the latter leading into the central cavity). f, vase-shaped sponge. g, varied spicules. h, radially arranged surface grooves in a stromatoporoid (redrawn from Lehmann). i, an archaeocyathid.

wall, some in a radial direction and others parallel to the outer surface
(fig. 143e); spicules tightly packed, of irregular shape; siliceous.
M Cretaceous–Tertiary

Ventriculites (fig. 143f).   The skeleton is roughly vase- to bowl-shaped,
varying from deep to shallow, and was attached by 'rootlets'; the pores lead to
simple radial canals; spicules six-rayed, crossing at right angles; siliceous.
M–U Cretaceous

Raphidonema (fig. 144).   The skeleton is an open vase- or funnel-shape, thick-
walled with rough outer surface; canal system rather indistinct; spicules one-
and three-rayed; calcareous.
Trias–Cretaceous

Sclerosponges

Sclerosponges, coralline sponges recently discovered living deep down in
coral reefs off Jamaica, have a skeleton of organic fibres, needle-like siliceous
spicules and aragonite. The latter is secreted as a compact base into the top of
which the soft body penetrates via minute pits. Traces of these pits are
preserved as narrow tubes with horizontal partitions, within the base, and
silica spicules may also be entrapped. The living tissue is a thin veneer over
the base and has a knobbly surface with many small incurrent pores lying in
depressions between domes where larger excurrent pores open. Water

**144   *Raphidonema*, a calcareous sponge, L Greensand, L Cretaceous,
Faringdon, Berkshire.**

drawn into the sponge is channelled along radially arranged canal systems, each converging to open in an excurrent pore.

Sclerosponges are unimportant as fossils. But chaetitids (Ordovician–Cretaceous), once regarded as tabulate corals (p. 110) resemble them closely and some, at least, are now classed as an order of sclerosponges.

Mode of life

Sponges are gregarious, living for the most part attached to a hard substrate in relatively clear water. Calcareous sponges live in reasonably sheltered warm shallow waters, less than 100 m deep. Fossil forms show similar preferences. The siliceous sponges range into deeper water, most occurring between 100 and 400 m. Those with six-rayed spicules, the glass sponges, are also found in deeper water, down to about 6000 m or more, living on a muddy substrate.

Geological history

Cambrian–Recent
Generally, sponges are abundant only locally and occur mainly in shallow-

145 An association of a sponge (*Siphonia*) and a worm in flint nodules from the Chalk, U Cretaceous. The worm lived within the sponge body in a spiral tube which opened to the surface at the margin of the osculum.

a, a polished median section of a sponge impregnated with silica; the circles are sections of the mould of the worm tube. b, a mould of the central cavity of a sponge and of the spirally coiled worm tube (× 0.6).

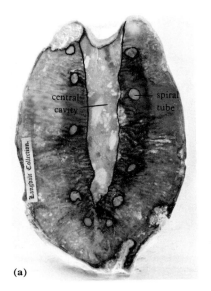
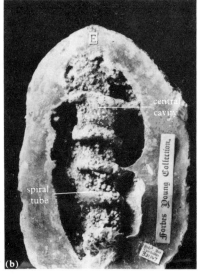

(a)   (b)

water rocks such as sandstones and limestones, including reefs. Palaeozoic sponges were predominantly siliceous forms and are commonly represented by isolated spicules in cherts and silicified limestones as, for instance, in the Lower Carboniferous. Chaetitids are common at this horizon in shallow-water reef structures and calcareous shales.

In Britain sponges are commonest in the Cretaceous, e.g. in the Faringdon Gravels (L Cretaceous, fig. 144) and also in the Chalk, where they are often enclosed in flint nodules (fig. 145).

Stromatoporoidea

? Cambrian, Ordovician–Cretaceous

Stromatoporoids, extinct marine organisms of uncertain affinities, secreted massive calcareous skeletons (fig. 146) forming irregular mounds and sheets varying from a few centimetres to over 2 m across. The structure is sometimes ill-defined but may show a closely reticulated arrangement of horizontal laminae and fine vertical pillars. Formerly they were often assigned to the cnidarian hydrozoa (p. 121). However, their structure is similar in some respects to that of the sclerosponges; for instance they may have radially arranged surface grooves (figs. 143h, 146d), and canal systems parallel to the growth laminae; also in some Mesozoic forms spicule pseudomorphs have been found. Such evidence indicates that some stromatoporoids, at least, should be referred to the Porifera.

Stromatoporoids are an important constituent of Silurian and lower to middle Devonian reefs associated with corals, calcareous algae and cyano-bacteria. They are rare in later rocks.

Archaeocyatha

L–M Cambrian

Archaeocyathans were marine animals with features reminiscent of sponges and corals but are distinct from either. The calcareous skeleton is typically an open inverted cone shape which is often flared into a holdfast for attachment to the substrate, and, usually, is twin-walled (fig. 143i). Both inner and outer walls are perforated by pores, and a series of radially arranged vertical plates lie between them. These, too, have pores. It is presumed they were filter-feeders. They were small, usually about 1–3 cm diameter and up to about 10 cm high, but sometimes larger. They occur in reef structures often associated with cyanobacteria and calcareous algae, in warm, shallow seas estimated to have been about 20–30 m in depth. They are confined to the calcareous facies of the Lower and Middle Cambrian with a world-wide, low-latitude distribution. They are useful stratigraphic guide fossils.

Bryozoa

Bryozoans are aquatic colonies of tiny individuals which typically secrete a protective skeleton. They are sessile, mainly marine and occur from tide level down to abyssal depths. They are encrusting (sea-mats) or freely-branching (sea-mosses), common today and abundant as fossils.

The bryozoan colony contains many interconnecting individuals (ZOOIDS). The skeleton may be chitinous but is calcareous in those species commonly occurring as fossils. The colony (ZOOARIUM) shows a range in form and size. It may be encrusting, stick-like, massive, or a delicate branching network. Usually it measures only a few centimetres across, but it may exceed 1 m. It consists of tubular or box-like chambers (ZOOECIA) which have an aperture (ORIFICE) at one end, and are rarely more than 1 mm in length (fig. 147a, b). Superficially, the zooid resembles a tiny 'coral' polyp but the body is relatively complex, generally similar in plan to that of the brachiopods. The mouth is surrounded by a ring- or crescent-shaped lophophore bearing ciliated tentacles. It leads to a U-looped digestive tract ending in the anus which opens near the mouth but outside the tentacles. During feeding the tentacles are extended to filter micro-organisms from the water; at rest they are retracted within the body. In some cases the zooids are modified for special functions.

An example

Fenestella (fig. 147c). The colony is fan- or cup-shaped arising from a calcified basal holdfast (seldom preserved); it consists of a network of delicate subparallel branches which bifurcate occasionally and are linked at regular intervals by cross-bars; the zooecia are short tubes bound together in calcareous tissue in which circular apertures form two rows along one side of each of the branches.
Ordovician–Permian

Mode of life

Living Bryozoa adhere to seaweeds, shells or stones and their form may be modified by turbulence. They are found mainly in clear, shallow water but will tolerate muddy conditions and have been found in abyssal depths. They are common in coral reefs, which encrusting forms help to bind. Fossil Bryozoa occur most abundantly in impure limestones and reef-limestones;

146 *Actinostroma*, **a stromatoporoid, Devonian.**
a, transverse section (× 0.8). b, c, vertical section, (b) is an enlarged view of part of (c). d, surface view showing star-shaped grooves. (Specimen in the Sedgwick Museum.)

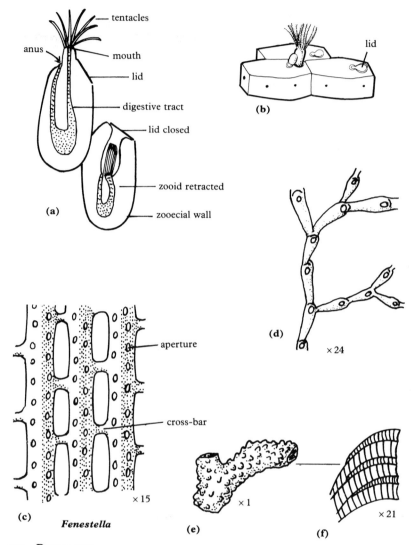

147 Bryozoans.
a, b, simplified diagram of the zooids in relation to the zooecia. c, part of the skeleton of *Fenestella*. d, part of an encrusting form. e, f, part of a stick-like bryozoan.

they may also be found in shales, but rarely in sandstones. The massive forms are often well preserved but the more delicate lacy zoaria tend to be fragmentary.

Geological history

L Ordovician–Recent

The Bryozoa range from the early Ordovician to the present day. They are abundant at many horizons and remain a flourishing group. A variety of

148 *Fenestella,* **L Carboniferous (× 2.6).**

forms are short-ranged, widely distributed and accordingly valuable for purposes of correlation.

Bryozoa were already diverse and widespread in the lower Ordovician, suggesting a prior, unrecorded, history of development. The predominant lower Palaeozoic forms were massive or stem-like. They played an important part (along with tabulate corals) in reef formation. They are common, for instance, in the Wenlock Limestone (M Silurian). Upper Palaeozoic bryozoans often formed lacy, net-like or branching colonies. *Fenestella* (fig. 148) is an example, occurring in limestones of the Carboniferous and Permian.

Most Palaeozoic orders did not survive the Permian but two groups, inconspicuous in the Palaeozoic, ranged on through the Mesozoic and Cainozoic to the present day. A new order with rounded box-like zooecia and the orifice closed by a hinged lid, appeared in the lower Cretaceous. Some were encrusting, others formed delicate branching stems or fronds. They were abundant and highly diverse in the upper Cretaceous (fig. 149) and are the dominant forms of the Cainozoic and the present day. The Pliocene Coralline Crag is particularly rich in bryozoan fragments.

149 *Castanopora magnifica*, **part of an encrusting bryozoan (cheilostome), U Cretaceous, Norwich (approx. × 60).** **(Courtesy of the British Museum (Natural History).)**

Annelida

Precambrian–Recent

Annelids are segmented worms with an elongate soft body. Behind the head region lies a series of mainly similar segments each carrying paired tufts of bristles, used to grip the substrate. Most are free-living, either pelagic or benthic, and some of these are predators with chitinous jaws. Others are sedentary, living within burrows or tubes made of detritus which may be mucus-bound (fig. 172a) or cemented by an organic matrix. Some secrete a calcareous shell (fig. 150a, b). These may be mud-eaters or filter-feeders with ciliated tentacles.

Annelids are widespread and abundant today. Marine forms are important as geological agents in scavenging and churning up soft sediment on the sea-floor which they ingest for its organic content, often extruding the used sediment on the substrate as casts. Their record is long but patchy and consists mainly of trace fossils such as tracks, burrows and casts. Burrows may be vertical tubes as in the Pipe Rock (L Cambrian, North Scotland) or U-shaped. Calcareous tubes on fossil shells are common and are either spirally coiled as in *Spirorbis* (Silurian–Recent, fig. 150a, b), or irregularly twisted as in *Serpula* (Silurian–Recent, fig. 150c). Finely preserved impressions of the

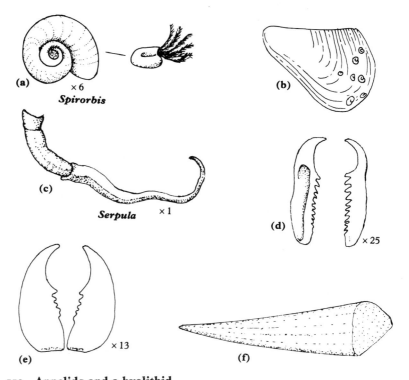

(a) ×6
Spirorbis

(b)

(c)
Serpula ×1

(d) ×25

(e) ×13

(f)

150 Annelids and a hyolithid.

a, b, spirally coiled calcareous tube; in (b) the worm-tube is shown on the posterior end of a bivalve shell where the worm would have benefitted from feeding-currents produced by the bivalve. c, an irregularly twisted form. d, e, scolecodonts (organic jaws). f, a hyolithid.

soft body are recorded from the Burgess Shale (M Cambrian) in British Columbia (fig. 151). Microscopic chitinous jaws (scolecodonts) are abundant in the Ordovician and Silurian (fig. 150d, e).

Giant 'tube-worms'

'Tube-worms' were first collected early this century from abyssal depths in the oceans and, in 1955, were identified as a unique phylum, the Pogonophora. Their geological interest lies in the discovery of giant forms living around hydrothermal vents at depths between 200 and 4000 m. They are associated there with a varied fauna including bacteria, bivalves, gastropods, arthropods, anemones and fish.

Tube-worms are thread-like, about 30 cm × 2 mm, with many tentacles on the 'head' region but without mouth or digestive system. They secrete chitinous living-tubes. The giant tube-worms (vestimentiferans) may measure up to 3 m × 3 cm. They are bathed by water issuing from the vents at a temperature varying between 2 and 23°C and containing dissolved minerals and hydrogen sulphide. Rich floras of sulphur-reducing bacteria live around

151 *Burgessochaeta setigera*, an annelid from the Burgess Shale, M Cambrian, British Columbia (× 10).

Part of a free-moving marine worm showing: (i) a series of segments, each with a pair of 'legs' bearing bundles of bristles; and (ii) the straight gut lying along the centre of the body. (Photograph, taken in ultraviolet light, of a fossil in the collection of the Geological Survey of Canada.)

the vents and form the base of the food chain for the varied fauna. The tube-worms absorb organic nutrients from bacteria living symbiotically in their tissues.

Moulds and casts of tube-worms have been described from fossil hydrothermal chimneys containing sulphides and other ore deposits of Carboniferous age in Ireland, and Cretaceous age in Oman. Several very long narrow tubes from the late Precambrian and Cambrian have also been assigned to the Pogonophora.

Hyolitha

Earliest Cambrian–U Permian

Hyolithids are an extinct group of unknown affinity. They are considered by some to be a separate phylum which might possibly be remotely related to ancestral mollusc stock.

They were marine, bilaterally symmetrical, and secreted a calcareous shell which, typically, was conical with its aperture closed by a lid. The shell was more or less subtriangular in cross-section (fig. 150f). Most were about 15–30 mm in length.

Hyolithids appeared in the earliest Cambrian, were most abundant during the Cambrian and rare after the Devonian.

11

Microfossils

Microfossils are the remains of organisms, whole or fragmentary, the study of which requires the use of a microscope. Most are unicellular, including members of the Kingdom Monera (i.e. the procaryotic cyanobacteria and bacteria) and of the Kingdom Protoctista (i.e. eucaryotes comprising the highly diverse phytoplankton and zooplankton). Calcareous algae are also included here following the present practice of treating algae as multicellular protoctists rather than as plants. There are also the minute remains of small multicelled animals, the ostracods and conodonts. A further category includes microscopic fossils of uncertain affinity, the acritarchs and chitinozoans. Pollen grains and spores, the study of which is known as palynology, are dealt with in the chapter on plants (chapter 15).

Kingdom Monera–procaryotes

Procaryotes are microscopic, one-celled, simple organisms distinguished by their lack of an organised nucleus and organelles (p. 5). They comprise cyanobacteria (or cyanophytes or blue–green algae) and bacteria. They have no skeleton and their cells are rarely preserved. Yet they are of major geological significance, especially in studies of the Precambrian and the evolution of early life.

Cyanobacteria

Cells of cyanobacteria are rounded (coccoid) (fig. 152a) to cylindrical in shape and range up to about 25 μm diameter. They occur singly or as groups of cells held within a mucilage-covered cellulose sheath; often cells are arranged in series to form single or clustered filaments (fig. 152b). They contain chlorophyll (p. 347) and make food by using light for photosynthesis releasing oxygen from water in the process. Some fix atmospheric nitrogen. Reproduction is by simple fission.

Cyanobacteria are tolerant of a wide range of conditions in water. They are resistant to extremes of temperature ranging from hot springs to pools in

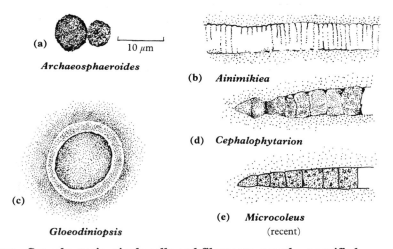

(a) 10 μm

Archaeosphaeroides

(b) *Ainimikiea*

(d) *Cephalophytarion*

(c)

Gloeodiniopsis

(e) *Microcoleus*
(recent)

152 Cyanobacteria, single cells and filaments, greatly magnified.
a, b, from the Gunflint Chert, Precambrian. c, d, from the Bitter Springs Chert,
Precambrian. e, a recent form to compare with d (redrawn from W.N. Stewart).

ice; some live in soils; some associate with other organisms as, for instance, in symbiosis with fungi in lichens; some are common in reefs. Their resting spores are resistant to desiccation. They tolerate low oxygen levels and ultraviolet radiation. They contain a blue pigment, phycocyanin, which enables them to photosynthesise at very low light levels: they have been found at depths of 1000 m in clear tropical waters of the Indian Ocean.

Cyanobacteria are among the most ancient known organisms dating back some 2800 my, possibly 3500 my. Over this period of time they show little change and some Precambrian fossils are remarkably like extant forms. They are involved today in the formation of various, aqueous sedimentary deposits, especially in the build-up of STROMATOLITES. These structures are of particular geological interest since they also occur in sedimentary rocks ranging in age back to early Precambrian times. Interpretation of these fossil stromatolites is based on our knowledge of the modern examples and the role of cyanobacteria in their formation.

Stromatolites

A small example of a stromatolite is a bun-shaped structure, a few centimetres across, built of numerous concentrically arranged laminae about 1 mm thick and usually consisting of calcium carbonate (fig. 153). Each lamina is formed by a sticky surface layer of filaments and cells of cyanobacteria with other bacteria and green algae, forming an algal mat which traps sediment washed over its surface. It may also precipitate calcium carbonate as the result of photosynthesis and the consequent abstraction of carbon dioxide from the water. As sediment accumulates, the organic layer grows through and over it to maintain its position at the

153   A stromatolite: polished section, Islay Limestone, Dalradian Supergroup, Precambrian, Argyll (× 2).

surface, thus building the laminated structure. Because of the role of organisms in their formation, stromatolites are termed biogenic structures. The process of their formation can be seen today at, for instance, Shark Bay in Western Australia, an area of salty, warm, intertidal shallow water (fig. 154). Stromatolites may also occur in brackish or fresh water, in alkaline lakes and in pools at hot springs.

Fossil stromatolites show considerable range in form and size: they may be small and bun-shaped, or be extensive thick sheets of complex structure formed by numerous adjacent dome-shaped growths. They occur in limestones or dolomites and sometimes in cherts. Modern examples are restricted, presumably by competition, to harsh environments hostile to most other organisms but in which the algal mat communities can flourish. In the Precambrian, however, they seem to have been more widely distributed, implying an absence of competition.

Laminated structures can, of course, be formed by purely physical processes and caution is needed in identifying them as true stromatolites in the absence of microfossils.

Fossils are extremely rare in Archaean rocks but stromatolites with microfossils are found in the Warrawoona Group in Western Australia. They are about 3500 my in age, one of the oldest examples so far discovered. The rock in which they occur is dark, carbonaceous chert which contains dark-

154 Present day stromatolites exposed at low tide, Shark Bay, Western Australia.
(Photograph courtesy of H.J. Hofmann and W.N. Stewart.)

walled coccoids and a variety of filamentous structures, some septate, which resemble bacteria and possibly cyanobacteria. Similar fossils are also found in cherty parts of stromatolitic limestones in the Fortescue Group, 2700 my old, in the same area.

In Proterozoic rocks, the lower limit of which is placed at 2500 my, stromatolites are of more frequent occurrence and authentic microfossils relatively common and well preserved. Examples occur in the Gunflint carbonaceous chert in Canada, about 2000 my old, which contains a diverse range of single cells and filaments (fig. 152) comparable with living cyanobacteria and bacteria. Still younger stromatolites in the Bitter Springs black chert in Australia, about 900 my old, contain a much greater diversity of microfossils, about 30 species, which have been identified as cyanobacteria, bacteria, green algae and other forms unascribed.

These examples cover a span of about 2600 my. The record of genuine fossils in this period is now such as to demonstrate a gradual increase in diversity of organisms associated with stromatolites. This is consistent with an increase in diversity of the stromatolites themselves, as shown by increasing complexity of the laminated structures, reaching a peak in the late Proterozoic. This development must reflect differences in types of algal mats, each reacting in a characteristic way with its environment.

It is generally held that the early atmosphere was anoxic and the earliest organisms were anaerobes. With the arrival of oxygen-releasing cyanobacteria came the start of a notable change in the atmosphere which led to a gradual build-up in the level of oxygen which it contained. Oxygen is, of course, essential for the life processes of metazoa which, appearing and diversifying in the late Proterozoic, competed with and disturbed the algal mats to an increasing degree. Invertebrates doubtless grazed the mats or burrowed through them. In the lower Cambrian, archaeocyathans built extensive reefs in shallow waters, thus usurping the procaryote niche. Later, in the Ordovician, other reef-builders, mainly sponges and calcareous algae, appeared and the stromatolites declined. They did not again become prominent in the marine environment.

Calcareous cyanobacteria

Calcium carbonate is deposited around the filaments of some cyanobacteria during photosynthesis, thus forming calcified tubes. One such is *Girvanella* (Cambrian–Recent) which occurs as small nodular masses. A thin-section shows a tangled mass of flexuous tubes 8–30 μm in diameter. *Girvanella* occurs in shallow-water marine limestones and reefs.

Bacteria

Bacteria are diverse procaryote cells occurring singly or as aggregates. They are rod-like, spherical or spiral in shape, measuring about 0.25–2 μm in diameter. Many are motile, moving by one or more lashing threads, flagella. They multiply by simple fission and some, in unfavourable conditions, form spores which can withstand desiccation and, for a time, high temperatures. This is known as a resting stage and will be followed by regeneration of the motile stage if conditions become favourable again.

Most bacteria are HETEROTROPHS, i.e. they live on organic matter, alive or dead. Some are AUTOTROPHS, i.e. they synthesise their organic constituents from simple inorganic materials.

Bacteria are abundant and ubiquitous on land, fresh water and sea. They are mainly aerobic but some are anaerobic. Most do not contain chlorophyll and do not photosythesise; they do not tolerate bright light or ultraviolet radiation. They are, however, tolerant with regard to salinity, temperature and pressure, and are found living, for example, at the high-temperature hydrothermal vents of the East Pacific oceanic ridge where the water pressure is about 250 atmospheres.

Bacteria form an important part of the food chain either as a direct food source for animals such as protozoans, worms, sponges, filter-feeders and mud-eaters; or indirectly, by breaking down organic matter into substances which can then be used by higher organisms. Some can fix nitrogen from the atmosphere thus making it available to plants.

Of great geological interest is bacterial activity in sediments which results in chemical and mineralogical changes. Organic matter in muds is partly decomposed by methanogens belonging to the primitive archaebacteria, producing methane; ferric hydroxide is reduced to the ferrous form, and sulphate is reduced to hydrogen sulphide, these two products reacting to form ferrous sulphide and, eventually, pyrites. In different circumstances bacterial activity may result in the formation of calcium carbonate from calcium sulphate; or of sulphur from calcium sulphate; or the direct precipitation of ferric hydroxide by 'iron' bacteria. Bog iron ore is formed in this way. In all these cases the presence of bacteria is recorded not so much by their presence as fossils, but by their products.

Fossil bacteria will obviously be extremely rare but forms similar to extant bacteria have been recorded from rocks, mainly cherts, ranging in age back to the Archaean. One example is the Gunflint Chert, 2000 my old; another is carbonaceous chert in the 3500-my-old Warrawoona Group which contains a variety of bacteria-like fossils. Other instances have been recorded from limestones, iron and manganese ores, oil shales, coals and in plant and animal tissues.

Kingdom Protoctista

Protoctists include a diversity of aquatic, mainly marine and aerobic eucaryotes (p. 5). It is convenient to distinguish two groups: plant-like AUTOTROPHS, 'algae', with chloroplasts, able to photosynthesise their food; and animal-like HETEROTROPHS, 'protozoans' which feed on other organisms. Only those which secrete a skeleton are considered here. The skeleton may be organic, calcareous or siliceous.

The algae range from the one-celled, largely planktonic forms which make up the phytoplankton to the many-celled benthic seaweeds. They are confined to the photic zone and are very important in marine ecology: they provide food, directly or indirectly, for all aquatic creatures and, by their photosynthesis, oxygenate their environment.

The protozoans are one-celled and include both planktonic forms, the zooplankton, and benthic forms.

The primarily unicellular forms are sometimes segregated in a more restricted Kingdom, Protista, which excludes the seaweeds.

Phytoplankton

Phylum Dinophyta

Silurian–Recent

Dinophytes or, more usually, dinoflagellates (fig. 155a, b) are plant-like by virtue of having cellulose in the cell wall, and chlorophyll. They also have reddish, light-absorbing pigment from which they take an alternative name:

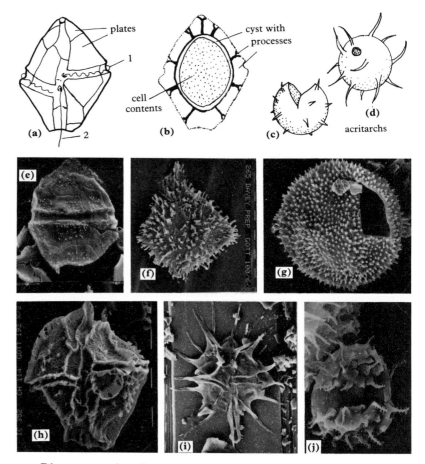

155 Dinocysts and acritarchs.
a, dinoflagellate, ventral view of motile cell showing 1, transverse and 2, longitudinal flagella. b, cyst formed inside cell wall which then decays. c, d, j, acritarchs (j, Cambrian, × 1600). e–i, dinocysts: e, *Subtiliophaera*, dorsal view (× 800); f, *Aptea*, dorsal view showing partly detached operculum of escape hole (archaeopyle) (× 400); g, *Trichodinium*, dorsal view showing escape hole (× 800); h, *Cribroperidinium*, ventral view (× 800); i, *Hystrichodinium*, ventral view (× 600). All from Barremian, L Cretaceous, north-west Europe. (Courtesy Dr I.C. Harding.)

pyrrophyte (fire plant). The shape is varied: it may be spherical, ovoid or polygonal and there may be horn-like projections. They measure about 20–200 μm. Some are fresh-water forms but typically they are marine phytoplankton. They are mostly motile, moving by means of two flagella which lie in furrows and work respectively transversely and lengthwise. Non-motile forms include ZOOXANTHELLAE which live in symbiosis with, for instance, corals (p. 115) and foraminifera (p. 258). The forms which occur as fossils are those which have a resting, encysted phase as CYSTS. The cyst, formed inside the cell, has a two-layered wall of tough organic material resistant to decay. Its surface may be smooth or sculptured with granules, spines, ridges or processes. Each has an 'escape' pore (archaeopyle) through which regeneration can take place.

Dinoflagellates live in the upper 50 m of the photic zone and are one of the major primary producers at the base of the food chain. They are commonest in warm, open ocean areas, preferring temperatures between 18 and 25 °C, though they tolerate a wider range. They have a roughly latitudinal distribution, being most diverse in low latitudes. Some genera can have a devastating effect on other organisms in shallow water when they 'bloom' in such vast numbers as to form a 'red tide' and their toxic wastes can kill fish and benthos such as mussels.

Fossil cysts occur mainly in clays and shales. Possible examples are recorded from the Silurian but their main history starts in the Permian. From late Trias times they were common and their diversity increased during the Mesozoic reaching an acme in the late Cretaceous. At the end of this period they declined sharply but were again important in the Eocene. Their numbers were very low in the Pliocene but, today, many species are known, though few of these encyst.

Group Acritarcha

U Precambrian–Recent

Acritarchs are minute, cyst-like fossils of uncertain origin (fig. 155c, d, j). They are single-celled and hollow; have a decay-resistant organic wall of one or more layers, which contain a substance similar to sporopollenin found in spores of vascular plants; and measure about 10–150 μm in diameter. The variable shape, reminiscent of cysts, spores or eggs, may be spherical, ovoid, triangular or polygonal. The surface may be smooth or sculptured with granules or spiny projections which may branch. There is an 'escape' hole in some; others must have opened by splitting. They possibly represent cysts of various sorts of planktonic algae such as dinoflagellates.

Acritarchs usually occur in fine-grained rocks such as shales and also in limestones, often in association with marine fossils. A few have been recorded from fresh-water Pleistocene peats. In marine rocks they are geographically widespread and seem to have been deposited away from shorelines, suggesting a planktonic mode of life. If they do represent phytoplankton they must have been an important food source.

Acritarchs are found in the late Precambrian and are stratigraphically useful there. They are mainly spherical forms. Spiny forms appeared in the early Cambrian and the group as a whole reached its acme in the Ordovician. They remained important and world-wide in the Silurian, but in late Devonian times a decline set in and they are generally rare in the latest Palaeozoic. They are insignificant in the Mesozoic and Cainozoic.

Phylum Chrysophyta–silicoflagellates

L Cretaceous–Recent

These are marine, flagellate, photosynthetic protists, about 20–50 μm in diameter. They secrete, within the cell, a delicate framework of opaline silica tubes with radiating spines. This framework may be ring-like, hexagonal or form a hemispherical lattice (fig. 156a, b).

They are minor constituents of the phytoplankton of upwelling silica-rich waters in high latitudes and other areas, their remains being found with those of the diatoms. Some species are temperature sensitive and their fossils are useful as indicators of temperature fluctuations.

They first appeared in the Lower Cretaceous and were most abundant in the Miocene.

156 Phytoplankton.
a, b, silicoflagellates (from Bolli *et al.*). c, d, diatoms: c, below, frustule of a simple centric form, surface view; above, sections showing cells before and after division; d, a pennate diatom. e, a coccolithophore.

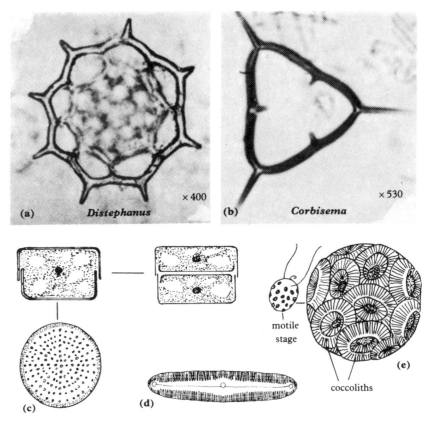

| (a) *Distephanus* | ×400 |
| (b) *Corbisema* | ×530 |

(c)   (d)

motile stage

coccoliths

(e)

Phylum Bacillariophyta – diatoms

L Cretaceous–Recent

Diatoms are minute, non-flagellate, photosynthetic forms, the cell wall of which is impregnated by opaline silica to form a delicately sculptured, reticulate skeleton, the FRUSTRULE (fig. 156c, d). They consist, pillbox-

157 Diatoms.

a, centric diatom from diatomite, Holocene, Kenya (× 2550). b, pennate diatom, Quaternary, Loch Droma, Scotland (× 11500). (Electron micrographs.)

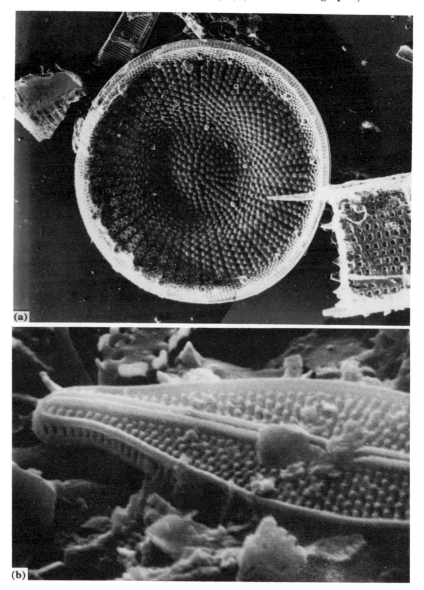

like, of two partly overlapping valves punctured by fine pores. They range in size between about 20 and 200 μm. They may be single or may join in mucus-bound filaments or chains. Two groups are distinguished by shape: CENTRIC diatoms, circular in plan with radial symmetry (fig. 157a); and PENNATE diatoms, elliptical in plan with bilateral symmetry (fig. 157b). They occupy many habitats in sea and fresh water. Most centric forms are marine and planktonic, living most densely in the upper 30 m of the photic zone. Most pennates are benthic and live in fresh water (and soil) and also in shallow seas.

Growth of marine, planktonic diatoms is seasonal with 'blooms' in spring and late summer, and they are of major importance as primary producers near the base of the food chain. They are enormously abundant in the seas of subpolar and certain other regions where upwelling deep waters provide the necessary nutrients and dissolved silica. On death the majority of the skeletons quickly dissolve but a small proportion accumulate on the ocean floor as diatomaceous ooze, subsequently lithified as chert.

Deposits also accumulate in fresh-water lakes in high latitudes, diatomaceous earths of Pleistocene age being found, for example, in northern Britain. Such deposits are of economic value in various ways, for example as fine abrasives. Deposits in dried-up lakes may become wind-borne as fine dust and widely redistributed on land and sea.

Centric diatoms first appeared in the Cretaceous and underwent a major radiation in the Palaeocene, reaching a peak in the Miocene, after which they declined in numbers. Pennate forms appeared in the Palaeocene and increased gradually through the Tertiary, being numerous today mostly as delicate fresh-water species.

Many living species have a fossil record and their changing geographical distribution in the rocks of the Cainozoic can be used to infer changes in oceanic conditions.

Phylum Haptophyta – Coccolithophorida

Trias–Recent

Coccoliths are minute calcite scales embedded in the cell walls of coccolitho-phores, marine, photosynthetic forms, spherical or ovoid in shape and about 5–20 μm in diameter (fig. 156e). They are motile with two flagella and a filiform appendage which may be coiled, the HAPTONEMA. The coccoliths, about 1–15 μm in diameter, are round, oval or stellate discs which show elaborate patterns of radially arranged calcite plates and rods (fig. 158). These are shed from, and replaced by, the parent cell as it grows.

Coccolithophores are found throughout the photic zone but are at their maximum numbers from a depth of a few metres to about 50 metres. They are most abundant and diverse in the warm, open oceans between latitudes 45 °N and S, but some species live in colder, temperate and subpolar waters.

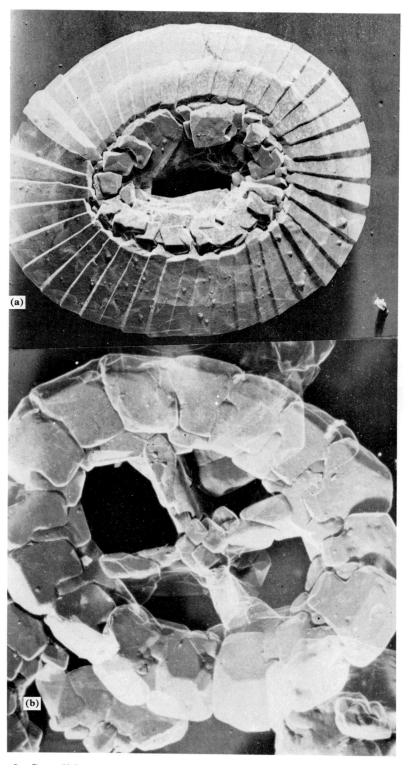

158   Coccoliths.
a, from the Miocene, Jamaica (× 11 300). b, from the chalk, U Cretaceous,
Bielgorod, Russia (× 22 500).

Most have a fairly narrow temperature range and some, found as fossils in the younger rocks, can be used as indicators of temperature fluctuations in the past. Like diatoms, they are a major primary food source.

Many warm-water species have a restricted geological range but are also widespread and hence of use in the biostratigraphy of the Mesozoic and Cainozoic rocks.

Coccoliths were rare in the late Trias. Their diversity and numbers increased through the Jurassic to reach a sharp high peak in the late Cretaceous, the Chalk owing much of its character to an exceptional abundance of coccoliths and coccolith debris. They were, however, almost wiped out in the end-Cretaceous extinctions and only a few species survived. New forms appeared in the Palaeocene followed by a rapid diversification and increase in numbers in the Eocene. In the later Eocene numbers again dropped and continued to decline apart from a temporary surge in the Miocene. Today, they are moderately abundant and their remains contribute to deposits of calcareous ooze on the ocean floors.

Seaweeds and other algae

Many-celled seaweeds are included by some taxonomists in the plant kingdom, but in the Protoctista by others. They are separated into phyla on the basis of their mode of reproduction and cell chemistry, including the type of photosynthetic pigments they contain. The forms considered here occur as fossils, often in fragmented state; they are identified by their microstructure.

Seaweeds have a simple body, the THALLUS, which in many-celled forms may be filamentous, branching or a flat membrane. They occur in fresh, brackish and sea water, and are benthic, anchored by a 'holdfast', or encrusting a hard surface. They live in the photic zone and are most abundant down to about 50 m. A number of calcareous algae secrete calcium carbonate and are potential fossils in which shape, tissue structure and sometimes reproductive structures may be preserved. The carbonate may encrust the thallus as the result of carbon dioxide being extracted from the water by photosynthesis; it may also form as an intercellular deposit by metabolic processes.

Phylum Rhodophyta – red algae

These are mainly marine, living in both warm and cold waters. They are coloured by reddish pigments which mask their green chlorophyll and, by absorbing blue light, enable them to live at depths to which longer wavelengths do not penetrate. They are most abundant just below low-tide level but have been found on sea mounts at a depth of 268 m where only 0.0005% of surface light remains. They use light much more efficiently than the shallow-water forms.

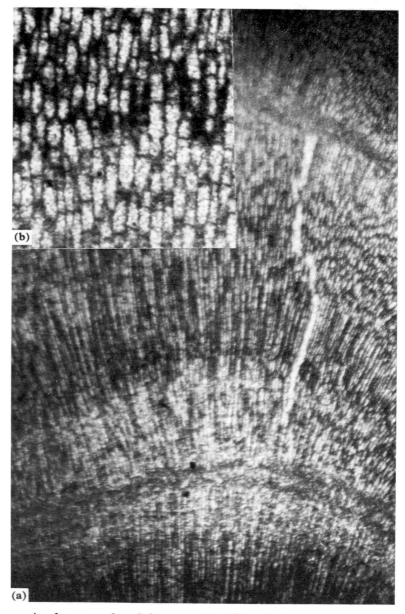

159 A calcareous alga, *Solenopora*, Bathonian, M Jurassic, Gloucestershire.
a, part of the alga showing banding (× 16). b, cell detail (× 60).

'*Solenopora*' (Cambrian–Cretaceous) belongs to a group of calcite-secreting algae which were common in the Palaeozoic and Mesozoic. It forms irregular, encrusting nodular masses of close-packed radiating filaments, more or less polygonal in cross-section, and with cell structure (fig. 159). Some species referred to '*Solenopora*' have been identified as unrelated organisms, e.g. cyanobacteria.

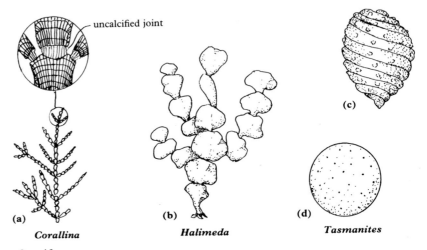

uncalcified joint

(a)

(b)

(c)

(d)

Corallina   *Halimeda*   *Tasmanites*

160 Algae.
a, coralline red alga with enlargement to show uncalcified joint. b, calcareous
green alga. c, calcified reproductive body of an Eocene charophyte (redrawn
from W.N. Stewart). d, a cyst, with perforate organic wall, found in some oil-
shales.

The other important group, the coralline algae, are found first in the
Carboniferous and are of increasing importance in the Mesozoic and
Cainozoic because of their association with coral reefs. *Porolithon* (fig. 161a)
and *Lithothamnion* are important components of modern reefs. They encrust,
helping to bind and cement their coral substrate, forming a purplish–red
algal ridge on the exposed, turbulent edge of the reef. *Lithothamnion*
(Cretaceous–Recent) is encrusting and may have irregular branches a few
centimetres high. It also grows in cold waters. Others, like *Corallina*
(Cretaceous–Recent) live in quieter waters such as back-reef areas. Here,
the thallus is erect, much branched, and consists of a series of calcified
segments with intervening non-calcified 'joints' (fig. 160a). It fragments
readily, adding to the sediment of the reef area.

Phylum Chlorophyta – green algae

These are very diverse and widespread. They are mainly fresh water
(including damp places) organisms but do occur in brackish and sea water.
Marine forms live in shallow water (to about 100 m) and are especially
abundant intertidally. The thallus is made of an interwoven tissue of non-
septate filaments, often freely branching. A number secrete aragonite which
is a factor in their preservation as fossils.

Green algae occur as fossils from Cambrian times onwards, often in
association with reefs. *Halimeda* (Cainozoic, fig. 160b) is commonly found
today in sheltered lagoons in coral reefs down to depths of about 60 m. It
forms erect, branching growths a few centimetres high, consisting of a series

of articulated, calcified discs. It fragments readily and adds greatly to the sediment flooring the lagoon.

Fresh-water green algae include the stoneworts (charophytes). An example is *Chara* (Eocene–Recent) with a calcite-encrusted thallus, forming a slender stalk with whorls of branches bearing male and female sex organs. The female organ, the oogonium, contains the egg cell; it is surrounded by spirally coiled filaments which may become calcified (fig. 160c). These are common fossils in fresh-water deposits in the Mesozoic and Cainozoic. Primitive forms are found in the Devonian.

Non-calcareous Chlorophyta. *Botryococcus* is a one-celled alga which secretes globules of reddish oil in the cell. It lives in fresh or brackish water and divides to form colonies about 10–100 μm across. The algae 'bloom' in season, quickly multiplying to form a dense cover over the surface of the water which becomes oxygen deficient and consequently lethal to fish. The algal remains may accumulate on the lake floor to a thickness of 5 or 10 m and such deposits in time are converted to a dark, compact, hydrocarbon-rich rock, thin-sections of which show amber-coloured masses of similar algae (fig. 161b). Such rocks are known in general as oil-shales but, in examples where fossil algae are conspicuous and abundant, special names have been given. One such is 'torbanite', examples of which are found in the Carboniferous of Scotland and the Permian of Australia.

A recent deposit of similar origin is known as 'coorongite'. It occurs in Australia where salt lakes and lagoons shrink or dry up in periods of drought.

161 Algae.
a, *Porolithon*, coralline red alga, Recent. b, *Botryococcus*, colonies from boghead coal, U Carboniferous. (Reproduced from W.N. Stewart.)

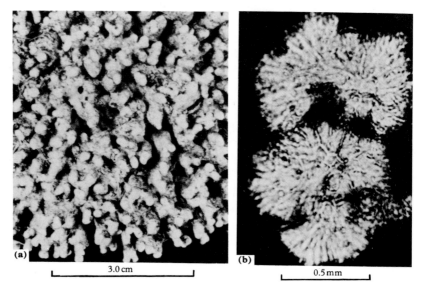

(a)
3.0 cm

(b)
0.5 mm

It is a soft, brown, rubbery substance formed by the dried-up accumulations of *Botryococcus* colonies in the resting stage, in which condition the algae contain up to 76% of hydrocarbons.

Other one-celled algae may be associated with coal and oil-shales. *Pachysphaera*, a living marine planktonic alga, forms a resting-stage cyst which is spherical and has a two-layered organic wall about 8–10 μm thick. It is similar to and may be related to *Tasmanites* Cambrian–Recent, (fig. 160d), cysts of which occur in large numbers in oil-shales (e.g. in Alaska, Jurassic and Cretaceous) and is prominent in Permian coal in Tasmania. The cysts are about 100–600 μm in diameter with a two-layered wall perforated by pores.

Zooplankton – Phylum Sarcodina

This group comprises animal-like protoctists alternatively classed as 'Proto-zoa'. Members of two groups, the radiolarians and foraminiferids, secrete skeletons and are of major importance as fossils because of their immense abundance, diversity and wide distribution in marine rocks.

Radiolarians (Subphylum Actinopoda)

M Cambrian–Recent

These are one-celled, free-floating, marine protoctists which secrete a delicate openwork endoskeleton composed, typically, of opaline silica; in one group, however, this is largely organic. They occur singly as a rule but may associate in colonies.

The term 'Radiolaria' includes members of two classes; the Polycistina with silica only in the skeleton; and the Phaeodorea with a mainly organic skeleton containing a small amount of silica. A third class, the Acantharea, is sometimes associated with the radiolaria and is worth mentioning because, unusually, its skeleton is made of celestine (strontium sulphate). The Polycystina have a substantial fossil record; the other two groups are insignificant as fossils, are confined to the Cainozoic, and are not considered further.

Polycystina. The protoplasm of the cell includes an inner endoplasm enclosed in a porous organic membrane, the central capsule (fig. 162a) and an outer ectoplasm. Long, straight, thread-like pseudopodia, some with a silica stiffening, radiate from the capsule to the outside. They are sticky and serve to trap passing food such as bacteria and diatoms. The food is passed in a stream of protoplasm into the capsule for digestion. The ectoplasm is frothy with vacuoles. It secretes the skeleton and, in forms living in the photic zone, contains symbiotic algae, zooxanthellae p. 244.

The skeleton is symmetrical, either spherical or discoidal (Order Spumel-

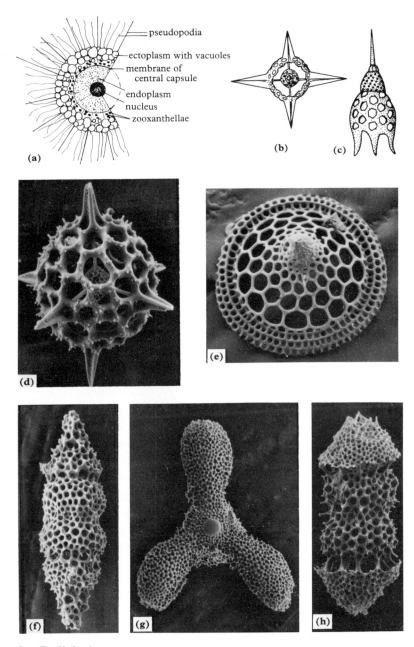

162 Radiolaria.

a, cross-section to show cell structure (skeleton omitted). b, spumellarid. c,
nasselarid. d–h, scanning electron micrographs of radiolaria from the Pliocene
(d–g) and Quaternary (h): d, a spumellarid showing inner concentric spinose
shell typical of this group (× 200); e, nassellarid (× 400); f–h, spumellarids
(f × 200, g × 165, and h × 200). (SEMs, courtesy Dr I.C. Harding.)

larida, fig. 162b) or bell-shaped (Order Nassellarida, fig. 162c). It is built of
spines, spicules and rods of silica which unite as a meshwork, from the surface
of which more spines radiate. The spumellarid skeleton has one or more
concentric, reticulate shells, frequently with spines projecting from the
surface of the outermost one. The nassellarid skeleton is like a spinose tripod
or a conical bell-shaped lattice, open at the base.

As planktonic animals, radiolarians are at the mercy of currents and
rough weather. It is essential for them to maintain buoyancy and the body is
designed to this end. Features which aid floating include the tiny size; the
spherical or bell-shape; the open-work skeleton, light but strong; the long
pseudopodia which increase resistance to sinking; secretion of oil globules;
and the frothy ectoplasm with vacuoles which can be collapsed or expanded
to vary their volume.

Radiolarians are mainly oceanic and are widespread from subpolar
regions to low latitudes. They are most diverse in warmer seas and most
profuse in nutrient-rich waters where their food supply is ample and the
necessary silica is available. They tend to 'bloom' seasonally as their silica
and food supply waxes and wanes. Different species inhabit distinct depth
zones in the ocean, most living in the photic zone but occurring down to
about 4000 m depth. They also show temperature preferences and, as with
foraminifera, these are useful in elucidating past movements of water masses.

Radiolarians range from the Middle Cambrian when spumellarids
appeared. They showed a gradual increase through the Palaeozoic becoming more diverse in the Carboniferous. Their numbers dropped sharply in
the Permian and remained low in the Triassic. The Jurassic saw a radiation
of new forms, including nassellarids, which led to a peak of numbers in the
Cretaceous. Numbers slumped at the end of this period and they were
generally less abundant in the Cainozoic. Today, however, they are very
prolific and include many forms with skeletons so delicate that they are
unlikely to be preserved in the fossil record, but dissolve immediately on
death. Only a small proportion survive as fossils, either as a minor
constituent of ocean sediments or, where the population was extremely large
and other sediments absent, as radiolarian ooze later lithified to chert.

Foraminifera (Subphylum Rhizopoda)

These are aquatic, mainly marine protoctists which secrete a shell, the TEST,
which consists typically of calcium carbonate but may be organic or built of
agglutinated detrital particles. Most foraminifera are tiny, less than 1 mm
diameter; a few are large and may reach several centimetres.

The soft tissue, PROTOPLASM, lies mainly within the test but some extends through one or more openings, APERTURES, to enswathe it and form
sticky threads, PSEUDOPODIA (fig. 163b). These are used for movement
and to trap and engulf food particles such as bacteria, diatoms, small

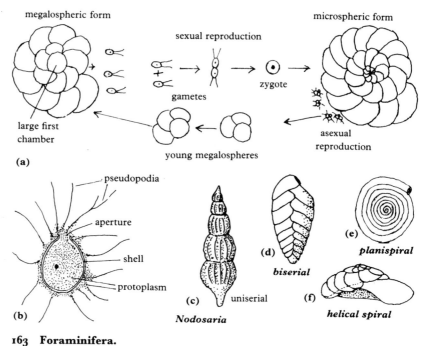

163 Foraminifera.
a, life-cycle of a foraminiferid. b, relationship of protoplasm and shell in a form with one chamber. c–f, varied arrangements of chambers.

metazoa and organic debris. Certain large, reef-dwelling and planktonic foraminifera enjoy a symbiotic association with photosynthetic algae.

The test consists either of a single CHAMBER or, more usually, of several chambers (MULTILOCULAR) which are separated by SEPTA. Each chamber communicates with the next one via an opening, the FORAMEN, in the septum between them. The opening to the outside, the APERTURE, is in the wall of the final chamber. The chambers are arranged in varied ways; they may lie in a single or double row or, commonly, are coiled in a plane or helical spiral as shown in fig. 163c–f.

In many species two forms are found which are known, from living foraminifera, to represent stages in a complex life history involving alternation of sexual and asexual generations (fig. 163a). In the sexual generation the first chamber is relatively large (MEGALOSPHERE) yet the whole test may be smaller, with fewer chambers, than that of the asexual generation which starts with a small chamber (MICROSPHERE).

Test wall

The composition and structure of the test wall is the basis for separating foraminifera into five suborders. Thus, the test may be entirely organic, consisting of a horny, proteinaceous substance (tectin); such forms (Allogromiina) are rare as fossils (range late Cambrian–Recent). In those with an agglutinated test (Textulariina), detrital grains such as sand or shell fragments are held in an organic cement (range Cambrian–Recent). Calcified tests, the majority, also have an organic constituent. There are three distinct wall structures: (i) microgranular (Fusulinina) with more or less randomly arranged, equidimensional grains of calcite giving a sugary appearance (range Ordovician–Trias); (ii) porcelaneous (Miliolina) with tiny calcite needles, most lying at random in an organic matrix but those at the outer surface lying parallel to give a smooth, shiny, milk-white surface (range Carboniferous–Recent); (iii) hyaline (Rotaliina) with crystals, usually of calcite, but aragonite in some, arranged in one or more layers, giving a glassy appearance; the walls may be PERFORATE, pierced by fine pores through which pseudopodia emerge (range Permian–Recent).

Foraminifera are mainly marine; a few live in brackish water (estuaries) where the salinity may fall to between 15‰ and 5‰ and a number tolerate hypersaline water. They are enormously abundant in the sea at all depths from the shoreline to the abyssal zone, in temperatures from 1 to 50 °C, and in high and low latitudes. They are prey to many creatures, yet in parts of the abyssal regions where little other detrital sediment accumulates, their remains may build up extensive deposits of calcareous ooze. Fossil forms constitute the bulk of some limestones, as in the nummulitic limestone from which the Egyptian Pyramids were, in part, built.

Mode of life

Most foraminifera are benthic but a number, like *Globigerina*, are planktonic. BENTHIC foraminifera may be sessile, or mobile and use their pseudopodia to crawl sluggishly. Diversity of species generally increases offshore, is high over the continental shelf but decreases in deeper water and lower temperatures. While some forms are cosmopolitan, many species are restricted to a narrow range of physical conditions. They may be sensitive to salinity, temperature, light, depth or to the nature of the substrate, e.g. whether hard, sandy or muddy. Organic-rich muds may be densely populated by foraminifera feeding on bacteria.

Foraminifera with agglutinated tests are characteristic of near-shore and brackish-water habitats, especially in colder waters, and they are the predominant forms below about 5000 m. Those with calcareous tests are commonest in warmer regions and shallow waters. Species with large tests are mainly restricted to well-lit shallow, warm waters (above 20 °C) including reefal habitats, probably because of their association with minute photosynthetic algae, zooxanthellae, which live embedded in the animal's protoplasm. This is a symbiotic situation similar to that of reef-corals and zooxanthellae (p. 115). The algae use waste products as nutrients and, in their uptake of carbon dioxide, may facilitate calcite secretion in the large foraminiferal test. The host may in part be nourished by soluble byproducts of photosynthesis and may also digest some of the algae.

PLANKTONIC foraminifera are widespread in the open ocean away from coastlines. There are relatively few species but these occur in enormous numbers. There may be a high seasonal peak in numbers as in the Arctic Ocean in late summer. Most live in the food-rich surface waters between 10 and 15 m and numbers decrease with increasing depth; few live below 1000 m. Different species live at different depths and, for each, the density of the test must be compatible with the preferred depth. While, in general, the test is relatively small and thus slow to sink, buoyancy may be improved by high porosity of the test wall or by a growth of long spines which increase resistance to sinking. In general, smaller forms live near the surface and larger forms in deeper water. Some species may migrate downwards as they mature; juveniles with small thin tests live near the surface and mature forms, with larger, thicker tests live and reproduce in deeper water. Some forms show a diurnal rhythm of vertical movement.

Diversity is highest in tropical waters and falls polewards. The distribution of many species is controlled by temperature and so there are 'warm'-water and 'cold'-water species which characterise faunal provinces. However, species characteristic of one province may also be found, in less abundance, in adjacent provinces. The faunal provinces are roughly parallel to latitude but surface currents and deeper water counter-currents, which influence temperature, complicate the pattern. In addition, cold-water species found at shallow depths in high latitudes also occur in deeper water of the same temperature in low latitudes.

In a few species the test is coiled, in some to the left and in others to the right. An example is *Neogloboquadrina pachyderma* which lives polewards of 25 °N and 25 °S. Over 90% of specimens in polar waters are coiled to the left while, in more temperate waters, most are coiled to the right. The temperature at which coiling direction is reversed varies from about 7 °C in some regions to 10 °C elsewhere. It is not understood why these changes occur.

Some examples of common benthic foraminifera

Saccammina (fig. 164a). Test agglutinated; simple; globular shape with aperture opening at the end of short, projecting neck. *Rhabdammina* (Ordovician–Recent), a related form, has tubular branches, each with a terminal aperture, radiating from a central area.
Silurian–Recent

Textularia (fig. 164b). Test agglutinated; multilocular and biserial with chambers in two alternating series; slit-like aperture is terminal.
U Carboniferous – Recent

Fusulina (fig. 164c, d). Test calcareous, microgranular, perforate; spindle-shaped, tapering at each end, with many whorls of chambers coiled

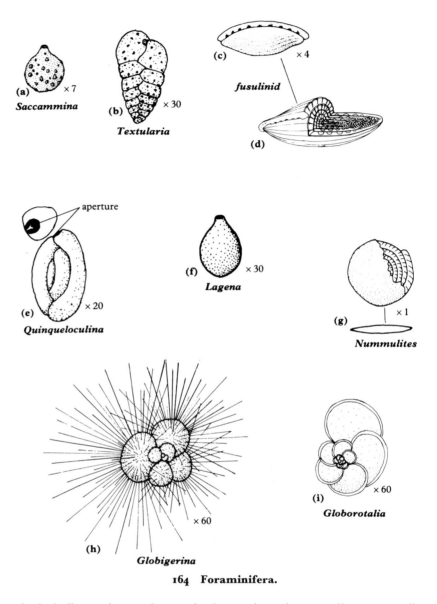

164   Foraminifera.

planispirally so that each completely envelops the preceding one; wall layered and cut by small pores; septa fluted transverse to the axis of coiling, in such a way that forward folds of one septum may meet backward folds of the next to form 'chamberlets'; aperture slit-like.
U Carboniferous

Quinqueloculina (fig. 164e).   Test calcareous, porcelaneous, imperforate; multilocular; coiled with sausage-shaped chambers added at angles of 144° so that five can be seen; aperture simple with a tooth-like projection.
Jurassic–Recent

Lagena (fig. 164f).   Test calcareous, hyaline, perforate; consists of one flask-shaped chamber with a long neck carrying the aperture. *Nodosaria* (Permian–Recent), a related form, has several chambers arranged in a series (fig. 163c).

Jurassic–Recent

165   Foraminifera.

a, b, internal structures: a, section of a Permian fusulinid showing chamberlets formed by folding of septa (× 7.5); b, section of a nummulite with spirally coiled whorls partitioned by septa, *Nummulites*, Bracklesham Beds, Eocene, Isle of Wight (× 4.2). c, *Ammonia* (a rotaliinid) (× 82). d, *Quinqueloculina* (a miliolinid) (× 55); c, and d, from recent shell gravel. (SEMs, c, and d, courtesy Dr I.C. Harding.)

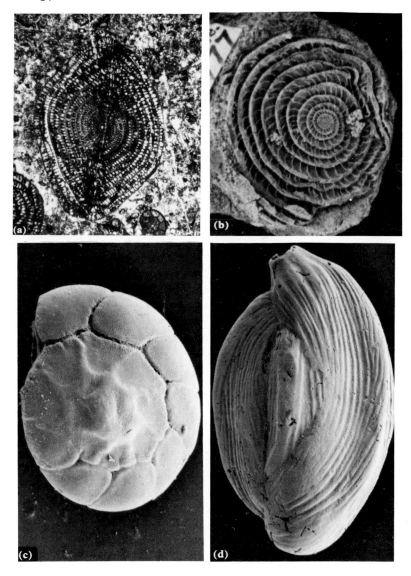

Nummulites (figs. 164g, 165b).   Test calcareous, hyaline, perforate; biconvex lens-shape with many simple chambers coiled planispirally, each whorl completely enveloping the previous one; the walls are thus layered; they carry a system of canals which open at the surface through pores; aperture slit-like and curved.
Palaeocene–Recent

Heterostegina (Palaeocene–Recent) is a reef-dwelling nummulite with the chamberlets divided into chamberlets. Study of this form has explained the role of the elaborate canal and pore system. It is a means of communication between the chamberlets and the sea water by which the varied biological processes, including locomotion, reproduction and protection are carried on. *Heterostegina* lives in shallow nutrient-poor but well-lit tropical waters. It contains symbiotic algae on which it depends for much of its nourishment; and its test, flattened and translucent, is well designed to allow easy penetration of light for the algae within to photosynthesise.

Some examples of common planktonic foraminifera

Globigerina (fig. 164h).   Test calcareous, hyaline, coarsely perforate; consists of several almost globular chambers coiled in a helical spiral (trochospiral); in life bears many very long fine spines; large basal aperture.
Palaeocene–Recent

Orbulina (Miocene–Recent) is a related form in which the final chamber is spherical and encloses the *Globigerina*-like earlier chambers. The spherical shape is an efficient design to promote buoyancy.

Globorotalia (Palaeocene–Recent, fig. 164i) is similar to *Globigerina* but chambers are smooth-surfaced, without spines, less inflated, and are "pinched'-out on the periphery into prominent keel. Species of all three have limited stratigraphic ranges and are, therefore, valuable for correlation purposes.

Geological history
Cambrian–Recent
Foraminifera appeared in the Cambrian. Early forms were benthic and typically agglutinated. By Silurian times they had become diverse and accompanied by forms with microgranular tests and these became the dominant forms in the Upper Palaeozoic, the fusulines being typical. Forms with porcelaneous tests appeared in the Carboniferous; and with hyaline tests in the Permian.

The fusulines include relatively simple lens-shaped forms, the endothyrids, which were important in the early Carboniferous; followed by giant fusulines, spindle-shaped forms with highly complex internal structure in the late Carboniferous and Permian, characteristic of warm, shallow seas and reefs. These are of great stratigraphic value and have been widely used in

North America, Russia and the Far East. They died out in the massive extinctions of organisms in the late Permian.

In the Mesozoic there was renewed diversification, mainly of forms with hyaline tests, but agglutinated and porcelaneous tests were also common. In the later Jurassic a new niche was occupied when planktonic forms appeared. A spectacular radiation occurred in the Cretaceous, notably of large porcelaneous, benthic forms living in shallow tropical seas; and of planktonic forms in the widespread Chalk seas. The Cretaceous ended with a large-scale disappearance of both benthic and planktonic forms.

A resurgence of numbers in the Palaeocene included survivors from the Cretaceous and newcomers like the planktonic globigerinids (still prominent today) and the large, benthic nummulites. The latter contributed massively to lower Tertiary warm-water limestones of the Tethys region, ranging eastwards from the Mediterranean area to the East Indies. Several species of *Nummulites* occur in the lower Tertiary rocks of the Hampshire and Paris Basins. Foraminifera are less diverse today though they are very abundant in warm seas.

Palaeoecology

The ecology of modern foraminifera is a direct guide to the living conditions of their fossil ancestors in the newer rocks; and an indirect guide to the palaeoecology of similar forms in older rocks. Data concerning temperature and depth of water, for example, can be used to establish fluctuating conditions in Quaternary sequences.

More general guides may also be used: today, large, calcareous foraminifera with symbiotic algae live in shallow, tropical waters and are likely to have done so in the past; open ocean waters have a high ratio of planktonic to benthic forms; extremes of salinity are indicated by low diversity; in brackish water mainly agglutinated forms with only an occasional hyaline form, such as *Ammonia* (fig. 165c), are found; hypersaline waters deter most foraminifera except porcelaneous forms such as the miliolines.

Uses of foraminifera

Foraminifera, like other fossils which show evolutionary changes with time and are geographically widespread, are of great value in establishing stratigraphical ages of rocks and making correlations on regional or continental scales.

In the oil industry, exploration drilling and oil-field development involve constant monitoring of the stratigraphic age of the rocks being penetrated; foraminifera are valuable for this purpose. As with other microfossils, they are particularly useful simply because of their small size, many specimens being obtainable from a small chip of rock. Such small samples are often the norm in drilling operations which comminute the rocks and flush the bits to the surface. Larger fossils are fragmented by this process but microfossils, if present, can be sampled adequately. Even when rock cores are obtained these are often only of very small diameter and the same considerations apply.

Exploration of the ocean floors by deep-sea drilling has involved the stratigraphic use of foraminifera, together with coccoliths (nannofossils) and radiolaria, on an extensive scale. These are the common fossils in widespread oceanic sedimentary deposits, normally up to a few hundred metres thick and ranging in age back to the upper Jurassic.

As one example we may take the testing of the sea-floor spreading theory. This postulates formation of new basaltic crust at mid-ocean ridges and its subsequent lateral movement away from the ridge as yet more new crust is formed. If the theory is correct, the age of the basaltic crust must increase away from the ridge. In practice, the most effective way to determine the age of the basalt is to determine, from the fossils, the stratigraphic age of the sedimentary rock resting immediately on it. Then, by referring to the stratigraphic–radiometric time scale, one can convert the stratigraphic age into an age expressed in years. To test the theory holes were drilled at increasing distances from the mid-Atlantic ridge in the South Atlantic, and the basal stratigraphic ages, hole by hole, over a total distance of 1300 km, proved to be: Upper Miocene (11 my); Lower Miocene (24 my); Upper Eocene (40 my); Middle Eocene (49 my); Upper Cretaceous (67 my). The results strongly support the theory and also allow an average spreading rate to be calculated: 2 cm per year.

As a second example we take the interpretation of a sedimentary sequence found at a hole drilled in the Pacific floor, about 41 °N and about 900 km east of Japan in a water depth of 5600 m. The hole was 300 m deep and penetrated the top of the basaltic crust. The stratigraphic age of the basal sediment is early Cretaceous (120 my) which dates the underlying basalt. Palaeomagnetic determinations give the latter a palaeolatitude of 6 °S at the time of its formation. The crust, therefore, has moved, like a conveyor belt, at a rate of a few centimetres a year, from that latitude to its present position at 41 °N in 120 my. This is consistent with the general motion of the plate in a northwesterly direction towards the trench in which it is being consumed. In the following account the different members of the sedimentary sequence are mentioned in turn, starting at the base.

The sediments immediately above the basalt are calcareous and siliceous oozes of planktonic foraminifera, coccoliths (nannofossils), and radiolaria, formed above the CCD (p. 23) from surface waters rich in nutrients, in the equatorial region. At this time the newly formed crust was located on the flanks of the ridge at a comparatively shallow depth. As it moved further away from the ridge it cooled and subsided below the CCD with formation of siliceous, radiolarian ooze only. Continued motion northwards then took it away from the nutrient-rich equatorial waters, formation of ooze ceased, and only thin clays, sparsely fossiliferous, accumulated, below the CCD.

Finally, nutrient-rich waters in the north-west Pacific were reached with the formation of another siliceous ooze of radiolaria and diatoms, below the CCD, and extending to the present ocean floor.

This combination of palaeontology and palaeomagnetics has proved a very effective technique in unravelling the history of the ocean floors.

Problematica

Chitinozoa

?U Precambrian, Ordovician–Permian

Chitinozoans are microscopic fossils of unknown affinities. They are hollow, flask or bottle-shaped vesicles (fig. 166a, b), open at one end and with a wall consisting of a chitinous organic substance, pseudochitin. The wall, the only part preserved, is two-layered, dark coloured and highly resistant to both chemical and physical changes. Chitinozoans have, for instance, been recognized intact in deformed rocks like slates. They are generally between 50 and 300 μm in size but may be longer.

The vesicle varies in shape and structure but in cross-section shows radial symmetry. Usually one end is drawn out into a neck with an aperture which is closed by a lid, the OPERCULUM; this may be recessed within the neck or be terminal. The surface may be smooth, striate, tuberculate or bear hollow spines. Sometimes the vesicles occur as clusters.

Chitinozoans are generally regarded as animal in origin because of the chitinous substance of the wall. Suggestions as to their nature include invertebrate egg cases and protective shells for animal protoctists.

Chitinozoans are confined to and widely distributed within marine deposits of Palaeozoic age. They occur commonly in shallow-water silts and shales and are also found in limestones, graptolitic shales and cherts, and in slates.

Possible chitinozoans have been reported from the late Precambrian but

166   Chitozoans and a pteropod.
a, b, chitinozoa. c, a recent pteropod.

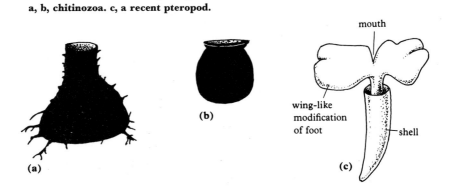

mouth

wing-like
modification
of foot

shell

(a)   (b)   (c)

none are known from the Cambrian. They became common after a major radiation in the Ordovician. In the middle and late Devonian they suffered extinctions and were rare in the Carboniferous. Only a few species have been reported from the Permian.

Kingdom Animalia ('animal microfossils')

Gastropoda – Group Pteropoda

Cretaceous–Recent

Pteropods are very small, marine, free-swimming gastropods some of which secrete an aragonite shell (fig. 166c). They swim using the foot which is modified to form two wing-like fins at the anterior end. The surface of the fin is ciliated and so directs currents of water with food, e.g. diatoms, to the mouth.

The pteropod shell is thin, translucent and either conical or planispiral. It measures about 2.5 to 10 mm or more.

Pteropods are most abundant and diverse in warm tropical oceans and their numbers fall polewards; only one species is found in polar waters. Most occur in the upper 500 m and they show diurnal rhythm, rising at dusk and falling at dawn. In places they occur in enormous shoals and their remains contribute to calcareous oozes in depths down to about 2500 m, below which aragonite, being more soluble than calcite, is unlikely to be preserved. An example of pteropod ooze is found in the mid-area of the South Atlantic.

Crustacea – Class Ostracoda

Ostracods are tiny, aquatic crustaceans which first appeared in the Cambrian, are common as fossils, and are still widely distributed today, mainly in the sea but also in brackish and fresh water. They are typically ovate or kidney-shaped in outline, and between 0.5 and 5 mm in size, though some are much larger, up to a few centimetres.

Ostracods have a laterally compressed body which is indistinctly segmented; the head and thorax are large, the abdomen rudimentary. The body is enclosed by a two-valved CARAPACE (fig. 167) which is periodically shed and regrown. This develops as chitinous outgrowths extending as VALVES on either side of the body. The carapace usually becomes calcified except along the dorsal margin where the chitin acts as an elastic ligament serving to open the valves. A hinge structure of teeth and sockets, or ridges and

167 Ostracods.

a, a recent ostracod showing appendages. b, c, phosphatised appendages preserved in primitive ostracods (phosphatocopids) from U Cambrian, Sweden. (See fig. 96 for comment on preservation of other arthropods from this horizon. SEMs courtesy Professor K.J. Müller.)

1–4 head appendages
5–7 thoracic appendages

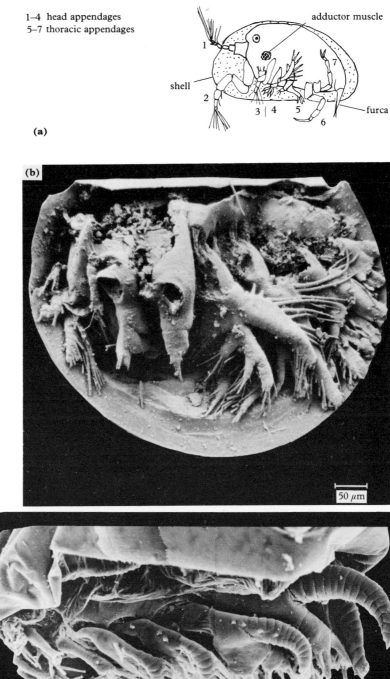

(a)

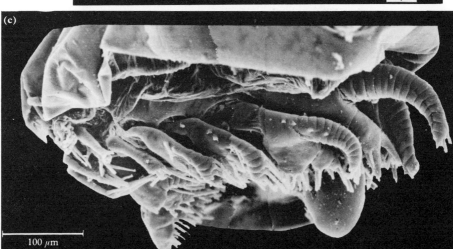

grooves, controls articulation. The valves are unequal in size and one overlaps the other. They are closed by adductor muscles which run transversely from one valve to the other; muscle scars on the inside of each valve mark their position. The carapace is pierced by canals which appear on the surface as pores through which tactile bristles, setae, pass to the outside. In addition, many, usually shallow-water forms, have eye spots or eye tubercles near the anterior end. The carapace may be smooth, or bear sculpture of tubercles or ridges, and often it shows dimorphism, the size and shape differing in the male and female.

When the carapace opens, the limbs extend. There are five to seven pairs of jointed and bristled limbs, either one-branched (uniramous) or two-branched (biramous). These are modified according to function. The head carries four pairs: two anterior uniramous pairs used in swimming or walking; and two biramous pairs beside and behind the mouth, concerned with feeding. The thoracic limbs serve for walking or digging.

Reproduction is sexual. The female may brood her eggs in brood pouches within the carapace, or may lay them separately in the water. The eggs of fresh-water forms are often resistant to desiccation and thus survive periods of drought.

Mode of life

Most ostracods live in normal sea water (salinity about 35‰) and may be found in temperatures ranging from 0 to 50 °C. They are most diverse in the neritic zone and in tropical waters; and are mostly benthic in habit though some are pelagic.

Benthic forms generally crawl or may burrow in sediment. They feed by filtering organic detritus or organisms like diatoms, bacteria and protozoans; some scavenge, e.g. on dead fish. The nature of the carapace may reflect the animal's habitat. For instance, in turbulent near-shore areas it is relatively stout, with eye spots and marked sculpture; and is streamlined in those forms which burrow in fine silts or muds. The small number which live in deep water, where it is uniformly cold (4–6 °C) and dark, have a relatively large, thin-walled, sculptured carapace. These species occupy conditions which are widespread in the oceans, and tend to be cosmopolitan.

Pelagic forms swim using their long, bristled antennae. They filter fine food particles and some capture small animals like copepods or tiny fish. Their carapace is weakly calcified and larger than average, up to 3 cm. They are most numerous in nutrient-rich currents, mostly in the upper waters, numbers and diversity falling with increasing depth. They show a diurnal rhythm, rising at night and falling by day.

Some ostracods are tolerant of a range in salinity and a few such species may colonise brackish water (up to 30‰) or hypersaline water (over about 40‰). The number of individuals tends to be very large in each case. Living forms with a long fossil record are thus useful as indicators of palaeosalinity.

Fresh-water ostracods live in lakes, rivers and ponds including some which may be ephemeral. Most belong to one family (Cyprididae) and tend to have a smooth, thin-walled carapace. Some are scavengers of plant or animal matter.

Some common examples

Leperditia (fig. 168a). Carapace stout and well-calcified. Purse-shaped with long, straight hinge margin; right valve overlaps the left; surface smooth; distinct eye tubercles; adductor muscle scars large. Widespread in shallow-water, marine deposits; benthic.
Silurian–Devonian

Beyrichia (fig. 168b). Carapace bean-shaped with long, straight hinge line and convex ventral margin; surface granular; three distinct lobes, the central

168 Ostracods.
a–f, the arrows point towards the anterior. g, from Chalk Marl, U Cretaceous, Barrington, Cambridgeshire (× 66). h, from shell gravel, Recent (× 82). (SEMs, g, and h, courtesy Dr I.C. Harding.)

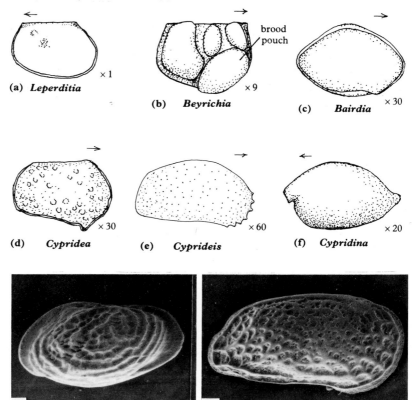

one being the smallest; anterior lobe is enlarged in the female shell to form a brood pouch. Widespread in shallow-water, marine deposits; benthic.
Silurian–Devonian

Bairdia (fig. 168c).   Long-ranging, marine, benthic, found today living in coarse sands in shallow waters. Thick, smooth carapace, strongly convex on the dorsal margin and with a pointed posterior end; left valve larger and partly overlaps the right.
Ordovician–Recent

Cypridea (fig. 168d).   Lives in fresh or slightly brackish water. Carapace thin and chitinous or weakly calcified, kidney-shaped to ovate with almost straight dorsal and ventral margins; there is a 'beak' at the antero-ventral angle (where the lower antennae emerged); surface pitted and usually tuberculate.
U Jurassic–L Cretaceous

Cyprideis (fig. 168e).   Is tolerant of salinity and may be found commonly in both brackish and hypersaline waters. Smooth, subovate carapace.
Miocene–Recent

Cypridina (fig. 168f).   Marine, pelagic form. Carapace smooth, thin-walled and ovate in shape with convex dorsal and ventral margins; prominent anterior process, the rostrum, accommodates the antennae which are used actively in swimming.
U. Cretaceous–Recent

Geological history
Cambrian–Recent
The ostracods appeared early in the Cambrian. The first forms had thin, chitinous valves but in the early Ordovician a major radiation brought a diversity of new forms with stouter, well-calcified valves. These have been separated into four orders, of which two were confined to the Palaeozoic and two were more long-ranging. One order, of which *Leperditia* is typical, died out at the end of the Devonian. *Beyrichia* belonged to a much larger, more important order which continued beyond the Devonian, though in lessening numbers, until it finally disappeared in the Permian extinctions. Of the remaining two orders, only a few representatives survived into the Mesozoic and their numbers did not increase until the Jurassic. The majority were benthic and these show great diversity in the Cretaceous. Numbers were low at the start of the Cainozoic but then increased. Ostracods are very abundant at the present day. The group to which most pelagic forms belong are not generally common as fossils perhaps because the rather thin carapace is not easily preserved.

Fresh-water ostracods appeared in the Devonian and were common in non-marine episodes from then on.

Phylum Chordata – conodonts

Conodont elements are microscopic, tooth-like, phosphatic fossils, brownish in colour, glassy in appearance and mostly measuring between 0.2 and 3 mm in length. They occur in marine rocks either singly or, rarely, in assemblages each of which represents the mineralised apparatus of an individual animal. Fossils of complete conodont animals have been found with conodont apparatuses preserved *in situ* in the mouth region where they were presumably part of a food-processing device. These animals are now referred to the phylum Chordata.

Conodont elements are translucent structures made up of calcium phosphate together with organic proteinaceous matter. They resemble teeth and are described using terms appropriate to teeth. The upper, oral, surface

169 Conodont elements.
a–c, types of conodont elements: a, coniform; b, ramiform; c, pectiniform. d, e, sections to show mode of growth of elements: d, protoconodont; e, euconodont.

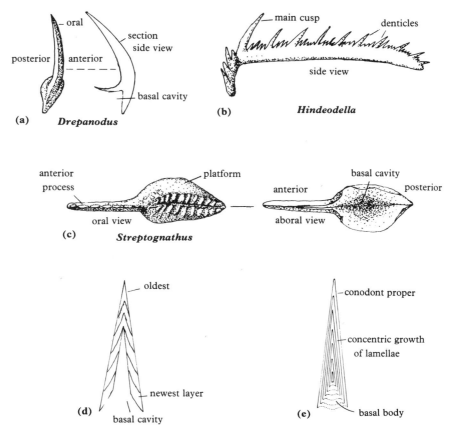

bears one or more cusps or denticles; the lower, aboral, surface is hollowed out to form a basal cavity. Sometimes the basal cavity contains an inner cone, the basal body.

Conodont elements are found in three basic shapes:

(i) Simple cones, or CONIFORM elements (fig. 169a), each shaped like a curved rose thorn with a sharp pointed cusp on the oral surface and an expanded base aborally. The convex edge is anterior and the concave edge posterior.

(ii) Compound, RAMIFORM, elements (fig. 169b) have the base extended to form bar-like processes bearing smaller, conical denticles in front of, or behind the main cusp, or on one or both sides of it. The processes may be directed downwards from the main cusp so that the element, as a whole, is arched. The basal cavity is deepest under the main cusp.

(iii) In compound PECTINIFORM elements (fig. 169c), the processes form a laterally flattened blade, or may be extended at the base to form a broad shelf or platform on one or both sides. The processes, with small denticles, form a central ridge, the keel-like carina; commonly the anterior process remains blade-like. Other, more complex shapes are produced by the development of lateral processes with platforms. The surface of the platforms may be sculpted by ridges or tubercles. The basal cavity may be large or very small; the basal body, if preserved, may be quite large.

While some compound conodont elements are bilaterally symmetrical in plan, most have a curved outline with a convex outer and concave inner side. They may be found in pairs which are mirror images and are regarded as right- and left-hand elements of a bilaterally symmetrical arrangement.

Mode of growth

Studies of the ultrastructure of conodonts show that they grew by accretion of very fine lamellae laid down in one of three growth patterns which define three conodont types. It is not known how these are related, if at all. Two types, protoconodonts and paraconodonts, were short-lived and numerically insignificant. The third, the euconodonts (true conodonts) was a major and long-lasting group.

PROTOCONODONTS (U Precambrian–L Ordovician, fig. 169d) are simple cones of translucent phosphate with much organic matter and with a deep basal cavity. The lamellae show a cone-in-cone arrangement with the newest layer at the base. They were obviously superficial structures and show close similarities in morphology to the grasping spines of chaetognaths (arrow worms, common marine pelagic animals). PARACONODONTS are similar but the cone-in-cone structure is modified by a slight overlapping of the newer lamellae up the sides of the cone, indicating that they were partly embedded in soft tissue.

EUCONODONTS (U Cambrian – U Trias, figs. 169b, c, e, 170) include coniform, ramiform and pectiniform elements, and comprise conodont and basal body. The lamellae, closely spaced, were added over the entire oral and aboral surface of each part, producing a concentric growth pattern which indicates that the entire element was enclosed in soft tissue. While generally translucent, most forms contain some 'white matter', vesicular

170 Conodont elements, scanning electron micrographs.
a, a bedding plane assemblage, U Carboniferous, Bailey Falls, Illinois, USA.
(University of Nottingham collection, × 25). b–j, conodont elements (all × 50): b,
***Corylodus*, Ordovician, lateral view; c, *Ctenognathus*, Silurian, inner lateral**
view; d, *Merrillina*, Permian, inner lateral view; e, *Pterospathodus*, Silurian,
upper view; f, *Ozarkodina*, Silurian, lateral view; g, *Palmatolepis*, Devonian,
upper view; h, *Gnathodus*, Carboniferous, upper view; i, *Icriodus*, Devonian,
upper view; j, *Neogondolella*, Permian, oblique upper view (SEMs, courtesy Dr
R.J. Aldridge.)

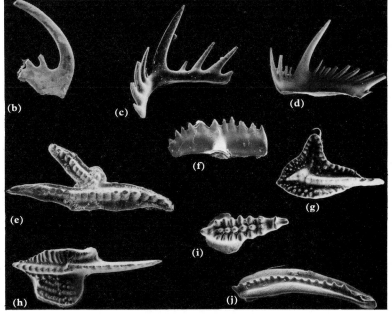

patches within older layers in the denticles. Only euconodonts are considered further.

The conodont animal

Fossils of conodont-bearing animals have been found in the lower Silurian of Wisconsin, USA, and in the lower Carboniferous of east Scotland. The latter were discovered in a thin limestone, associated with varied fossils such as shrimps, ostracods, nautiloids and fish. Three specimens contain conodont elements *in situ*. The fossils are of a soft-bodied, elongate and very slender animal rather like a miniature eel. The best preserved is about 4 cm long by 1.8 mm wide. The head end is slightly wider (1.95 mm) and is bilobed. The assemblage of conodonts present in the mouth region includes ramiform elements in front and pectiniform elements at the back. Towards the posterior end there are faint traces of segmentation with segments which 'Vee' forwards as in amphioxus (a cephalochordate). Traces of fin-rays at the tail end suggest the presence of caudal fins. The picture thus presented is of a tiny, elongate animal, bilaterally symmetrical, perhaps laterally flattened, and designed for swimming.

Affinities of conodont animals

The shape and segmentation of conodont animals, together with the phosphatic composition of the elements, suggests that their affinities may lie with primitive chordates, possibly with the jawless agnathans. Conodont elements themselves have no counterpart in other chordates, but their nature and location suggests that they had a grasping, cutting and grinding function concerned with feeding.

Living agnathans are eel-like forms. They include lampreys (with a record going back to the upper Carboniferous, p. 289) and hagfish (a fossil of which has recently been found in the Mazon Creek fauna (U Carboniferous) of Illinois, USA. Hagfish differ from lampreys in many ways. They have a feeding mechanism which suggests a model for understanding how the conodont teeth may have operated (fig. 171a,b).

The hagfish, up to 40 cm long, is marine, living in a burrow from which it can emerge and swim. It feeds on animals such as polychaete worms and fish, alive or dead. Its sucker-like mouth has an extensible tongue, supported by cartilage and bearing a many-toothed horny plate along either side. The hagfish attaches to its prey and extends its tongue with the plates spread out against the soft flesh. As the tongue is retracted, it folds so that the 'teeth' interdigitate and tear off flesh. The hagfish is not, of course, an exact model. Its 'teeth' are horny and lie on the tongue whereas the phosphatic conodont elements were embedded in tissue which carried out growth and repair but which was not sufficiently robust for preservation.

(a) *hagfish*   gill openings

nostril — vertical section of hagfish head

tentacles —

tongue with horny teeth   cartilage plate supporting tongue

(b)

(c)

171   Hagfish and conodont animal.
a, b, Hagfish (used as a model to explain how conodont elements may have been used). c, Conodont animal, Lower Carboniferous, Granton, Edinburgh. (Courtesy Dr R.J. Aldridge.)

Geological occurrence

Conodonts are widespread in marine rocks including limestones from which they can easily be extracted by chemical means since they are resistant to dilute organic acids. They are stratigraphically valuable since they show rapid evolutionary change. Some were cosmopolitan and may occur in a wide range of facies, including relatively deep-water shales, perhaps with limestones and cherts. Here they are associated with pelagic fossils: e.g. graptolites or goniatites according to the geological age of the rocks concerned but, usually, with no benthic fossils. Such conodont animals were probably also pelagic, perhaps nektonic, a life-style borne out by the shape of the body and the presence of fin-rays in the fossils. Other conodonts were relatively restricted in distribution. They may occur in near-shore deposits associated with benthic fossils of various sorts. Often such elements were larger and more robust than those occurring in deeper water facies.

Geological history

Euconodonts (U Cambrian–Trias) had a long and varied history. They underwent a major radiation in the early Ordovician, reaching a peak of diversity and widespread distribution by the mid-Ordovician. At this time the elements were predominantly of ramiform and coniform types. In the late Ordovician numbers fell and were at a reduced level until late Devonian when there was a moderate resurgence of ramiform and pectiniform elements which continued into the Carboniferous. Coniform elements are not found after the Devonian. In the late Carboniferous a decline set in, and elements are generally smaller and less common in the Permian and Trias. They survived longer in the Tethys region than elsewhere, finally disappearing in the late Trias.

12

Trace fossils

Trace fossils are signs of animal activity preserved in rocks: a footprint is the obvious example; the bones of the foot would form a body fossil. They are often the only evidence for the presence of soft-bodied animals not otherwise preserved. Their study, ICHNOLOGY, involves identifying the activity and making deductions about the animal responsible for it.

Trace fossils are made by both skeletal and non-skeletal organisms. They include signs of locomotion (footprints and crawling trails); resting and dwelling (surface depressions, burrows and borings); feeding processes (tooth marks, grazing furrows, 'mining' burrows, stomach contents, gastroliths), faeces (coprolites, faecal pellets); reproduction (nests); and use of tools (flint implements). They are found in marine and non-marine rocks, and are best preserved in sandy sediments, preferably calcareous, and mudstones or shales. In some rocks they may be the only evidence of life at the time, and have then a special value.

A trace, such as a footprint, made on a soft but firm substrate, is preserved when covered, soon after, by further sediment, preferably different in texture and colour. The eventual lithified fossil may be exposed both on the original substrate as an impression, and as a natural cast on the underside of the covering rock.

Rarely, the animal responsible for a trace is found as a body fossil, e.g. a crawling invertebrate at the end of its trail. Usually, however, the trace-maker cannot be directly identified and circumstantial evidence is used. This is based largely on studies of modern organisms.

Marine trace fossils

In the modern sea the various animals in an area of the sea-floor constitute a community, and the component members change with depth of water and type of substrate. The sea-floor can, therefore, be divided into broad zones, each characterised by particular types of potential trace fossils. In shallow water, the traces are often those of infaunal filter-feeders, whereas in deep water they are made by deposit feeders, mainly surface dwellers. Generally, conditions are well oxygenated but there are some organisms which can tolerate relatively low levels of oxygen.

Marine environments fall broadly into four regions: the intertidal, the sublittoral (extending over the continental shelf), the bathyal and the abyssal. Each grades into the next without any sharp demarcation; lithology rather than actual depth is the controlling factor. Similarly, there is overlap in the distribution of organisms and the traces they make.

The intertidal area is a high-energy zone with a variety of substrates ranging from rocky foreshores and sheets of well-sorted, shifting sands to the more sheltered bays and inlets where tidal flats of muddy sands and silts are found. It is not well represented in the fossil record. Organisms leaving traces are not particularly diverse though individuals may be numerous.

On hard substrates some organisms, such as bivalves and echinoderms, may carve out cavities, sometimes with a constricted entrance, e.g. *Pholas* (p. 37). In some cases the boring organisms make their dwellings in shells,

172   Invertebrate animal traces, recent and fossil.

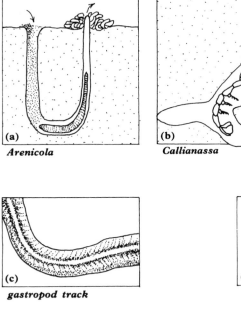

(a) *Arenicola* (b) *Callianassa*

(c) *gastropod track* (d) *spiral feeding trace*

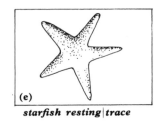

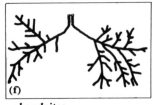

(e) *starfish resting/trace* (f) *chondrites*

e.g. the polychaete *Polydora*, which lives in tubes, about 1–2.5 mm in diameter, bored in oyster shells. The common boring sponge, *Clionia*, makes a network of tunnels with openings about 2 mm in diameter in shells and soft rocks. Feeding traces abound: shells which have been bored by predatory gastropods such as *Natica* (fig. 36) and *Murex* (neogastropod), or gastropod shells with the aperture damaged by predating crabs, are common.

On sandy substrates traces are most likely to be of infaunal filter-feeders living in burrows, vertical or U-shaped. A well-known fossil example is found in the Cambrian 'piperock' of north-west Scotland which takes its name from the abundant vertical burrows known as *Skolithus*. A variety of living 'worms' occupy vertical burrows; some line it with a stabilising mucus, others bind detritus to form a tube or, like *Sabella*, secrete a membranous tube.

On tidal flats some of the sediments have a high content of organic matter, a source of food for deposit feeders which may be present in large numbers. An example is the polychaete worm, *Arenicola*, which constructs a U-shaped burrow (fig. 172a). About two-thirds of this is lined with mucus, and the remaining third, lying above the worm's head, is filled by sand and organic matter being drawn into the burrow and ingested by the animal. The unwanted sand is defaecated at the entrance to the tail shaft as a spiral faecal casting. Periodically the worm pumps water into the burrow via the tail shaft, and this performs several functions: it provides oxygen, introduces an additional food source in the form of suspended micro-organisms, and is finally used to mobilise the sand in the head shaft. These burrows are marked at the surface by the spiral casting and also by a funnel-shaped depression above the head shaft.

The sublittoral region is one of moderate to low energy. It passes, as depth increases, to a region of quiet water unaffected even by storms, and finally merges into the bathyal zone. Similarly, changes in trace fossils are not sharply marked. Deposits are variable but generally finer than in the intertidal area: fine sands, silts and muds, usually rich in organic matter which has settled from higher levels in the water. The fauna is diverse and abundant, including surface and infaunal suspension and deposit feeders. Where sedimentation rates are very low the surface layers may be worked and reworked by burrowers over a long period of time producing a chaotic pattern of structures. This process is known as BIOTURBATION. Conversely, rapid sedimentation may result in better preservation of traces.

Traces include crawling trails, resting traces, and vertical and oblique feeding burrows. Vertical pipes and U-shaped burrows are abundant and made by many different organisms including worms and arthropods. They retain their form readily if they have been plastered with mucus or if a tubular covering has been secreted by the organism. Sometimes a funnel-like depression is present or faecal deposits such as worm castings are found. Modern examples include *Chaetopterus*, a polychaete worm which lives in a parchment-like U-shaped tube in mud or sand and potentially might be preserved as an entire U-shaped trace. U-shaped burrows, complete or otherwise, are found throughout the Phanerozoic. Some burrows are inclined or more-or-less horizontal. *Callianassa*, a type of crab living and feeding in muds and sands at the present day, constructs a complex system of cross-connected branching tunnels, the walls of which are supported by mucus (fig. 172b). It keeps open a pipe to the surface through which it circulates water with oxygen. Similar burrows in Mesozoic rocks are called *Thalassinoides*. Deposit feeders include certain echinoids, e.g. *Echinocardium* (p. 183), which excavate horizontal tunnels, backfilling as they advance.

Surface markings are usually preserved under quiet water conditions. They are essentially grooves which are semicircular in cross-section and which were impressed in soft but firm sediment. Some are food-searching trails made by free-living, mobile organisms such as snails, trilobites and worms, and may be random, winding, sinuous or a regular meandering pattern (fig. 172c). Some snails have a groove along the length of the foot and, as they crawl, leave a double trail. Signs of their feeding are not usually obvious. *Cruziana* is the name given to another type of double trail common in Lower Palaeozoic rocks. This consists of two parallel grooves, separated by a median ridge, and showing numerous scratch marks made by the ventral appendages of an unknown animal (probably an arthropod) as it ploughed into a firm substrate of sand overlying mud. Resting traces, called *Rusophycus*, are thought to have been excavated by trilobites; they are more or less bilobed depressions with scratch marks.

The sublittoral region merges into the bathyal region, an area of low energy where muddy fine sands, silts and muds rich in organic matter are laid down. In places the oxygen level may be low. The diversity of organisms is low though individuals are abundant. Most are deposit feeders which graze the surface or mine along shallow tunnels parallel to or inclined to the substrate as if the animal was following an organic-rich seam within the sediment. Feeding traces may be quite simple or moderately complex, ribbon-like or spiral patterns (fig. 172d).

The bathyal region merges into the abyssal region where the deposits are very fine muds. Conditions are quiet though at times disturbed by turbidity currents flowing down the continental slope bringing sediment with organic matter which augments that which continually rains down from the upper waters. It is dark, pressure is in excess of 400 atmospheres, and it is uniformly cold, about 4–5 °C. There is adequate oxygen. A wide range of different, sometimes unusual, organisms lives here and may be abundant, but they live at a slow tempo. They are mainly deposit feeders or scavengers. Traces include crawling, grazing and shallow feeding-cum-dwelling traces.

Burrows are often elaborate and complex in design and sometimes three-dimensional, indicating a systematic searching for food. Oxygen supply is ensured by short breathing tubes reaching to the surface. Patterns include ribbon-like meanderings, intricate spirals and honeycomb structures, and dendritic systems. *Chondrites* is a dendritic system of descending and radiating tunnels common in the Cretaceous and Tertiary flysch in Europe (fig. 172f). It was made, perhaps, by a thin worm-like animal which made an initial tunnel from which it excavated side branches alternately on either side starting at the far end and backtracking towards the entrance. The tunnels are back-filled with excreted sediment.

Trace fossils connected with feeding processes include COPROLITES. These represent the fossilised faeces of fish and other aquatic vertebrates and are generally phosphatised. Faecal pellets are smaller and produced by invertebrates of many types. Coprolites and faecal pellets are abundant and widespread. They may contain remains of food which has resisted digestive juices. Examples are belemnite hooks in coprolites; and foraminiferal tests or coccolith plates in faecal pellets made by copepods feeding in the plankton. In the latter case the calcareous material is partly protected from dissolution in the deeper waters to which it falls.

Continental trace fossils

On the continents there are few environments in which trace fossils are likely to be preserved. Low-lying areas where sediments can accumulate in river flood-plains, deltas and lakes are the most favourable. Animal populations are likely to be high in such areas, especially near communal watering places. Volcanic areas are also a possibility: imprints in damp volcanic ash, quickly covered by another ash-fall, are now well known.

The traces are most commonly of locomotion but traces of feeding and dwellings are sometimes preserved. The main animal track-makers include invertebrates such as snails, insects and other arthropods: and vertebrates (fish, amphibia, reptiles, birds and mammals). Plant traces are usually of root systems.

Animal tracks are instructive. Over a period of geological time they record the anatomy of the foot and the increased efficiency of locomotion as indicated by changes in foot and limb structure. In the case of vertebrates, the record is of a change from sprawling, undulating movement on four legs, sometimes with tail trails, to the development of speed such as is shown by the horse or the cheetah.

Early vertebrate signs include markings apparently made by lobe-finned fish in shallow-water rocks and found, for instance, in the middle Devonian of Orkney. Amphibian tracks appeared in the upper Devonian, for instance in Victoria, Australia, where tracks, possibly of an ichthyostegid (p. 302), show left and right 'hand' and 'foot' marks, set wide apart, each with splayed-out short digits. Some also show an undulating tail mark between the left- and right-hand tracks. Amphibian tracks are common in the Carboniferous but after the Permian are rarely found.

Reptile tracks are found in the upper Carboniferous and, while rare at

173 Reptile footprints from a quarry in Purbeck Limestone at Langton Matravers, Dorset, and now on display in the Hunterian Museum, Glasgow.
The track, attributed to *Iguanodon*, was found a few feet above the Middle Purbeck Cinder Bed and is thus of basal Cretaceous age.

first, become increasingly common and widespread in the Mesozoic. The early tracks are of a sprawling, clumsy gait. Later forms indicate more efficient striding with quickening movement: the trackway becomes narrower and the individual prints more widely separated. To a varying extent the characteristics of the track-makers can be deduced from the trail and the individual imprints: quadrupedal or bipedal (figs. 173, 174); walking plantigrade (full foot) or digitigrade; shape of foot and number of digits; whether claws or hoofs were present; general indication of size. The tracks show, too, whether the animal was moving slowly or running, alone or in herds. The size of footprints is a rough guide to the size of the animal which made them; and the distance between successive prints made by the same foot, the STRIDE, may give some idea of the speed at which the animal was moving. Estimates are based on the relationship between length of stride and length of limb to the hip. For a given size, the longer the stride the faster the movement. For instance, a large sauropod which made huge, deeply impressed footprints arranged in a narrow trackway with a stride of 2 m is judged to have been walking at about 3.6 kph.

(a)

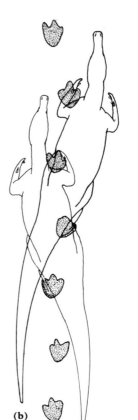

(a) The footprints shown in fig. 173 were made by a large bipedal dinosaur, probably *Iguanodon*. Diagram (a) shows how closely the right foot skeleton fits into one of the footprints.

(b) The disposition of the footprints sheds some light on the way *Iguanodon* walked. The impressions, though clearly alternate, are almost in line. This suggests that at each stride the body swayed from side to side, the head and tail region remaining parallel to the direction of movement. This kind of movement involves some rotation of the foot relative to the body which would normally take place at the ankle. The absence of a tail groove indicates that the tail was held clear of the ground to counter-balance the body.

(b)

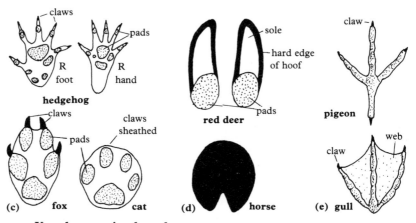

174 Vertebrate animal tracks.

a, b, a three-toed reptile (redrawn from a photograph. The explanation, by Dr J.K. Ingham, accompanies the exhibit of the foot tracks shown in the previous figure, 173). c, d, mammals: c, above, a primitive insectivore; and, below, two carnivores; d, above, a cloven-hoofed and, below, a one-toed herbivore. e, bird tracks: above, a perching bird, below, a swimming bird.

Tracks in lacustrine deposits may have been made by animals moving across sediment temporarily exposed by drought, but some appear to be connected with movement in water. Repeated impressions of front feet only, in a brontosaur trail, suggest that the animal was buoyant in shallow water, propelling itself forward by its front legs and with its rear afloat.

Bird footprints are found in the middle Cretaceous and are widespread in the Cainozoic (fig. 174e).

Mammal tracks (fig. 174c, d) are numerous and varied throughout the Cainozoic. They are best preserved when the animal was moving slowly and the feet went down in regular order: left hind-foot, left fore-foot, right hind-foot, right fore-foot. An animal moving fast puts two or three feet down simultaneously and, when jumping, all four feet come down close together with long gaps separating the groups of prints.

Among the most interesting mammal tracks recorded are those found at Laetoli in Tanzania. About 3.6–3.75 my ago a surface of fine volcanic ash was dampened sufficiently by rain to take the impressions of a multitude of varied footprints. Later these dried and were covered by more ash and blown sand. They include clear hominid tracks made by three upright bipedal individuals strolling along. A larger one is followed, perhaps later, by a smaller person who was accompanied by a child walking alongside. The feet are broad with clear impressions of the heel, the ball and the outer edge of the foot, indicating a well-developed arch. The toes are rather long, the big one being the longest. A clear picture of the fauna existing at the time is given by other prints. Some 20 taxa are represented including monkeys, elephants, hares, cats, a sabre-toothed tiger, hyaenas, horses, rhinoceros, giraffes, bovids, deer of various sorts, a guinea fowl and a centipede. The evidence provided by these tracks is supported by fossil bones and teeth.

Evidence of feeding, faeces and dwellings is restricted. Examples include fossil leaves with edges chewed by insects or caterpillars (on occasion with associated faecal droppings); and bones which have been gnawed by rodents or carnivores. Hyaenas, for instance, crunch bones for their marrow content in a characteristic fashion and leave tooth marks on them. The bones, which may be accompanied by distinctive high-calcium coprolites, have been found in caves and occur in later Cainozoic deposits.

Gastroliths are smooth, rounded 'stomach' stones, sometimes found within a fossil reptile. They appear to have been swallowed by some dinosaurs as an aid to grinding their food (which was swallowed whole rather than chewed) thus speeding digestion. Some modern grain-eating birds swallow grit for the same purpose.

Generally, dung is unlikely to be preserved under subaerial conditions. It is attacked by a variety of organisms, mainly bacteria and insect grubs (e.g. dung beetles) which break it down for food. Dried dung has been found, however, associated with an extinct sloth. Coprolites and bird food-pellets may contain signs of the animals' diet such as bone fragments, teeth or beetle elytra.

Dwellings are occasionally found. They are subsurface excavations subsequently infilled by sediment. They may be simple, like worm burrows, or a more complex system of tunnels.

Precambrian trace fossils

Indications of soft-bodied metazoans are found in Precambrian rocks from about 1000 my onwards. The earliest trace fossils, which are very rare, are signs of animal activity such as a string of faecal pellets, about 0.4 mm across. In the later Precambrian (Vendian, age about 670–570 my) metazoan fossils are found more commonly and sometimes in abundance. In the Ediacara Group, for instance, a variety of fossils of soft-bodied organisms are preserved as impressions and a small number of trace fossils also occur. These include chains of pellets made of sediment, interpreted as 'worm' casts, and a variety of grooves, one gently sinuous and smooth, up to 25 mm wide; and another a more systematic meandering groove. These traces were made by benthic, worm-shaped animals while searching for food either on the surface or, in some cases, at a very shallow depth in the substrate. It has not been possible to relate these traces either to the body fossils or to known phyla. It is probable that they were made by worm-shaped animals moving either by the use of appendages (of which there is no trace) or by PERISTALSIS. The latter is the mechanism used by modern, coelomate, segmented worms and is very effective in burrowing. The coelom is a fluid-filled cavity lying between the gut and the outer wall and divided into segments by transverse septa. Each segment is equipped with two sets of muscles (circular and longitudinal) which, by alternate contractions, send muscular waves along the body, a process known as peristalsis. Since the fluid in the coelom is incompressible the effect of the muscular action is that each segment is narrowed and elongated, then swollen and shortened. The action is similar to that of a wedge, displacing sediment in a forward motion. Anchorage is provided by swollen segments as the rest of the body is moved forwards.

Surface crawling traces could have been made either by coelomates or acoelomates (flatworms, animals without a coelom). It is probable, however, that burrowing traces were the work of coelomates only.

From Vendian times onwards trace fossils are found in increasing numbers and diversity: simple dwelling burrows, straight or U-shaped, appear in the lower Cambrian and complexity increases thereafter. While many traces must have been caused by the activity of worm-like animals, it is likely that some were made by ancestral members of groups such as molluscs and arthropods which became skeletalised in the early Cambrian.

13

Vertebrata – introduction and fish

Introduction

The vertebrates, or backboned animals, have a jointed internal skeleton of bone or cartilage, the cardinal feature of which is the brain-case (cranium), a box-like structure enclosing the brain. Vertebrates comprise the largest and most varied section (subphylum) of the Chordata. This great diversity of animals can be grouped broadly into aquatic forms, i.e. fish; and terrestrial forms, i.e. amphibians, reptiles, birds and mammals, all of which have two pairs of 'walking' limbs, the 'tetrapods'. Their extensive fossil record documents the modifications to the primitive fish body (adapted for aquatic life) which culminated in the tetrapod, fully equipped for life on land.

A brief classification is as follows:

Phylum: Chordata
Subphylum: Vertebrata
Class: Agnatha   –fish without jaws

       Acanthodia (extinct)
       Placodermi (extinct)   – fish with jaws and
       Chondrichthyes (sharks)   paired limbs
       Osteichthyes (bony fish)

       Amphibia
       Reptilia   – tetrapods
       Aves
       Mammalia

As chordates they show bilateral symmetry, and their distinctive features include a notochord, gill slits (which may be present in the adult or may occur only at an early stage in the development of the individual), and a dorsal nerve cord. The NOTOCHORD is an axial rod composed of soft tissue encased by a tough sheath which primitively forms a flexible support for the body. It persists as such in some lower chordates; in the vertebrates it is the basis of the backbone by which it may be replaced in development. The gill slits, paired openings in the pharynx (throat), form part of a food-filtering device in the lower chordates, but in vertebrates they are either concerned with breathing as in fish, or occur only in the embryo as in tetrapods. The nerve cord is contained lengthwise within the backbone and lies above (i.e. dorsally to) the notochord.

The vertebrate fossil record begins with fragments of bony plates of jawless fish found in the upper Cambrian of North America and, more widely, in the lower Ordovician of, for example, North America, Spitzbergen and Australia. The main expansion of fish history, however, did not start until the upper Silurian when more complete fossils are found.

175 Basic structural plan of the vertebrate skeleton.
a–f, fish: a, distribution of the main parts of the skeleton; b, enlarged view of head region to show arrangement of the gill arches; c, d, vertebrae from the tail region seen from the front (c) and from the side (d); e, f, longitudinal sections of ring-shaped centra with unconstricted notochord (e) and biconcave centra with constricted notochord (f). g, the disposition of the limb bones in a tetrapod.

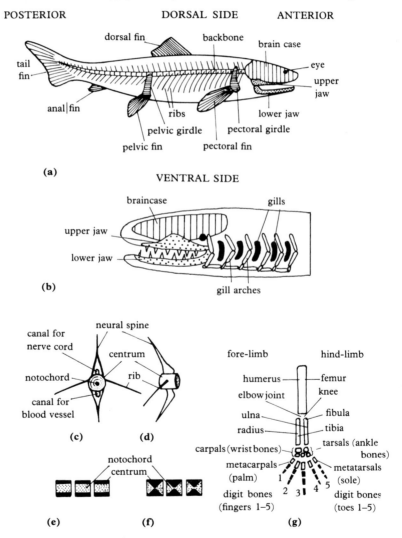

POSTERIOR   DORSAL SIDE   ANTERIOR

dorsal fin   backbone
                                          brain case
tail
fin   eye
                                                          upper
                                                          jaw
anal|fin
            ribs   lower jaw
    pelvic girdle   pectoral girdle
  pelvic fin   pectoral fin

(a)   VENTRAL SIDE

braincase   gills

upper jaw

lower jaw

(b)   gill arches

neural spine
canal for
nerve cord
              centrum
notochord   rib   fore-limb   hind-limb
canal for
blood vessel   humerus   femur
                                      elbow joint   knee
(c)   **(d)**   ulna   fibula
                                      radius   tibia
                                      carpals (wrist bones)   tarsals (ankle
          notochord   metacarpals   bones)
          centrum   (palm)   1   metatarsals
                                                    4 5   (sole)
                                      digit bones 2 3   digit bones
                                      (fingers 1–5)   (toes 1–5)
(e)   **(f)**   **(g)**

Skeleton.   The skeleton may consist of cartilage, a translucent organic tissue which readily decays, or of bone (p. 12), which, since it is about three-fifths mineral salts, is harder and more resistant to decay. Teeth (p. 12), which are an inconspicuous part of the body, are the most resistant part of the skeleton, being composed almost entirely of calcium salts, and may often be the only parts preserved.

The skeleton is largely internal (endoskeleton) and consists of many bones which are grouped into the AXIAL SKELETON, the SKULL and the PAIRED LIMBS and LIMB GIRDLES (fig. 175a). There is also an external skeleton of scales, or bony plates, in fish, amphibians and reptiles; of feathers in birds and of hair in mammals.

The axial skeleton is made up of a series of articulating bones, the VERTEBRAE (fig. 175c, d), forming a flexible support, the BACKBONE, which is situated dorsally along the length of the body (fig. 175a). Each vertebra has a central part, the CENTRUM, which primitively forms a ring around the notochord (fig. 175e), and in higher forms constricts (fig. 175f) or replaces it. Two processes on the dorsal side unite above the nerve cord forming a spiny projection (neural spine). Similar processes on the ventral side of the tail vertebrae protect blood vessels, and articulating outgrowths from the sides form the ribs. The vertebrae are all similar in fish but in tetrapods they differ according to their position in the body, e.g. they have ribs in the chest region but none in the waist region. The skull articulates with the front vertebra of the backbone. It includes the brain-case and jaws.

Most vertebrates have two pairs of appendages used for steering or movement, FINS in fish (fig. 175a) and LIMBS in tetrapods (fig. 175g). These have skeletal supports in the body wall with which they articulate, the fore-limbs with the shoulder or PECTORAL GIRDLE, and the hind-limbs with the hip or PELVIC GIRDLE.

Fish

Most fish have scales covering the body, paired fins, and 'breathe' by paired gills. The gills lie within slits in the pharynx. They are richly endowed with blood vessels and control the uptake of O_2 and release of CO_2 between blood and water. They are supported by jointed skeletal bars which meet the corresponding bars ventrally forming inverted arches (fig. 175b).

Typically, the fish body is a pliant, streamlined, spindle-shape, tapering tailwards and slightly flattened laterally. It is an ideal design for active swimming, the tail being the main swimming organ. The fins, which can pivot, control stability and direction of movement. Paired fins act as foils to control pitch, and also act as brakes. Unpaired fins act as stabilisers. Forms which are adapted to a sedentary life on the sea-bed, e.g. the skate or sole, have a depressed, flattened body with the eyes placed dorsally on top of the head instead of on each side as in active forms like the herring.

Jawless fish

Agnatha

Agnathans include the simplest 'fish' which have no jaws, and typically lack paired fins; they have many paired gill openings. They are represented today by only a few species including the lamprey (a degenerate, eel-like parasite) and the hagfish, with a skeleton of cartilage and naked of scales. Extinct forms, referred to broadly as OSTRACODERMS (shell-skinned), were mostly heavily armoured forms encased by a box-like carapace of thick, bony plates and scales. The endoskeleton, of cartilage, is not usually preserved. They were bottom-living forms, generally lacking paired appendages, and rather like tadpoles in shape, probably swimming in a similar way by undulating the tail. They became extinct at the end of the Devonian.

The earliest remains, small, bony plates, occur in shallow-water marine rocks (often sandstones) of late Cambrian and Ordovician age in many parts

176 Jawless fish (f, modified from Moy-Thomas and Miles).

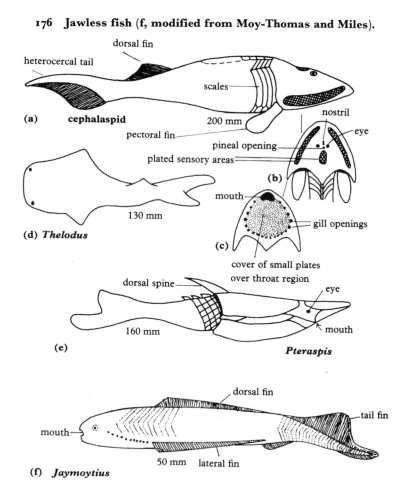

of the world. The plates were discrete elements in a mosaic arrangement and separated after death. They are referred to a group, HETEROSTRACANS, of which better preserved fossils, with the plates more regularly arranged, are found in the Silurian. In the lower Devonian they occur in fresh-water deposits. *Pteraspis* (L Devonian, fig. 176e) is a common form with much of the anterior body encased by a shield of several plates. The shield projects forwards beyond the mouth as a 'snout' and, at the back, forms a sharp dorsal spine. Later, related forms developed lateral stabilising projections (horns). *Cephalaspis* (U Silurian–M Devonian, fig. 176a, b), one of the best known ostracoderms, represents a different group, OSTEOSTRACANS. The head region is covered by a rigid, bony shield and the rest of the body by elongated scales. It has, unlike most ostracoderms, a pair of fins lying just behind the head shield. The depressed shape of the head shield with the eyes close together on top is typical of a bottom-dwelling fish. The position of the mouth on the underside, with paired gill slits ranged behind it, is shown in fig. 176c. Dissection of the cranium in some particularly well-preserved fossils has shown that they possessed a brain structure generally similar to the

177 Cephalaspids.
a, *Cephalaspis*, Lower Old Red Sandstone (Devonian), Scotland (× 0.5); much of the bony plating of the head shield is absent, and fine tubules can be seen radiating towards the left margin. b, *Aceraspis robustus*, Downtonian, U Silurian, Roingerike, Norway (× 0.8), lateral view of a nearly complete fish with scales and fins (pectoral, dorsal and tail) preserved.

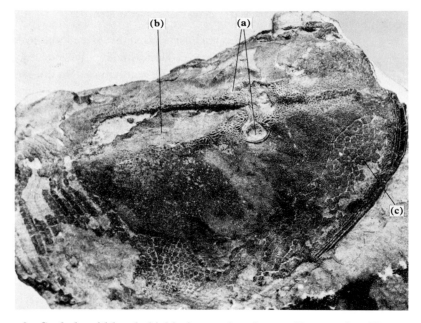

178 Cephalaspid head shield: *Aceraspis robustus*, Downtonian, U Silurian, (× 1.5).
Position of the eyes (a), and of the dorsal sensory field (b); c, small plates covering the lateral sensory field.

modern lamprey. Also of interest are traces of tubules (fig. 177) that run towards three areas covered with small plates lying one behind the eyes, and the other two near the lateral margins of the head shield (figs. 176b, 178). It has been suggested that these areas had some special, sensory function. Possibly they were pressure or vibration receptors. *Cephalaspis* is thought to have lived in fresh-water pools or streams, feeding on organic matter filtered in the gill pouches from the sediment on the stream bed.

Two groups lacked heavy armour. One, which lived in coastal waters, is typified by *Thelodus* (U Silurian–L Devonian, fig. 176d). Its body was studded with tooth-like scales (denticles). Similar scales are found dispersed in Silurian rocks and possibly also in the upper Ordovician. The second group, the ANASPIDS, is found only in fresh-water deposits (U Silurian–M Devonian). These were small, minnow-like forms, about 10–15 cm long, with the body covered by thin, narrow scales, and with a small terminal mouth. Their shape suggests they were active swimmers. It has been suggested that one of these, *Jaymoytius* (U Silurian, fig. 176f) was possibly related to the ancestral stock of the lampreys.

The potential for preservation of a fish like the naked lamprey, with its cartilaginous skeleton, is not good. However, fossils such as *Mayomyzon* (U Carboniferous) found in the Mazon Creek fauna are very similar, indicating a prolonged, conservative history.

Fish with jaws and paired limbs

The development of jaws in fish was a notable event in vertebrate evolution. Together with paired limbs and lungs it widened the possibilities of water as an environment, and eventually led to the emergence on land of the tetrapods. The bones which form the jaws originated, on embryological and other evidence, from a pair of jointed skeletal bars which supported an anterior set of gills (fig. 175b). The development of jaws implies predatory habits and consequently a need for increased mobility. This was met by the development of fins, and these are of a characteristic type in each of the four classes of fish with jaws (p. 286).

Acanthodia

The name 'Acanthodia' refers to the stout, thorn-like spines which form the leading edge of their many fins. *Climatius* (U Silurian–L Devonian, fig. 179a) is typical with paired fins ventrally, and single, median-dorsal and anal fins. It has the fusiform body of an actively swimming predator. The axis of the tail is tilted upwards, overlying the tail fin (heterocercal condition). Head and body are covered with small, rhombic scales, and larger scales cover the gill openings. The eyes and mouth are large, and the jaws well-furnished with sharp teeth.

Acanthodians were mainly small fish, about 20 cm long, though some reached 2 m. They lived in both fresh and marine waters. They were the earliest jawed fish to appear. Scales and spines are found early in the Silurian and possibly in the Ordovician. These fish were most diverse in the lower Devonian and persisted through the Carboniferous and Permian, showing little divergence from the basic plan seen in *Climatius*.

Placodermi

The extinct placoderms were a heterogeneous group of fish with primitive jaws carrying teeth, not of normal fish type, but modified, tooth-like bony plates with slicing edges. They included heavily armoured forms, reminiscent of the ostracoderms, and others, with a less rigid scaly covering, which were probably more agile predators. Most were relatively small forms but some, like *Dunkleosteus*, reached 10 m in length. They included fresh-water and marine forms.

Coccosteus (M–U Devonian, fig. 179c) represents a group (the arthrodires) in which the head shield was hinged to a second shield covering the shoulder region so that the head as well as the lower jaw may have moved to open the mouth. This fish must have had a very wide gape which, in conjunction with sharp biting bones on the jaws and a relatively streamlined body, is indicative of predatory habits.

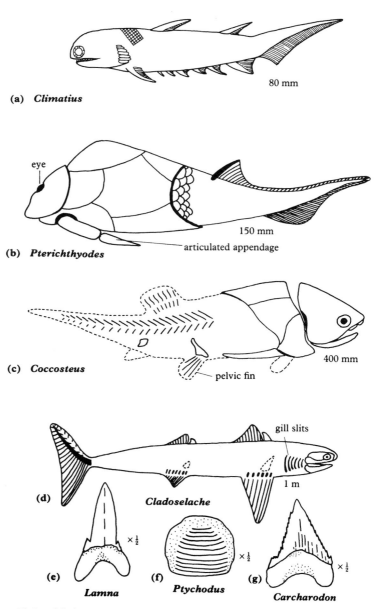

(a) *Climatius*   80 mm

(b) *Pterichthyodes*
eye
150 mm
articulated appendage

(c) *Coccosteus*
D
pelvic fin
400 mm

(d) *Cladoselache*
gill slits
1 m

(e) *Lamna*   ×½

(f) *Ptychodus*   ×½

(g) *Carcharodon*   ×½

179 Fish with jaws.
a, an acanthodian. b, c, placoderms. d–g, sharks: d, an upper Devonian form; e,
g, teeth of predatory sharks; f, tooth adapted to crushing shells.

Pterichthyodes (M Devonian, fig. 179b) belongs to another group of placoderms (the antiarchs) in which a pair of spiny, jointed appendages articulated with a box-like shield of fused plates enclosing the head and trunk. Scales covered the rest of the body. Impressions of soft parts in a related genus show a pair of sacs, connected with the pharynx, which have been interpreted as lungs.

Placoderms were already highly diversified when they appeared in the upper Silurian, and for a time, in the upper Devonian, they were the supreme vertebrates with a world-wide distribution. Their numbers then fell abruptly and few survived into the early Carboniferous.

Chondrichthyes – the sharks

The sharks and related forms, the skates and rays, form a group in which the endoskeleton consists of cartilage (p. 12). Their gill slits are open, and their skin may be studded with tiny tooth-like scales, DENTICLES. Like teeth, these are made of dentine covered with enamel. They have neither lungs nor air bladders with which to adjust buoyancy. Typically the sharks are predators. They are strong swimmers and their two pairs of fins are used mainly in steering. The mouth opens on the underside of the head, and the well-developed jaws are typically armed with sharp pointed teeth which are replaced in succession as they move forward and drop out (fig. 179e). The skates are bottom dwellers with the body flattened and greatly extended sideways by enlarged (pectoral) fins. They feed mainly on molluscs, and their teeth are modified to a blunt, smooth or grooved surface adapted to crushing shells (fig. 179f).

Since cartilage is rarely preserved in fossils, the shark record consists mainly of teeth and fin-spines and, apart from some early fresh-water forms, is found mainly in marine deposits. *Cladoselache* (U Devonian, fig. 179d), a marine shark, has been found as finely preserved fossils showing traces of soft tissue, such as gill filaments. It had a typically shark-like body up to 2 m in length with broad-based fins comprising dorsal fins; paired pectoral and pelvic fins; and a forked tail fin. Its teeth were sharply cusped with a large central cusp flanked by smaller ones. It was probably a relatively fast-swimming predator.

Sharks appeared in the lower Devonian and increased greatly in variety during the Carboniferous and Permian. With renewed diversification in the Mesozoic, forms of modern aspect appeared, including bottom dwellers with shell-crushing teeth, e.g. *Ptychodus* (Cretaceous, fig. 179f). Diversity diminished in the Cretaceous with the rise of teleost bony fish. Today there are few species though individuals are numerous and widespread. They include *Carcharodon* (Cretaceous–Recent, fig. 179g), the great white shark.

Osteichthyes—the bony fish

These are the most complex fish, being distinguished by their bony skeleton and scales, and by the bony flap covering the gills. The skull is covered by a complex arrangement of plates, and the body by scales. Most possess an air bladder or, more primitively, lungs. The air bladder functions as a buoyancy control. It contains gas which is adjusted to vary the specific gravity of the body so that the fish can maintain effortlessly its depth in the water. It arises, during development, as a sac-like outgrowth from the pharynx. Originally, it appears to have functioned as a lung, used for breathing when oxygen in the water was reduced by, for instance, a rise in temperature. Such conditions are encountered today by some tropical fresh-water fish in periods of drought and, probably, in the past by, for example, Devonian fresh-water forms. Traces of lungs are found in some early fossil bony fish and also in placoderms.

Bony fish far outnumber other types of fish and are enormously varied in form and habit. Today, most live either in the sea or in fresh water, but a few, like the eel and the salmon, migrate from the one environment to the other. Their early fossils occur in near-shore, marine and non-marine rocks in the lower Devonian when they were already clearly defined into two groups: the ray-finned ACTINOPTERYGII, and the fleshy-finned SARCOPTERYGII. Subsequently, the ray-finned fish have had an expansionist evolution and comprise the bulk of living fish like the herring, perch and salmon as well as many fossil forms. By contrast, the fleshy-finned fish have, since the Devonian, remained numerically restricted and only four forms survive. Their interest lies primarily in their role in tetrapod evolution.

Ray-finned fish
In these fish, the actinopterygians, the paired fins each resemble a fan consisting of a web of skin supported only by horny rays, and with the skeletal bones lying inside the body (fig. 180d). The fins are flexible foils used for braking as well as steering, and there is a single, dorsal stabilising fin. Most living members have an air bladder and typically breathe by gills although lungs are present in two primitive, surviving forms and have been found as traces in some fossil forms. The early ray-finned fish, chondrosteans, were strong-swimming predators. They are found in fresh-water deposits in the mid-Devonian, and later they spread into the sea. They were small, up to 25 cm in length with the head covered by bony plates, and the body by thick, shiny, enamel-covered bony scales of rhombic shape (GANOID scales). The tail axis was tilted upwards overlying the fin (HETEROCERCAL, fig. 180a). A lung was present and is retained in the two extant genera. *Cheirolepis* (M Devonian) is typical and *Palaeoniscus* (Permian, fig. 180a) differs little. Chondrosteans showed some variety of body form in the course of the Carboniferous and Permian but declined in numbers as a rich variety of more advanced forms, holosteans, appeared.

Holosteans occur in Mesozoic fresh and sea-water deposits (fig. 181) and show modifications connected with improved feeding and locomotion. Changes include shortening of the jaw and the tail axis; reduction of the fin skeleton; thinning of the bony scales, though the enamel cover was retained; and modification of the lung to form the air bladder.

By the later Cretaceous the holosteans in their turn were largely replaced

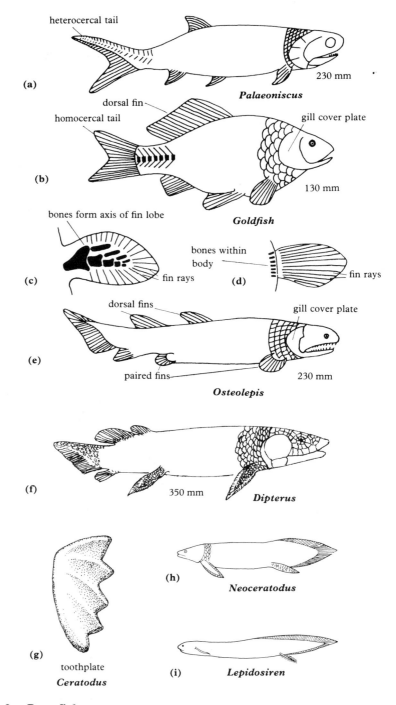

180 Bony fish.

a, a Permian ray-finned fish with heterocercal tail. b, a modern teleost with homocercal tail. c, d, paired fins of a crossopterygian fish (c) and of a ray-finned fish (d). e, f, a crossopterygian fish (e) and a lungfish (f) from the Middle Old Red Sandstone (Devonian). g, a lungfish tooth, Trias. h, i, extant lungfish.

181 A holostean fish, *Dapedium*, L Lias, L Jurassic (× 0.5).

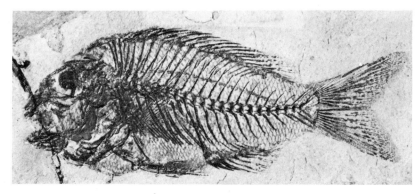

182 A teleost fish: *Sparnodus*, U Eocene, Verona, Italy (× 0.7).

by the 'modern' type of bony fish, the teleosts, which are the dominant bony fish in both sea and fresh-water today (figs. 180b, 182). In the teleosts the skull bones lie below the skin; the body scales are very thin, bony plates lacking enamel on the surface; and the tail axis is shortened so that the fin is symmetrical and terminal (HOMOCERCAL, fig. 180b). The air bladder is fully developed. The adaptive radiation of the teleosts brought a vast array of modifications in body form and behaviour, ranging from the bottom-dwelling, inactive flatfish to forms capable of remarkable bursts of speed: for instance, swordfish and marlins, which reach speeds of 65–80 kph. A few species of chondrosteans and holostean fish survive today in fresh-water and are invaluable aids to interpreting fossil forms.

Fleshy-finned fish

Fossils of these fish, the sarcopterygians, are first found in the early Devonian
and are similar in appearance to the co-existing ray-fins with bony plates on
the skull, heavy scales on the body and a heterocercal tail. But they were
generally larger, up to about 70 cm, and show a variety of distinctive
features, two of which are of particular significance: the paired fins are each
supported by a scale-covered, fleshy lobe with an axial bony skeleton
(fig. 180c), and there are nostrils opening in the roof of the mouth. Other
characteristics, best seen in Devonian fossil forms, include two dorsal fins
(fig. 180e), and bony scales with only a thin enamel layer. Two groups are
distinguished: the Dipnoi (lungfish) and the Crossopterygii.

Dipnoi.   A typical lungfish like *Dipterus* (M Devonian, fig. 180f) is distinguished by, for instance, the mosaic arrangement of the skull bones, and by the
nature of its specialised teeth. These lie, not on the marginal jaw bones, but
on the palate and floor of the mouth. They are broad, ridged plates with
denticles, formed by fusion of many teeth, and well suited for crushing hard
food such as shells (fig. 180g). *Dipterus* occurs in both marine and fresh-water
deposits, but later forms are restricted to fresh water. Fossils, mainly of teeth,
are relatively common in the upper Palaeozoic but rare after the Trias.

   Three surviving genera are found today, one in each of Australia, South
Africa and South America. They have broadly-ridged tooth plates
(fig. 180g) similar to Devonian forms, but they show a great reduction of
bone in the skeleton and the fins are extremely slender. They are eel-like
creatures, with lungs, adapted to life in shallow water which, seasonally, is
stagnant and liable to dry up. The Australian form, *Neoceratodus* (fig. 180h),
is little different from its Trias relative *Ceratodus*. When stressed for oxygen it
gulps air at the surface but cannot live out of water. The other two,
Protopterus and *Lepidosiren* (fig. 180i), have paired lungs and die if kept
submerged. They survive dry periods by burrowing in mud and breathing
air until the rainy season returns. Similar burrows, some with tooth plates,
have been found in upper Palaeozoic rocks, so this adaptation dates from an
early stage in lungfish history.

Crossopterygii.   Devonian crossopterygians differed from co-eval lungfish in,
for instance, the pattern of the skull bones (fig. 184d), and the nature of the
teeth. The latter lay along the margins of the jaws and were sharply pointed,
grasping teeth appropriate to an active predator. Their surface is grooved
vertically as a result of radial infolding of the enamel layer into the dentine
(fig. 184a). In transverse section this gives a tortuous, labyrinthine pattern
(fig. 184b) and the teeth are accordingly termed labyrinthodont. Such teeth
are unique to crossopterygians and to certain amphibia, the labyrinthodonts
(p. 304). In some Devonian crossopterygians the centrum of each vertebra
consists of two bony elements instead of the single ring- or spool-shaped

183   A fossil shoal of crossopterygian fish, *Holoptychius flemingi*, U Old Red Sandstone (U Devonian), Dura Den, Fife (× 0.4).

element typical of other fish. A similar vertebral structure is found in the early amphibians (p. 301) and the pattern persists in a modified, more complex form in higher vertebrates. Crossopterygians appeared in the early Devonian and forms like *Osteolepis* (M Devonian, fig. 180e) and others (fig. 183) were common in shallow fresh-water until the early Permian. *Osteolepis* and related forms, rhipidistians, show many skeletal characters which suggest that they may be the stock from which the tetrapods diverged during the Devonian.

The coelacanths were an offshoot which developed in the mid-Devonian and migrated into the sea. They were common in the Mesozoic until the end of the Cretaceous. Later fossils have not been found but an extant form, *Latimeria*, lives in relatively deep water off the Comoro Islands in the Indian Ocean. It is unusual for a fish in giving birth to live young.

14

Vertebrata–tetrapods

Tetrapods are essentially terrestrial vertebrates which have lungs and, typically, two pairs of limbs, each with five digits. They are divided into four classes: Amphibia, Reptilia, Aves (birds), and Mammalia.

Origins of the tetrapods

Tetrapods appeared in the late Devonian as fully developed amphibians able to function on dry land: e.g. to breathe air direct; and, as predators, to move with agility and catch prey. They resemble the fleshy-finned fish of that time in so many respects that, despite the lack of intermediate fossils, there is little doubt as to their relatedness. These fish had, in their lungs and in the structure of their bony axial skeleton and paired limbs (fig. 180c), the necessary potential for life out of water. The choice of ancestral stock lies with either the lungfish, then already specialised and adapted to a diet of hard food, or the crossopterygians. These, like the early amphibians, were active predators, and the traditional view is that they gave rise to the amphibia via rhipidistian (p. 299) stock. The basic arrangement and structure of their skeletal parts is similar. Details of particular interest which may be closely matched include the pattern of the skull bones, and the nature of the teeth (fig. 184a–d).

Structural contrasts in the skeleton of the fish and the amphibian reflect mechanical adjustments needed to support the extra weight of the body when it was no longer buoyed up by water. This involved a modification of the vertebrae into stout, interlocking pieces (fig. 184f) to form a strong, articulated backbone acting as a bridge between the limbs; and the development of ribs in the chest region to support the lungs. The limbs and girdles, too, were modified and strengthened to support the backbone and carry the body clear of the ground (fig. 184j). The girdles are larger, and the pelvic girdle is united with the backbone; the limb bones are longer and fewer in number, and there are additional bones, the digits, which support the fingers and toes. Other modifications include those concerned with seeing, hearing and smelling in air instead of water.

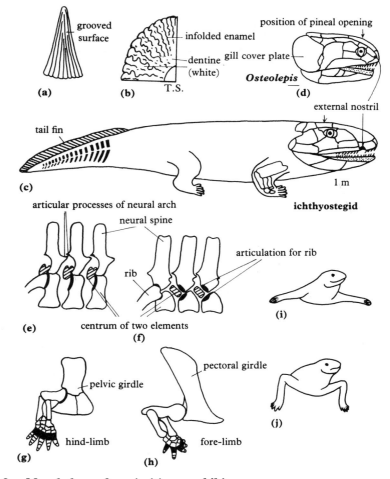

184 Morphology of a primitive amphibian.

a, b, labyrinthodont tooth, lateral view (a) and part of a transverse section (b). c, restoration of a primitive amphibian; the skull bones are sketched in to show the similarity in their arrangement with that in (d), d, crossopterygian skull. e, f, lateral view of vertebrae of an ichthyostegid (e) and of a later, more advanced amphibian (f). g, h, limb structure of a labyrinthodont amphibian. i, j, sketches to show the change in attitude of the limbs from the sprawling fish-like position in (i), to the condition in (j) typical of the early tetrapod, with the body supported clear of the ground.

Amphibia

The amphibia are the oldest and simplest class of tetrapods. Living forms are cold-blooded and have a moist, permeable skin used, in addition to the lungs, for respiration. They are terrestrial only in part. Typically they must go back to the water for breeding, since their eggs lack protection against desiccation. Living forms, like newts, frogs and toads, are highly specialised and show little outward resemblance to their Palaeozoic ancestors.

Perhaps the most familiar amphibian is the frog. It feeds on insects and

185 Restoration of an ichthyostegid (× 0.1).
(Painting by M. Wilson.)

worms. Its young are completely aquatic tadpoles, with feathery gills and a long, finned tail for swimming. In growth the tail and gills are resorbed, lungs and paired limbs develop and, when its metamorphosis is complete, it comes out on land as a tiny frog.

Fossil amphibia are found in non-marine rocks but are rather rare. The earliest known are the ichthyostegids, of which several genera occur in fresh-water deposits of late Devonian age in Australia and Greenland. They were sprawling creatures (fig. 185) showing a mixture of fish-like and amphibian features. They had a finned, fish-like tail, scales on the body, and vertebrae similar in pattern to those of crossopterygians; but their limbs were of tetrapod type with five digits. The skull was flattened but the arrangement of the bones differed in only a few details from that of the crossopterygians: the eye openings were closer together and set higher on the skull; and they had no gill cover-plates. Ichthyostegids, though the earliest known, are regarded as an offshoot from the main line of amphibian evolution.

Recent finds from the eastern part of the Midland Valley of Scotland have greatly increased our knowledge of lower Carboniferous amphibians. These include about six different aquatic forms (one specimen almost complete) from one locality; and at least four land amphibians from another. The latter are associated with other land-living animals (such as millipedes) and include the earliest known ancestors of frogs and salamanders (fig. 186). Amphibians are relatively common in the coal measure facies of the upper

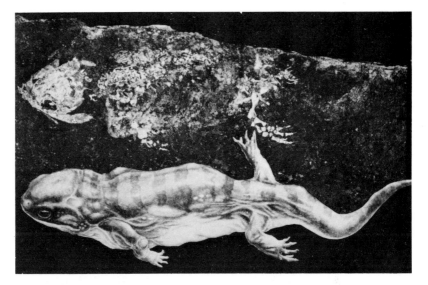

186   A land-living amphibian: a temnospondyl amphibian, L Carboniferous, Bathgate, Edinburgh.

It represents the ancestral stock from which frogs and salamanders arose. (Specimen in the Royal Museum, Scotland; photograph Dr J.K. Ingham; reconstruction by Michael Coates.)

187   A Permian labyrinthodont amphibian, *Eryops*, Red Beds, L Permian, Texas, USA (× 0.06 approx.).

Background: a restoration by Dr J.K. Ingham; foreground: a replica of a skeleton in the American Museum of Natural History, exhibited in the Hunterian Museum. Length of original, about 2 m.

Carboniferous in Europe and North America. Presumably, they found the moist, warm climate congenial. They include the labyrinthodonts, so called because of the complex infolded pattern of the enamel layer of their teeth as seen in thin-section (fig. 184b). These were notable for the heavy bony plating of the skull (fig. 187). They disappeared in the early Mesozoic, but during their heyday had provided a variety of offshoots. These included the true reptiles, and also a number of reptile-like forms whose affinities are difficult to resolve. These include *Diadectes* (L Permian), the first known tetrapod herbivore. Some forms had external gills in the early stages of development, and are therefore classed as amphibians.

Fossil amphibians are rare in the Mesozoic and Cainozoic though modern forms are abundant and varied. Their structure has changed little since the Jurassic.

Reptiles

Reptiles are 'cold-blooded' tetrapods whose body is protected by a dry, horny, often scaly skin. Their major innovation, which sets them apart from amphibia, is the amniotic egg with its protective shell designed to hatch on dry land. In becoming independent of water for breeding purposes they were able to spread over land to occupy new habitats. The egg, of course, requires internal fertilisation. The embryo, as it develops within the shell (fig. 188), lies in a fluid-filled sac (amnion); it is connected to a large yolk (energy food store); to water and protein (albumen); to a sac-enclosed waste disposal system (allantois); and, through the porous shell, to air. Eggs are found fossilised but are rare; the earliest known is of Permian age.

'Cold-blooded' means simply that the reptiles' body temperature is more or less that of the ambient temperature. Reptiles have a low metabolic rate and produce little internal heat. Instead, they are ectothermic, deriving their energy from an external source, for instance by sun-basking. Once warmed to their preferred temperature they become active. But lacking insulation they cool quickly as heat is lost from the body surface. Smaller reptiles, having a greater surface area/volume ratio than larger forms, cool

188 Embryo of reptile growing within egg.

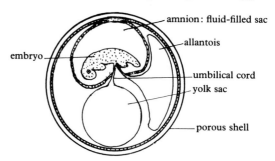

more quickly. Thus the larger the reptile the longer it retains body warmth. It is likely that the very large Mesozoic reptiles, living in an equable climate, were able to maintain a relatively stable body temperature.

Living, and most fossil, reptiles are easily distinguished from amphibia by many skeletal details, especially in the skull, vertebrae and limbs, as well as in the soft parts. As mentioned earlier (p. 304), a variety of upper Carboniferous and Permian amphibia show reptilian features, but these cannot be the ancestral stock of the true reptiles which were living alongside them. They do, however, stress the close relationship and leave no doubt as to the amphibian origin of reptiles.

Reptiles provide a classic example of adaptive radiation. Initially scarce, they diverged to exploit and adapt to many different habitats thereby becoming successful and abundant. This cycle was repeated many times in the course of their history as new 'improved' forms appeared and diversified.

The earliest known reptiles occur in the mid-Carboniferous. They mark the start of the first great radiation during which they dispersed widely throughout the world in Permian times. It was to culminate in the later, vast and varied array of Mesozoic reptiles of which the extant snakes, lizards, turtles, tortoises, crocodiles and the unique New Zealand tuatara are the few relics. It also led to the emergence of birds and mammals.

In the later Palaeozoic there were great changes in climate and vegetation forming new habitats on land to which the reptiles adapted and diversified. This involved changes in locomotion and diet which are reflected in their skeletal details, especially in their jaws and limbs.

The early reptiles were probably insect-eaters using an abundant and diversifying food source. Later forms exploited alternative foods. Some adapted to aquatic life and developed sharply-pointed grasping teeth, suited to fish-eaters; or broad, flat-surfaced teeth, suitable for crushing hard shells. Most, however, remained on land, some adapting to feed on plants and others to prey on animals.

Plants form food of low energy value, must be eaten in bulk and are slow to digest. Accordingly, the plant-eaters tended to become large, bulky, slow-moving creatures with heavy jaws, and teeth modified for cropping and grinding. They had also to evolve a new digestive process. Flesh, by contrast, has a high energy yield and less of it need be eaten at a time. But live prey has to be caught: so the carnivores had to become more agile and light-bodied. In the process, the sprawling gait of their ancestors was altered and the legs were swung under the body, lengthened and straightened so as to lift the animal clear of the ground. They had to develop powerful jaws to cope with struggling prey, and sharp teeth to kill and to tear off flesh.

Classification

Skull structure provides a basis for separating reptiles into four subclasses: Anapsida, Synapsida, Diapsida and Euryapsida (fig. 189). Basically, the

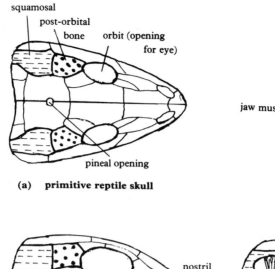

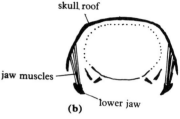

189   Basic skull patterns in reptiles, a basis for separating four major groups.

a, surface view of a skull without temporal openings to show the arrangement of bones. b, cross-section behind eyes. c, anapsid skull; no temporal openings. d, synapsid skull; one temporal opening in lower positions. e, euryapsid skull; one temporal opening in upper position. f, diapsid skull; two temporal openings.

primitive reptile skull is a 'box' enclosing the brain, and overlying the upper and lower jaws. The 'box' has an inner part around the brain and an outer cover made up of bones lying just below the skin and fused together to form a roof. This roof, with paired openings for the eyes and nostrils, extends down over the brain-case forming the cheeks. The muscles which work the lower jaw are attached on the inner surface of the cheeks in the temporal region, i.e. just posterior to the eyes. Early in reptile history new and stronger systems of jaw muscles evolved and new paired openings developed between bones in the temporal region. These openings provided a larger, more secure anchorage for the muscles and gave them space to bulge as they contracted. There are four basic patterns of temporal openings which are shown in fig. 189. These may be greatly modified in later reptiles.

Early reptiles

Anapsids

One of the oldest reptiles is *Hylonomus* (M Carboniferous), a small insect-eater, about 20 cm in length, and lizard-like in appearance. It marks the start of the anapsid reptiles, the captorhinomorphs. They became widespread and dominant in the Permian with offshoots including carnivores and the massive plant-eating pareiasaurs (U Permian, fig. 190), which reached 2–3 m in length. In Britain they are known mainly by their footprints, but

190 A restoration of *Elginia*, a pareiasaur reptile from the upper Permian, Elgin, Scotland, Length 760 mm.
(Painting by L. Kennedy and J.K. Ingham.)

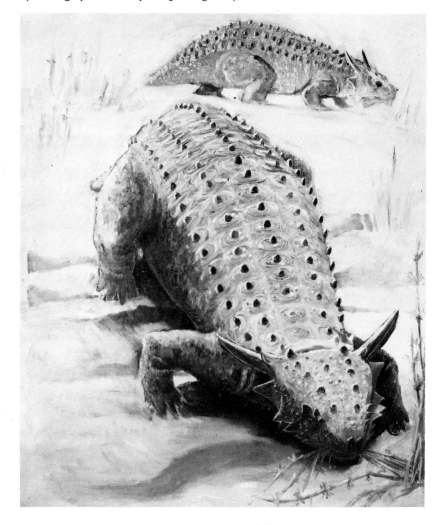

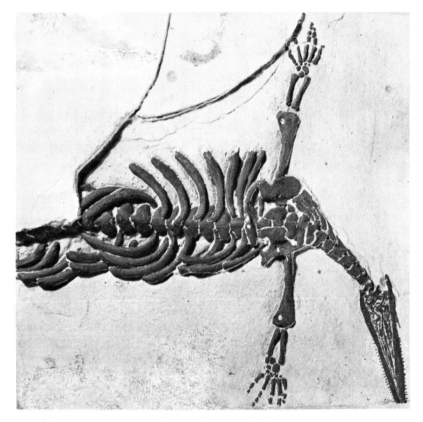

191 *Mesosaurus*, an aquatic reptile of early Permian age from the Karroo Supergroup, South Africa. Replica (× 0.7).
The first evidence for continental drift was largely geological and included the distribution of certain fossils such as *Mesosaurus*, which is known only from fresh-water deposits occurring on opposite sides of the South Atlantic in parts of South Africa and South Brazil.

occasional bones or moulds thereof are found, for instance, in sandstones (U Permian/L Trias) near Elgin in north-east Scotland.

Mesosaurus (L Permian, fig. 191), usually classed as an anapsid, was an aquatic fish-eating form restricted to rocks occurring on opposite sides of the South Atlantic in parts of South Africa and Brazil.

A few anapsids occur in the Trias; and the tortoises are extant forms (p. 326).

Synapsids

The synapsids or mammal-like reptiles were the dominant land animals of the late Palaeozoic and early Trias. During this period of time the skeleton became increasingly mammal-like, and in the late Trias a new class, the true mammals, emerged. By then the synapsids were in decline and only a few lingered into the Jurassic.

There are three orders of synapsids: pelycosaurs (U Carboniferous–M Permian); therapsids (M Permian–L Trias); and therosaurs (U Permian–M Jurassic).

Pelycosaurs

These were the most primitive synapsids. The first known, *Archaeothyris* (early U Carboniferous; N America) was found, along with other reptiles and terrestrial arthropods, in hollow lycopod stumps. It was lizard-like and probably fed on insects. Other Carboniferous forms were generally small, under 1 m in length; some were semi-aquatic fish-eaters. Permian forms were larger. They included some herbivores, but most were carnivores with sharp biting teeth. One of these, *Dimetrodon* (L–M Permian, fig. 192d) is notable for the elongated spines projecting from its backbone. These apparently supported a sail-like membrane which was well supplied with blood vessels. This structure greatly increased the animal's surface area and was probably a temperature regulating device which, it is estimated, may have increased the rate of heat exchange by $2\frac{1}{2}$ times. Turned side-on to the sun there would be net gain in heat to the body; turned end-on there would be a net loss.

The mid-Permian saw the start of a series of radiations of more advanced synapsids. Therapsids, the first of these, evolved from and replaced the pelycosaurs, and they were joined in the late Permian by the therosaurs. Each group included herbivores and carnivores. They became numerous and spread world-wide, being especially prominent in the Karroo deposits of South Africa. The Permo–Trias transition marked a drop in their numbers from which they did not fully recover.

Therapsids

Plant-eating therapsids were varied and abundant. They were heavy-limbed sprawlers, some reaching 4 m in length. The front teeth were reduced or lost and, in some, replaced by a horny beak with sharp cutting edges. One such form, *Lystrosaurus* (L Trias), was semi-aquatic. Its discovery in Antarctica, as well as South Africa and India, is useful evidence for the timing of the break-up of Gondwanaland.

Therosaurs

The therosaurs, derived from therapsid carnivores in the late Permian, gave rise to progressive forms, the CYNODONTS, which in the course of the Trias became increasingly mammal-like. They included a range of herbivores, a few of which persisted into the Jurassic, and carnivores, less numerous, from which the mammals emerged in the late Trias. Later forms were generally smaller and more agile than their predecessors with longer, straighter limbs which held the body higher off the ground. Their teeth, in contrast to those of a typical reptile which are all alike, were differentiated for biting and chewing (fig. 192b). They probably cut their food into small, more easily

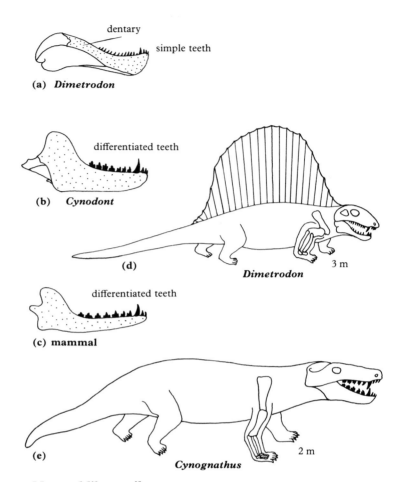

(a) **Dimetrodon**

(b) **Cynodont**

(c) **mammal**

(d)   **Dimetrodon**   3 m

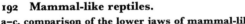

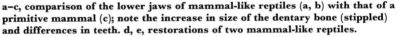

(e)   **Cynognathus**   2 m

192 Mammal-like reptiles.
**a–c, comparison of the lower jaws of mammal-like reptiles (a, b) with that of a
primitive mammal (c); note the increase in size of the dentary bone (stippled)
and differences in teeth. d, e, restorations of two mammal-like reptiles.**

digested bits instead of swallowing it whole in reptile fashion. They had, too,
a separate nasal passage formed by the growth of a secondary palate under
the original roof of the mouth. This allowed breathing while eating. The
lower jaw, as in other reptiles, contained several bones on each side; but one,
the DENTARY, was enlarged at the expense of the other bones (fig. 192b). In
mammals the lower jaw consists of the dentary alone (fig. 192c). By the late
Trias little carnivores like *Diarthrognathus* had become so mammal-like that
they are classed as reptiles only by the form of the jaw hinge (p. 327). They
co-existed with the earliest true mammals and it may be speculated that
they, too, may have been warm-blooded and possibly also hair-covered.
There is, however, no direct evidence for this, though it may be noted that, in
some fossils, there are small pits in the snout bones similar to those seen in
mammals with whiskers.

Mesozoic reptiles

From the human point of view the emergence of the mammals from the mammal-like reptiles might seem to be the culmination of the primary, early reptile radiations. It was, none-the-less, an aberrant and relatively minor event in reptile history. The major part of their chronicle is concerned with some of the most spectacular animals of all time: the dinosaurs, pterosaurs, ichthyosaurs and plesiosaurs of the Triassic, Jurassic and Cretaceous; and also with those lesser reptiles which, overshadowed in the Mesozoic, have survived to the present day. These Mesozoic reptiles radiated into new niches, adopting modes of locomotion which involved highly modified limb and body structure in the flying pterosaurs, and the marine plesiosaurs and ichthyosaurs. The dinosaurs, the most numerous, remained earth-bound, achieving greater agility by evolving erect posture.

During the Mesozoic these reptiles enjoyed a climate which, over wide areas, was warm and equable. Food was plentiful and varied. For the land-based herbivores there was a luxuriant diversity of plants, especially gymnosperms and, in the Cretaceous, the emergent angiosperms. The herbivores, in turn, provided flesh for carnivores. In the seas the marine reptiles preyed on an abundance of invertebrates and fish.

Terrestrial reptiles

Most Mesozoic reptiles, including lepidosaurs (lizards), rhynchosaurs and archosaurs, were diapsids which originated in the late Palaeozoic, the first known being *Petrolacosaurus* (U Carboniferous). This was a lightly built, lizard-like, insect-eating form which in many respects resembled its contemporary anapsid captorhinomorphs (p. 307). It represents the stock from which, during the Permian, two major lines diverged: the Lepidosauria (lizards, p. 325) and the Archosauria (which includes dinosaurs, pterosaurs and crocodiles). By the late Trias the archosaurs had become the dominant tetrapods. Another important diapsid group, the rhynchosaurs (p. 316) appeared in the late Permian but became extinct before the end of the Trias.

Archosaurs

The archosaurs were a very varied group which shared certain distinctive features: for instance, in the skull, an extra pair of openings lay in front of the eyes; and the fore-limbs were shorter than the hind-limbs. Primitive forms in late Permian and early Trias times included carnivores with sharp, slightly recurved teeth set in sockets in the jaw. They showed improved, though not fully erect, posture, walking with the body held off the ground. Subsequent radiations of archosaurs during the Trias, produced both short- and long-ranged forms adapted to a variety of habitats. Some were confined to the Trias; another, the crocodile line, still survives today; the pterosaurs, the first flight specialists, ranged to the end of the Cretaceous, as also did the dinosaurs, the dominant land tetrapods of the Jurassic and Cretaceous. A further innovation resulted in the birds (Class Aves), an offshoot of the dinosaurs.

Dinosaurs

Dinosaurs were a highly varied group of archosaurs which evolved erect posture and greatly improved their efficiency of gait. In the process the limb structure was remodelled. The limbs were straightened and held in a vertical position under the body and thus brought closer together. This allowed direct fore and aft limb motion and increased the length of stride. Speed of movement was furthered by placing, not the whole hands and feet, but only the digits on the ground. While some dinosaurs were quadrupedal, walking on all fours, other were bipedal and walked on the hind legs. In these, the weight of the body was pivoted on the hip bone which became fused to the backbone, giving more rigid support. The fore-limbs in many were short and adapted for grasping.

DINOSAUR is a collective term for two separate orders, the Saurischia and the Ornithischia. They are distinguished by a variety of characters including the nature of the pelvis. The saurischian pelvis is the normal reptile type, and the ornithischian pelvis resembles that of a bird (fig. 193). Both orders appeared in the late Trias, diversified, and existed in large numbers until late in the Cretaceous.

193   Dinosaur hip structure.

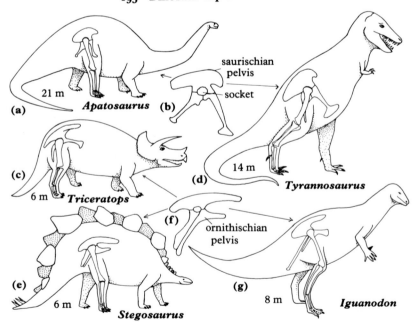

Saurischia. The saurischians include bipedal carnivores, the Theropoda, and quadrupedal herbivores, the Sauropodomorpha. The theropods (U Trias–U Cretaceous), when walking, left bird-like tracks made by three-clawed toes. A further toe, which did not reach the ground, pointed backwards. There are several different groups. For instance, the coelurosaurs (U Trias–Cretaceous), ranged from quite small forms, like *Composognathus* (U Jurassic) which was about 65 cm long, up to about 3–4 m in length. They were active, lightly built predators with small heads and flexible grasping hands. Some Cretaceous forms were ostrich-like with toothless jaws covered by a horny bill. The carnosaurs (Jurassic–Cretaceous), more heavily built and with large heads, were important in the late Jurassic. Some of the largest terrestrial killers known to have existed, the tyrannosaurs, had a limited distribution in the late Cretaceous. *Tyrannosaurus* (U Cretaceous, fig. 194), for instance, was about 12 m long and stood 5 m tall. It had a massive skull, over 1 m in length, and its jaw held serrated, dagger-like teeth about 15 cm long. Its massive, powerful hind-limbs, with great clawed toes, contrasted with its arms, which were tiny with two-fingered hands and too short to reach the mouth.

The sauropods (Jurassic–Cretaceous) include not merely the largest dinosaurs, but also the largest known land animals. Some, it is estimated, may have weighed 30 to 80 tonnes or even more. The record for length (about 27 m) is held by *Diplodocus* (U Jurassic), though its skeleton was less massively built than that of *Apatosaurus* (U Jurassic, fig. 193a). Typically, in the sauropods, the neck and tail were extremely long, and the body relatively short and bulky. They had stout, columnar legs with short-toed feet rather

194 *Tyrannosaurus*, **reconstruction of a skeleton from the upper Cretaceous of Montana, USA. Length overall about 12 m.**

like an elephant's. The design of the body has been compared to a cantilever bridge. The backbone had strong, tensional ligaments running dorsally from skull to tail-tip. The vertebrae, while massive, had hollow spaces which reduced the weight without losing structural strength. The skull was very small, only about 60 cm long in *Diplodocus*, and the brain capacity unusually small. There was, however, an enlargement of the spinal cord cavity in the pelvic region which may have accommodated an enlarged nerve cord to co-ordinate movement of the hind-quarters. The jaws were small in *Diplodocus*, containing a few simple, peg-like teeth in front.

Views differ as to whether sauropods inhabited swamps or dry land. In some respects, especially in the form of the body and limbs, the nearest living analogy is the elephant. Certainly, their eggs were laid on land, and their fossils are associated with those of undoubted land animals. Fossil trackways show that, like elephants, they moved in herds. In feeding, however, they differed, for they lack the elephant's grinders (p. 343). Instead, their teeth form rake-like structures with which, giraffe-style, they could have stripped leaves from high trees. The function of grinders may have been served by gastroliths (p. 284) or by microbes. It is known that they could not have lived in deep water where the pressure would have inhibited breathing. But sauropod 'hand prints', made when swimming and making intermittent contact with the bottom, are recorded (p. 284); so perhaps they enjoyed both habitats, retreating to shallow water to counter excessive heat.

Ornithischia.   The ornithischians were herbivores with a unique, single bone forming the tip of the lower jaw. Typically, the jaws were toothless in front and probably horn-covered, while at the back they were equipped with a great battery of grinding teeth. These were continually replaced as they were worn down and shed. Ornithischians were the dominant herbivores in the Cretaceous. They were highly varied, and included bipedal forms, the ORNITHOPODS; and quadrupeds, such as the STEGOSAURS, ANKYLOSAURS and CERATOPIANS, by saurischian standards, none were of gigantic size.

The ornithopods (Jurassic–Cretaceous) appeared in the early Jurassic and were varied and numerous in the Cretaceous. *Iguanodon* (L Cretaceous, fig. 195) is one of the best known. It was a large, kangaroo-like creature, about 5 m high, standing on feet with three broadly tapering toes, the blunt ends of which were encased by hoofs. It had sturdy fore-limbs, the fingers of which were also hoof-clad, and a spike-shaped thumb (fig. 195b). While usually bipedal, the hoofed fingers indicate that it also moved on all fours at times. Its teeth, when freshly erupted, were oval and blade-like with grooved surfaces and serrated edges (fig. 195a) but were worn, with use, to flat, grinding surfaces. Fossils of, and footprints made by *Iguanodon* are found in the non-marine, lower Cretaceous Wealden Beds of south-east England (fig. 173), together with the remains of plants on which they must have fed.

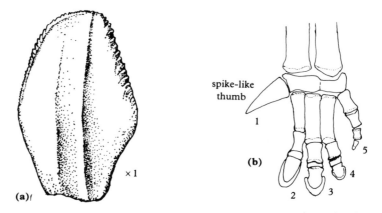

195 *Iguanodon.*
a, unworn tooth. b, structure of hand, with hoof-clad fingers.

In the upper Cretaceous a range of ornithopods show skull modifications. For instance, the hadrosaurs had the front jaw extended as a duck-like bill, while at the back were hundreds of closely packed grinding teeth which could have dealt with tough plant material. The stomach contents of mummified fossils include pine-needles and twigs. Some hadrosaurs developed a variety of bony processes, often hollow, on the skull. In some cases the nasal passages were diverted up and backwards through these.

The four-footed ornithiscians are largely confined to the northern

196 *Stegosaurus,* **a restoration of the life appearance. Length of original about 9 m.**

hemisphere. They are mainly Cretaceous in age though the stegosaurs appeared in mid-Jurassic.

Stegosaurus (U Jurassic, fig. 196) was about 6 m in length. It had two rows of large, triangular, bony plates along the backbone, and paired spikes on its stumpy tail. The plates bear traces of numerous blood vessels. This implies they possibly acted as a heat-exchange device (cf. *Dimetrodon*, p. 309).

The ankylosaurs are mainly Cretaceous forms. They were heavily armoured with bony plates, often spinose, set in leathery skin over the back, much as in the modern mammal, the armadillo.

The ceratopians appeared in the upper Cretaceous and occur mainly in North America and some in North Asia. *Triceratops* (fig. 193c) was about 7 m in length and reminiscent of a rhinoceros. The skull, which was over 2 m long, had bony horns on the nose and forehead, while an enormous, solid, bony frill projected backwards over the neck and shoulders. The frill doubtless served to protect the neck against attack. *Protoceratops* occurs in slightly earlier upper Cretaceous rocks in the Gobi Desert of Mongolia and is known in unusual detail. Various stages of its development have been found, from fragmentary embryos inside eggs to the fully developed adult (fig. 197).

Rhynchosaurs

Rhynchosaurs (Trias), numerous and widespread in mid and late Trias, were large and heavy-bodied diapsids with specialised dentition adapted to

197   Dinosaur eggs: a nest of eggs found along with the remains of a ceratopid dinosaur (*Protoceratops*) in Mongolia.

plant-eating. They had powerful jaws, modified in front into a beak, and, behind, bearing rows of deep-rooted, heavy toothplates.

Marine reptiles

Several different groups of Mesozoic reptiles were adapted in varying degrees and in quite different ways to life in the sea. They showed changes in body shape and limb structure which enhanced their ability to swim but left them recognisably reptiles. To illustrate some of these adaptations two groups, the plesiosaurs and ichthyosaurs, which are relatively common as fossils, are dealt with here. Both have a euryapsid skull structure but show little else in common. Their origins are uncertain but recent finds indicate that both groups may have originated from diapsid stock.

Plesiosaurs (fig. 198). These were relatively large reptiles ranging from about 3 m to over 12 m in length in later forms. Two different lineages are distinguished: forms with small heads and long necks, such as *Plesiosaurus* (U Trias–M Jurassic); and forms with big head and short necks like *Pliosaurus* (Jurassic). In both the body was short, almost barrel-like, and ending in a short tail. The limb girdles had strong, plate-like bones ventrally for the attachment of the limb muscles. Both front and hind-limbs were long, flattened and tapering towards the tip. The number of bones (phalanges) in the digits was greatly increased, in some cases to over 12 in each of the five digits (fig. 199b). The long-necked plesiosaurs reached their acme in the upper Cretaceous with *Elasmosaurus* which had over 70 vertebrae in the neck, more than double the number found in *Plesiosaurus*. These forms had sharply pointed teeth, suitable for catching fish.

The short necked *Pliosaurus* (Jurassic) had a 3 m-long skull and an overall length of 12 m. It was a more streamlined creature than *Plesiosaurus* and probably a more efficient swimmer. Its teeth were short and broader-crowned, and often show signs of wear. It is thought to have fed mainly on cephalopods. Gastroliths are sometimes associated with plesiosaur remains and may have acted as a digestive aid or, it has been suggested, to adjust their buoyancy.

Plesiosaurs probably spent most of their time at sea. Their mode of swimming has been compared with that of the sea-lions. It has been suggested that they swam, without undulating the body, moving both pairs of limbs in unison to give the main thrust by a down and backwards sweep followed by 'feathering' for the recovery stroke. An alternative view is that they swam like sea turtles (p. 326). Almost certainly they returned to the shore for breeding. Plesiosaurs appeared in the mid-Trias and were widespread in the Jurassic and Cretaceous. They occur throughout marine rocks of this age in Britain.

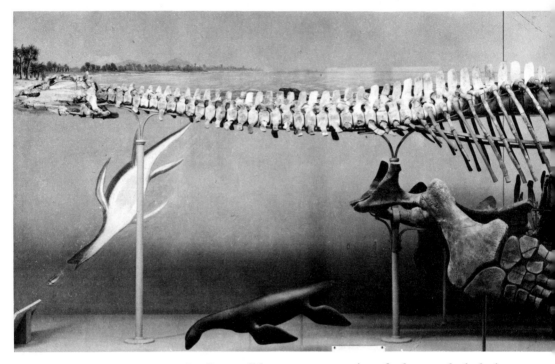

198 *Cryptoclidus*, a reconstruction of a long-necked plesiosaur.

Ichthyosaurs.   These were perhaps the most specialised of all aquatic reptiles. *Ichthyosaurus* (Jurassic), a typical example, was about 3 m in length. Well-preserved fossils of a related genus from dark shale in the lower Jurassic of Germany show the outline of the body (fig. 200). It was spindle-shaped with a dorsal fin, and a tail fin above the down-turned axis of the tail. The tail was the main propulsive organ, moving side to side, fish-wise, and the limbs were reduced to short, steering and braking 'fins', each with many small, disc-like bones (figs. 199c, 201). Later forms tend to show both an increase in the number of digits and of the digit bones (phalanges). The skull had long, pointed jaws with many sharp conical teeth; the eyes (or more precisely the orbits) were large, and the nostrils were placed just in front of the eyes (fig. 199a).

The dolphin-like shape of ichthyosaurs suggests they were relatively fast-swimming predators. They are believed to have fed on fish and cephalopods since associated coprolites contain fish remains and belemnite hooklets. One form, *Omphalosaurus* (M Trias), with blunt, button-like teeth, apparently fed on shellfish.

Ichthyosaurs, like modern whales, are thought to have given birth to live young at sea. Occasional fossils have been found showing very small ichthyosaur skeletons within the body cavity.

Ichthyosaurs range from lower Trias on through most of the Mesozoic. They were at their peak in the Jurassic and in decline during the Cretaceous.

Their remains are common in some marine rocks in Britain, from the Jurassic to the mid-Cretaceous. Occasionally, more or less complete skeletons are found, but isolated bones, such as vertebrae or teeth, are more common.

Flying reptiles and birds

The power of flight was developed independently in the flying reptiles, or pterosaurs, and the birds. They represent separate offshoots, from diapsid stock, which appeared in the upper Trias (pterosaurs) and the upper Jurassic (birds). They have a number of features in common. For example, both groups are bipedal; the fore-limbs were modified to form wings; the breast-bone was developed as a surface for attachment of the flight muscles. The pterosaurs had delicate bones, hollow and thin-walled to achieve a light skeleton without sacrificing strength. Birds, too, except for the earliest one known, have hollow (pneumatic) bones which, in living forms, contain air-sacs. Some pterosaurs and the earliest birds had long, pointed jaws with sharp teeth set in sockets.

However, despite these, and other reptilian features, the birds are a very distinct group, sufficiently so to merit their being assigned to a separate class, Aves. In particular, they differ from reptiles in wing structure (fig. 202), in having an insulating cover of feathers, and in maintaining a constant, warm,

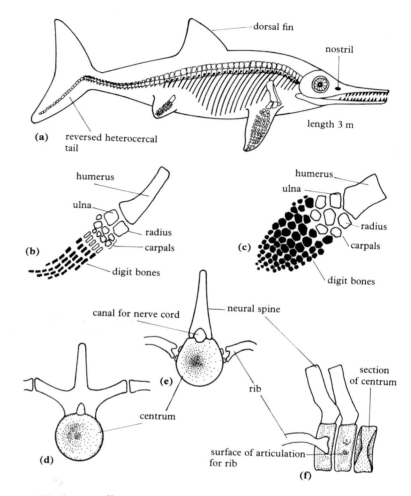

(a)   reversed heterocercal tail

dorsal fin

nostril

length 3 m

humerus

ulna

radius

carpals

(b)

digit bones

humerus

ulna

radius

carpals

(c)

digit bones

canal for nerve cord

neural spine

(e)

rib

section of centrum

centrum

(d)

surface of articulation for rib

(f)

199 Marine reptiles.
a, restoration of an ichthyosaur. b, c, right fore-limbs of a plesiosaur (b) and an ichthyosaur (c). d–f, dorsal vertebrae of a plesiosaur (d) and an ichthyosaur (e, f,).

200   An ichthyosaur with the outline of the soft body preserved:
Stenopterygius, **U Lias, L Jurassic, Holzmaden, West Germany (× 0.1).**

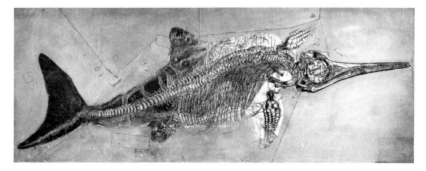

201   Head and fore-limbs of an ichthyosaur (× 0.15).
(Specimen in Sedgwick Museum.)

body temperature. They have, too, a unique respiratory system, with extensions from the lungs into the air-sacs in the bones. It has long been conjectured that pterosaurs, also, might have been warm-blooded because of the great expenditure of energy involved in flight. Evidence supporting this view comes from the discovery of *Sordes pilosus* in very fine-grained rock which has retained impressions of a covering of hair-like structures on the body, an insulating feature.

Pterosaurs
Rhamphorhynchus (U Trias, fig. 202a) is typical of one group of pterosaurs, the rhamphorhyncoids, found from upper Trias to upper Jurassic. It had a relatively small body with a long tail ending in a rudder-like membrane, and an elongated skull with large eyes and forwards projecting pointed teeth. It had a wing span of about 70 cm–1 m. The wings, long and relatively narrow, were formed by a membrane of skin which was apparently strengthened by an arrangement of collagen fibres. This membrane was stretched between body and fore-limbs, being supported along its forward edge by the bones of the fore-limb and the greatly elongated fourth digit. The first three digits, which were short and clawed, remained free. A small fore-wing is thought to have extended from a small splint-like bone on the wrist to the neck. This would lower the stalling speed when landing. At rest the wings were probably folded birdwise. The hind-limbs, which were not attached to the wing membrane, are broadly similar to those of a small theropod dinosaur. There are four walking toes and a short fifth toe at the back of the foot.

  In a later group, the pterodactyls (U Jurassic–Cretaceous), the tail was

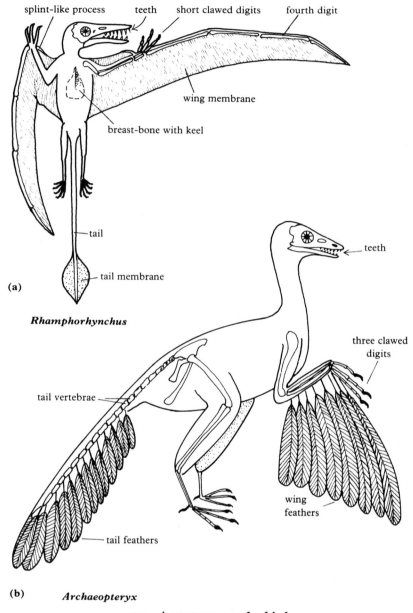

splint-like process   teeth   short clawed digits   fourth digit

wing membrane

breast-bone with keel

tail

tail membrane

(a)

Rhamphorhynchus

three clawed digits

teeth

tail vertebrae

wing feathers

tail feathers

(b)   *Archaeopteryx*

202   A pterosaur and a bird.

much reduced; and advanced members, like *Pteranodon* (U Cretaceous, fig. 203) lacked teeth.

Pterosaurs ranged from sparrow-sized forms like *Pterodactylus* (U Jurassic) to large forms like *Pteranodon* with a wing span of about 7 m, and *Quetzalcoatlus* (U Cretaceous) with an estimated wing span of over 10 m. Analysis of their flight potential has indicated that while the wing was quite unlike that of the bird, it was suitable for flapping flight. Also, the sternum,

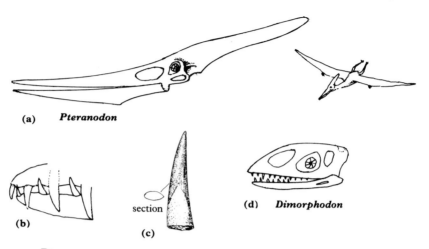

203 Pterosaurs.
a, pterodactyl, toothless form. b, fragment of a pterosaur jaw, teeth suited to catching fish. c, a pterosaur tooth. d, pterosaur skull, L Jurassic (possibly an insect-eater).

though often lacking the keel to which the flight muscles in birds are attached, was well developed. The small forms were probably the better, more active fliers while larger forms, perhaps, depended more on the use of thermals to give lift for soaring, as do some larger birds today like the albatross and the vultures. Pterosaur fossils are often found in marine rocks and it is possible that they spent much of their time at sea, snatching fish at the surface. Other forms may have fed on insects, or may have scavenged.

Pterosaurs had a long, successful and varied history. First known in the upper Trias with *Eudimorphodon*, they were already well developed and must have originated before that time. Initially they occur in Europe; by the early Cretaceous they had spread world-wide, and they became extinct at the end of that period. A British example, *Dimorphodon*, comes from the lower Lias of Southern England.

Birds (Aves)
The birds are a highly diverse group. Some, like the swift and albatross, are almost incessantly in the air, while others have lost the power of flight. For instance, the penguin has wings which are modified to swimming flippers; and the rhea, with much reduced wings, is a fast running bird. Birds are widespread today but their fossil record, especially of the early forms, is scanty.

The earliest known fossil bird, *Archaeopteryx* (figs. 202b, 204), from the Solnhofen Limestone (U Jurassic), is avian by virtue of the impressions of feathers splayed-out around its bones. Without these it would have been identified as a dinosaur. The general view now is that it was an offshoot from bipedal running stock related to the coelurosaurs (p. 313). Its main reptilian

204 *Archaeopteryx* in 'flight'.
(Model by Mr F. Munro.)

characters are the teeth, placed in sockets in the jaws; the nature of the wing
bones with three, clawed fingers on each wing; the lack of air spaces in the
bones; and the long tail consisting of a chain of reptile-like vertebrae. The
tail, with its row of feathers down each side, is a unique feature. Avian
features are the flight feathers, which are asymmetric like those of modern
flying birds; and the presence of a wishbone, boomerang-shaped and formed
from the fused collar bones. These two characters place *Archaeopteryx* firmly
in the bird class, but not capable of strong flight since it lacks the keeled
breast-bone to which flight muscles are attached in modern birds. The
strong wishbone, however, may have afforded some support for such
muscles. *Archaeopteryx* was probably not on the main line of bird descent but,
rather, a short-lived side-shoot.

Bird fossils are rare and, after *Archaeopteryx*, are next found in the lower
Cretaceous. By the end of that period some 35 species are known. They occur
in marine rocks, including the Chalk, and are of varied types of sea birds with
characters typical of modern forms, such as pneumatic bones, no tail, and,
with two exceptions, no teeth. The two forms with teeth are *Hesperornis* (U

Cretaceous), a diving form, measuring up to 2 m in length, with vestigial wings and no keel on the breast-bone; and *Ichthyornis* (U Cretaceous), a small, strong-flying tern-like bird with keeled breast-bone.

The fossil record is much better in the Cainozoic and includes land as well as sea birds. A major radiation in the Eocene introduced many of the kinds of birds now living. Later, in the Miocene, a further radiation brought still more diversification. Some forms were gigantic: for instance a vulture-like bird with a wing span of 7.5 m; flightless carnivores standing 2 m high; and, in historic times but now extinct, the moa, standing 3 m high.

Surviving types of reptiles

Living reptiles include members of lepidosaurs (p. 311) (*Sphenodon*, lizards and snakes); crocodiles; and turtles and tortoises. They are varied, none are particularly large, and most live in warm to tropical regions. Fossils of these forms are found in the Mesozoic and Cainozoic but are relatively insignificant in number.

Lepidosaurs. In these forms the diapsid condition of the skull has been greatly modified except in *Sphenodon* (the tuatara). *Sphenodon* is little different from its late Triassic forbears which had diverged from the main lizard line. It is a nocturnal, lizard-like creature living (and protected) on islands off New Zealand. It moves sluggishly but can operate when its body temperature is only 11 °C. A point of interest is its single pineal eye, an eye-like structure with lens and retina which lies below the skin on top of its head. This pineal eye explains the purpose of an opening in the skull in a number of fossil forms, e.g. the crossopterygians, some amphibians and reptiles.

Lizards and snakes are the commonest and most diverse of living reptiles. They are rare as fossils in Mesozoic rocks apart from a marine group, the mosasaurs, which were widespread predators in upper Cretaceous seas.

Snakes are basically legless lizards. They appeared in the Cretaceous.

Crocodiles. Crocodiles are semi-aquatic, mainly non-marine, predatory reptiles which today are restricted to warm regions. Of surviving reptiles they are the most nearly related to the dinosaurs and, like them, are archosaurs.

The skull is primitive with a diapsid pattern of skull openings. The jaws are elongated and equipped with numerous sharply conical teeth which interlock as the jaws close. The skin is armoured by horn-covered, bony plates (scutes, fig. 205) along the back and tail. They are sprawlers with hind-limbs longer than fore-limbs. They can, however, raise the body off the ground when moving fast. Untypically, for reptiles, they show a degree of care for their young.

Crocodiles appeared in the upper Trias. They became numerous and

205　Crocodile scute: one of a series of protective bony plates from the back of a crocodile, *Diplocynodon*, Oligocene, Hampshire (× 1.5).

varied in the Jurassic and were widely distributed. A short-lived group became highly specialised for life at sea during late Jurassic to early Cretaceous times; for instance, they developed paddle-like limbs and a finned tail. Crocodiles of modern type appeared in the Cretaceous and they have shown little change since then. In England crocodile remains are found in Cretaceous Wealden and Purbeck Beds, and also in the Eocene and Oligocene.

Turtles and tortoises.　These rather primitive reptiles are members of the chelonians, an anapsid order. The living members fall in three categories: tortoises, land-living and herbivorous; terrapins, fresh-water and mainly carnivorous; turtles, marine and also mainly carnivorous. The aquatic forms return to land for egg-laying. The main chelonian feature is the box-like shell within which the head and the limbs can be retracted. The shell consists of an inner layer of bony plates which is covered by horny material. The sturdy legs of the land tortoise are modified in the sea-going turtles into flippers. In swimming these act as underwater 'wings'. The earliest forms had teeth but these are absent in later members in which the jaws are covered by a horny beak.

Chelonians appeared in the upper Trias and their essential structures have changed little since then. The earliest were land-living, but aquatic forms soon developed and they diversified in the early Cretaceous to become widespread. Most fossils are of marine and fresh-water forms. In England their fossils occur at various horizons, for instance in the Purbeck, the Wealden, the Chalk and the Eocene London Clay. Some were very large such as the marine turtle, *Archelon*, which was over 3 m in length. Fossil tortoises are known from the Eocene onwards.

Mammals

Two diagnostic features of mammals are unique: the presence of fur or hair, and the secretion of milk as food for the young. As in birds, the body temperature is kept at a uniform level. Living varieties are adapted to a wide range of habitats on land, in water and in the air. They are assigned to three groups: the majority, including man and the hedgehog, to the placentals; a relatively small number, including the kangaroo, to the marsupials; and six species, comprising echidnas and platypuses, to the monotremes. Fossil mammals include quite rare Mesozoic forms and an extensive array of Cainozoic forms which document the changes leading to the modern mammals.

Mammals differ greatly from the main body of reptiles in their anatomy, physiology and behaviour. All parts of the skeleton show differences, the foremost being the larger brain-case in the mammals and its indications of a more intricate brain. The simplest diagnoses are provided by the nature of the lower jaws and by the teeth; these, fortunately, are the parts most commonly preserved. Each half of the lower jaw consists of one bone only, the DENTARY, in mammals (fig. 192c), but of several bones in reptiles (fig. 192a). A new joint developed in mammals, enabling the jaws to articulate directly with the skull. In the process, two small bones, which form part of the reptile's jaw hinge, became incorporated in the mammal's earbones. A mammal has dissimilar biting and chewing teeth, the latter having divided roots. In its lifetime only two sets grow: a juvenile or milk set, and an adult or permanent set. A reptile, on the other hand, has uniform, simple teeth, each with a single root, which may be replaced indefinitely. Other points of difference include the form of the articulating surface between the skull and the backbone, which in mammals is a double-headed knob, but in the typical reptile is a single knob.

There is convincing evidence that mammals are derived from advanced mammal-like reptiles, the cynodonts (p. 309). For example, there are transitional stages, preserved in the fossil record, between the fully reptilian and the fully mammalian state of the lower jaws and of their articulation with the skull.

The evolution of warm-bloodedness, or ENDOTHERMY, gave mammals a major advantage. Their stable, warm body temperature is maintained by an internal source of energy derived from a high metabolic rate. Thus, they can sustain activity, even in low ambient temperatures, for long periods but must eat frequently to do so. Insulation by fur and fat is an essential part of endothermy.

On fossil evidence, the early mammals, found in the Mesozoic, were very small, about shrew to small domestic cat in size. They are thought to have been nocturnal in habit (thus avoiding encounters with contemporary reptiles which were probably diurnal), and insect-eaters. It has been suggested that their life-style paralleled that of the modern form, *Tenrec*, a primitive nocturnal insectivore found in Madagascar. *Tenrec* has a body temperature around 28–30 °C, much lower than that of most diurnal, placental mammals which is around 38–40 °C. Presumably this is related to its nocturnal mode of life and the cool ambient temperatures which it consequently encounters. By analogy, therefore, it may well be that the early mammals had a similar low body temperature and that this increased as they invaded diurnal niches during their initial radiation at the close of the Cretaceous. *Tenrec* depends on well-developed senses (hearing, smelling and tactile) in seeking its diet of insect grubs and small invertebrates. A similar development of improved senses in the early mammals may be correlated with the increase in size of brain suggested by their fossils.

Mesozoic mammals

Of the early mammals which appeared in the upper Trias and Jurassic only one group, the multituberculates (U Jurassic–L Oligocene), survived into the Cainozoic. The three groups of living mammals (monotremes, marsupials and placentals) appeared in the Cretaceous and diversified during the Cainozoic. These are dealt with in the next section, on Cainozoic mammals.

Mammal remains occur only sporadically and as fragments for much of the Mesozoic. They consist mainly of teeth (fig. 206), jaws or parts of skulls; post-cranial remains are rare. All were very small. The nature of the chewing teeth, each with several sharp peaks (cusps), is the basis of classification. There are two main lines: therians, with teeth showing a triangular arrangement of three main cusps (fig. 206a–c); and non-therians, with other patterns of tooth cusps (fig. 206d, e). The early therians included the eupantotheres (M Jurassic–L Cretaceous), the ancestral stock giving rise to the marsupials and placentals during the Cretaceous.

The sharply cusped teeth, typical of most of these early mammals, were suited to eating insects or flesh; but the non-therian multituberculates, with broader teeth bearing many small cusps in rows, were probably plant-eaters.

The main horizons in Britain in which fossils of Mesozoic mammals have been found include upper Triassic infillings of fissures in Carboniferous Limestone in the Mendips and Glamorgan; the middle Jurassic Stonesfield 'slate'; the upper Jurassic Purbeck 'dirt' beds; and the lower Cretaceous Wealden Formation.

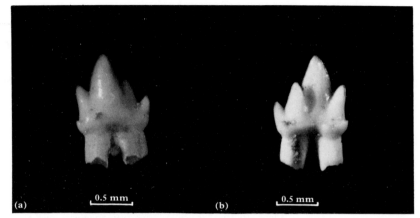

Kueneotherium

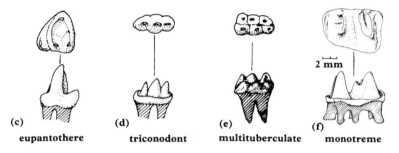

(c) eupantothere   (d) triconodont   (e) multituberculate   (f) monotreme

206 Mesozoic mammal teeth.

a, b, a many-cusped mammal tooth of pre-Rhaetic, late Trias age, possibly the oldest known fossil mammal; found in fissure deposits in the Mendip Hills, Somerset; a, cheek view; b, inner (tongue) side of lower left molar (photograph courtesy of Dr N.C. Fraser). c–e, lower molars from three types of Jurassic mammals; above, biting surface; below, side views showing twin roots. f, upper molar tooth of *Obdurodon*, Miocene; above, biting surface; below, inner side view (redrawn from Woodburne).

Cainozoic mammals

Monotremes

The duck-billed platypus and the spiny ant-eating echidna are the only known surviving monotremes. They are confined today, and as fossils, to the Australian region. They are the most primitive mammals alive and have retained a number of reptilian features in the skeleton and soft parts. They have fur and secrete milk, but they lay eggs. Their body temperature fluctuates somewhat and is rather below that usual for mammals. They are highly specialised and, as adults, lack teeth (teeth are present in the young platypus). The platypus, which is semi-aquatic (in streams), uses electroreceptors in its 'bill' to detect muscular movement of animals within sediment and thus find its prey of crustaceans.

The fossil record of monotremes is meagre. Prior to the Pleistocene it consists of mid-Miocene teeth (fig. 206f) and a recently found jaw fragment from the lower Cretaceous (Albian). The latter, referred to *Steropodon*, is of 'platypus' type; associated fossils include lungfish, plesiosaurs and crocodiles. The relationship of monotremes to other mammals is uncertain.

Marsupials

The marsupial mother carries her young in her pouch, a fold of skin over her belly, which is supported by special bones. The young are born while still very immature, and crawl into the pouch where, attached to teats, they suck milk. Marsupials form a minor group today and, apart from an occasional form like the American opossum, they are confined to the Australian region. In the past, however, they were more widespread. They diversified into a wide range of habitats, often paralleling adaptations found elsewhere in placental mammals. They include flesh-, plant-, and insect-eaters.

Marsupials and placentals, both characterised by live birth, share a common ancestry in eupantothere stock. They diverged, possibly in the early Cretaceous, along separate evolutionary paths. A variety of marsupials are found in the upper Cretaceous of North America where, for a time, they were common and diverse. In the early Cainozoic they spread to Europe and also to South America where a wide variety of fossils is recorded. In Antarctica only a single specimen has so far been found, in the upper Eocene. The first fossil marsupial recorded in Australia is of upper Oligocene age; in the Miocene they became common and diverse, as is the case today.

Australia became isolated from Antarctica and South America during the Eocene. At that time monotremes were already there (above). It is not known precisely when, or by what route, the marsupials arrived and established themselves as the dominant mammals. Placentals did not arrive till later and have occupied a lesser role, which is in contrast to their dominance elsewhere in the world. They spread south-eastwards from Asia, first bats (Miocene), followed by rodents (Pliocene) and, about 30 000 BP, man and *Canis dingo* ('yellow dog dingo').

Placentals

In placental mammals the developing embryo is nourished in its mother's uterus by a special growth, the placenta, and when born it is relatively mature. In some cases, like the whale, it may be able to follow its mother within hours of birth. Placentals differ from marsupials in many aspects of both soft body and skeleton; the brain-case, for instance, is larger. Placentals include the vast majority of living mammals, and most (about 95%) of Cainozoic forms.

A variety of placental fossils are found in the late Cretaceous. They comprise isolated teeth, jaws, skulls and some post-cranial remains. They represent small, long-snouted insect-eaters and give a picture of the possible rootstock from which the vast array of Cainozoic placentals arose. A major radiation, starting at the close of the Cretaceous, produced, in the Palaeocene, an archaic fauna dominated by varied insect-eaters, condy-larths (plant-eaters), and primates. In the Eocene, a sudden influx of new forms, some of great size, included the forerunners of surviving orders and other, now extinct orders. The early Oligocene saw the demise of the Mesozoic order of multituberculates (p. 328), until then a thriving group. Their decline coincides with, and may in part be a consequence of the arrival of the rodents to occupy a similar niche. Today over a third of living placental species are rodents.

The adaptive radiation of the placentals was accompanied by modific-ations of the skeleton which are related to three aspects in particular: development of the brain, locomotion and diet.

The brain in primitive placentals is small and relatively simple. In more

207 Limb form and brain size in placental mammals.
a, diagram showing the limbs of a typical four-legged mammal, aligned vertically below the body. b, d, adaptation of the fore-limb for flight (b) and for swimming (d). c, e–g, different types of foot structure in walking and running mammals. h, comparison of brain size of a primitive placental mammal (stippled) with that of a reptile (cross-hatched).

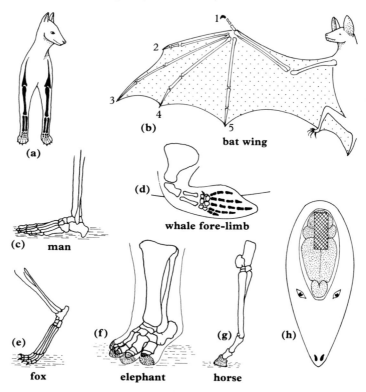

advanced placentals the brain-case is, relatively, larger (fig. 207h) and the brain itself is more complex. The latter implies improved co-ordination of bodily functions and development of sense organs like the eyes or ears.

Limbs of terrestrial mammals are hinged so as to move fore and aft in a vertical plane below the body which is thus held clear of the ground (fig. 207a). The primitive forms have short legs and walk with the five-toed feet flat on the ground (plantigrade) as found in hedgehogs. In more advanced forms the limbs and the way in which the body is carried are modified so that they become more agile. Usually there has been an elongation of some of the limb bones, resulting in increased length of stride; and there may also be some alteration in the arrangement of the foot bones. Thus an agile form like the fox may walk on the digits (digitigrade) with the sole of the foot raised off the ground (fig. 207e); or a fast runner like the horse may walk on the tip of its digits (fig. 207g). The most complete structural change of the basic limb plan is seen in a flying form like the bat (fig. 207b), or an aquatic form like the whale (fig. 207d).

The nature of the teeth is of particular importance in the study of placentals and is a guide to the relationship between different forms. It has to be remembered, however, that animals which are unrelated to one another but which eat the same food, may develop a similar tooth structure. The teeth are fully formed when they erupt from the jaws. With use, the chewing surface becomes worn.

The maximum number of teeth in placental mammals, 44, is found in the more primitive forms, which have three biting teeth, the INCISORS, one piercing tooth, the CANINE, and seven chewing or cheek teeth on each side of both upper and lower jaws (fig. 208d). The incisors are simple round pegs, with single roots inserted along the front margins of the jaws. The canine, which is longer and pointed, lies between the incisors and the cheek teeth. The latter, lying towards the back of the jaw, have two or more roots; and the visible part, the crown, bears sharp projections, the cusps, on the chewing surface. The first four cheek teeth, the PREMOLARS, are of simpler structure than the three rear teeth, the MOLARS, which typically are the most informative about diet (fig. 208k). In primitive forms the molars are triangular in plan with a cusp at each corner, but in the lower jaw each one has an added small platform jutting out on the rear side (fig. 208b).

The dentition described above is found primarily in insect-eating animals and it is modified in more advanced forms which feed on other types of food. The modifications include both a reduction in the number of teeth (e.g. to 32 in man; or, as an extreme case, the whalebone whale, to none) and an alteration in their structure. The most profound changes are seen in the cheek teeth. Thus carnivorous forms, like the cat, eat an easily digested food which need only be cut into bits small enough to swallow, and have fewer, less massive cheek teeth. In these forms, a pair of opposed teeth (i.e. one in each lower jaw and the other immediately above it in the upper jaw) have the cusps compressed and united to form a jagged cutting edge which can

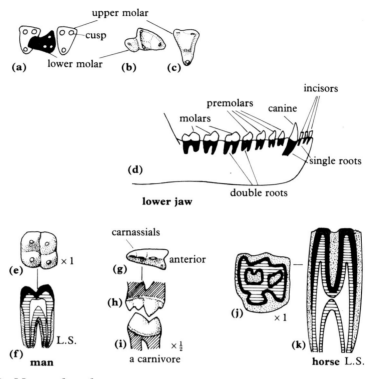

208 Mammal teeth.
a–c, teeth and d, lower jaw of a primitive (insectivorous) placental mammal (a, shows how the cusps of opposed upper and lower molars alternate in position in the process of chewing; b, c, chewing surface of molars). e–k, adaptation of a molar for a general diet (e, f,); for a carnivorous diet (g–i); and for a herbivorous diet (j, k); e, g, j, show chewing surface; h, i, side views of an upper and lower molar; f, k, sections of unworn molars to show the arrangement of enamel (black), dentine (lined), cement (stippled) and pulp cavity (unshaded).

slice through flesh and sinew, and crush the bones of their prey. These teeth are known as CARNASSIALS (fig. 108g–i). The cat has one pair on either side of the jaws. Herbivorous forms, like the horse, however, are dealing with a low-energy food which must be eaten in bulk and thoroughly chewed. Such forms have enlarged 'prismatic' cheek teeth with a high crown and complex grinding surface in which bony cement (which in other types of teeth is usually restricted to the roots) is incorporated (fig. 208j, k). With use, the tooth surface wears unevenly forming ridges which improve the grinding process. The tooth is highly resistant to the abrasive action involved in grass chewing. It may be noted that some grasses contain a small amount of silica.

Some examples to illustrate the adaptive radiation of the placental mammals

The placentals, being extremely diverse, are separated into a large number of orders. The following examples describe some of the structural variations involved.

Insectivores

The insectivores (*s.s.*) are represented today by small, often nocturnal forms like the hedgehog, shrew and mole, which feed on insects, worms or slugs. Although some forms are highly specialised for particular modes of life, they retain many of the primitive features of the skeleton shown by the early fossil insect-eaters. They have, for instance, a small, relatively simple brain, and primitive dentition. They have short legs and walk with their clawed feet flat on the ground. The group has a fossil record going back to the Palaeocene, and its ancestry is rooted in the earlier insect-eating placentals of the Cretaceous.

Bats

The bats, an offshoot of insectivores, are highly specialised for flying, primitively, in pursuit of insect prey. Their fossils are understandably rare. However, even in the earliest known, *Icaronycteris* (early Eocene), the power of flight was already well developed. In addition, details of skull structure suggest that some degree of echo-location may have been developed. The form of the wing makes an interesting contrast with that of the pterosaurs and birds. It consists of a fold of skin extending from the body as far back as the tail. It is supported by the long fore-arm and by greatly elongated finger bones (excepting the clawed thumb) which are arranged rather like the spokes in an umbrella (fig. 207b).

Primates

Primates are typically arboreal (tree-living) animals which, in some respects, are relatively unspecialised. They show a range of characteristics related to their adaptation to climbing, leaping and swinging in trees. This mode of life requires a very precise co-ordination of limbs and sight. Typically, they use all four limbs, apart from a few ground-living forms which are bipedal. The hands and feet are adapted for grasping with opposable thumbs and large toes (fig. 209f). All five digits are retained and their tips are protected with flat nails. In the more advanced forms the teeth are reduced in number and are modified for a mixed, omnivorous diet. The most significant modifications, however, are expansion of the brain with consequent enlargement of the brain-case; and improvements in eyesight resulting in binocular vision.

Living primates are widespread in warm climates. They include the relatively simple PROSIMIANS, the lemurs and tarsiers; and the more advanced ANTHROPOIDS, monkeys, apes and man. Fossil primates are quite rare.

The group is thought to have diverged from insectivorous stock. Various small primitive primates appeared in the course of the Palaeocene. The best known of these is *Plesiadapis*, a long-snouted, almost squirrel-like form with clawed digits (fig. 210i). Further variety was added in the Eocene with forms

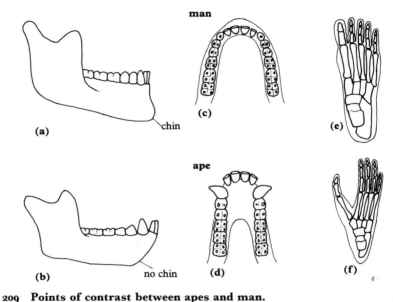

209 Points of contrast between apes and man.
a, b, profile of lower jaw. c, d, alignment of teeth in the lower jaw. e, f, form of
the foot (e, no grasping power and the toes are shortened; f, large toe opposable
and the foot can grasp).

much like the present day, nocturnal lemurs and tarsiers (fig. 210g). Lemurs
are long-tailed creatures with pointed, fox-like head, and eyes directed
sideways (fig. 210j). In tarsiers the snout is shorter, and the large eyes face
forwards (fig. 210h). Both have sharp-cusped cheek teeth suited to eating
leaves, fruit, insects, etc.

The higher primates, or anthropoids, show a number of specialised
features. For instance the skull is modified by an increase in size of the brain-
case, and the direction forwards of the orbits (which house the eyes). The
face is shortened too, by the reduction of the snout (due to diminished sense
of smell) and jaws. The smaller size of the jaws is linked with a reduction in
the number of teeth to 32 in most members. The teeth may be somewhat
modified in form: the molars for instance are quadrate and have four or five
blunt cusps, which are ideal for a generalised diet (fig. 209a–d).

In the course of time monkeys, apes and man have diverged to follow quite
different modes of life. Monkeys are the more remote from man. Typically
they are arboreal, using both hands and feet (sometimes the tail too) as they
run and leap in the trees. They comprise two groups: Old World and New
World monkeys, which were already separated in the Oligocene. Since then
they have followed parallel lines of development, the New World forms
being the more primitive, for instance in dentition and having a prehensile
tail. Apes are tail-less and semi-erect, with long arms and short legs. They are
not exclusively tree-living but range from the gibbons, expert in arm-
swinging from branches, to the chimpanzees and gorillas, which are more at

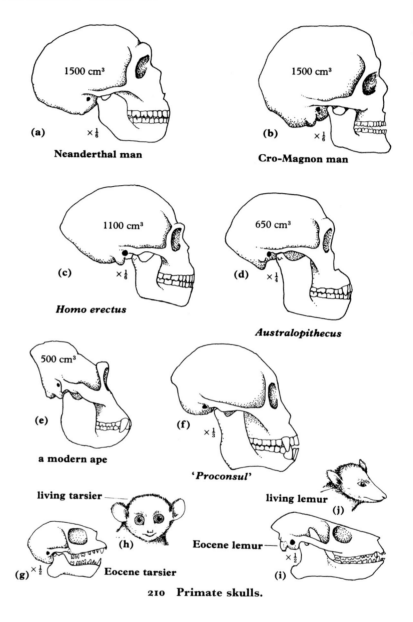

210 Primate skulls.

home on the ground where they usually move on all fours. Man, in contrast, is ground-living, fully erect and bipedal.

Man and the great apes (i.e. orang-utans, chimpanzees and gorillas) show marked similarities in their general anatomy, physiology and molecular chemistry; accordingly, they are grouped together as HOMINOIDS. However, there are significant differences in detail which serve to define two families: PONGIDS or apes, and HOMINIDS or man. They differ mainly in the development of the brain; in posture and locomotion; and to a lesser extent in diet.

The brain capacity in man is about $1400\,cm^3$ compared with about $500-600\,cm^3$ in a gorilla. Some parts of the brain, too, are more complex, especially in the frontal region, and this is reflected in the higher forehead in man (fig. 210b). Man has a chin (fig. 209a) and his jaws are less massive, with the teeth lying in a gently rounded arch (fig. 209c); whereas apes have no chin (fig. 209b), and the teeth lie parallel on either side of the jaw (fig. 209d). Man is distinguished, too, by his erect posture with the skull balanced on top of the neck. The backbone has an S-shaped curvature in man but is more C-shaped in apes. In walking, man places the sole of the foot on the ground, and the large toe is not opposable (fig. 209e); while apes continue, to some extent, to move on all fours, and their feet retain a grasping function (fig. 209f).

The earliest anthropoids, recorded from the late Eocene in Burma (about 44–40 my ago), indicate that the group had split away from the prosimians. *Amphipithecus*, for instance, just under 1 m high, had molars with low, rounded cusps suited to a generalised diet. Later, in the Oligocene of the Fayum, Egypt, diverse and unspecialised tree-living forms occur. Features in some of the later forms suggest that the divergence of Old World monkeys and hominoids was under way. One of these, *Aegyptopithecus*, around about 29 my ago, with ape-like jaws and teeth, may represent the sort of stock from which, in the Miocene, a range of hominoids arose.

The radiation of hominoids was initiated in the early Miocene, the earliest forms occurring in East Africa about 19–14 my ago. *Proconsul* (fig. 210f), for instance, was a small, tree-climbing form with a number of unspecialised features and a brain capacity of $167\,cm^3$. About 17 my ago a land-link formed between Africa and Eurasia enabling hominoids to spread from tropical towards cooler regions. Two of the varied forms may be mentioned: *Dryopithecus* (related to *Proconsul*), a forest dweller found in parts of south Europe; and *Sivapithecus*, a larger form living in more open woodland and found widely in Asia. They probably fed on a varied diet including leaves, and fruit. The divergence of apes and man began during the late Miocene, about 7.5 my ago. Apes specialised for life in tropical forest regions; man adapted to ground living in open savannas and grasslands. Skeletal and molecular evidence from the apes suggests: (i) that the orang-utan split off first, about mid-Miocene (16–15 my ago), *Sivapithecus* showing similarities with this form; and (ii) that, of the African apes, the pygmy chimpanzee is closest to man. No fossils of intermediate stages have been found.

The first known hominids are found in the Pliocene of east Africa and later in south Africa. They are scattered and fragmentary; many show a mix of characters that cause difficulties of interpretation. They date from about 4 my ago or possibly, 5.5 my ago, and include skeletal remains, footprints such as those at Laetoli (p. 284) and, from about 2 my ago, tools. Early hominids are assigned to several species of *Australopithecus*. These were small, 1–1.5 m high, short-legged, erect and bipedal. Their teeth were essentially human in type but the face was ape-like, and the brain capacity-small, about 400–650 cm³. They fall into two groups with different life-styles: earlier 'gracile' forms of slender build, about 3.75–2.3 my ago; and larger 'robust' forms about 2.5–1.2 my ago, partly overlapping in time. The most primitive is the gracile *Australopithecus afarensis*. A partial skeleton, dating from about 2.9–3.2 my ago, is of a female ('Lucy') only 1 m high. The robust forms were larger, about 1.5 m, and had massive jaws with thickly enamelled, worn molars which suggest a coarse vegetarian diet.

A variety of fossils suggest that true man, or *Homo*, was present about 2 my ago or possibly earlier, co-existing for a period with robust *Australopithecus*. Tools, about 2 my old, are associated with fossils referred to *Homo habilis*. He was rather like gracile *Australopithecus* but had a larger brain capacity, 700 cm³ or more. *Homo erectus* (fig. 210c) was a more advanced and larger form, 1.6 m high, who appeared about 1.6 my ago. He spread widely into Europe and the Far East, surviving until 250–220 000 BP. Fossils show an increase in brain size from about 800 cm³ to 1100 cm³ or more in later forms. He had large, human teeth but retained ape-like features in the skull such as pronounced brow ridges, a flattened receding forehead and absence of a chin. Primitive implements (Acheulian culture) and fire ash are commonly associated with his remains. Use of fire may go back to about 1.4 my but this point is debated.

The first appearance of *Homo sapiens*, 'modern' man, is uncertain. Skulls may show a mosaic of *erectus/sapiens* characters which are difficult to resolve. It may be that an archaic form of *sapiens* co-existed with *Homo erectus* some 400–300 000 BP, or his arrival may have been later, about 120–100 000 BP.

A distinct type, Neanderthal man (fig. 210a), was in existence about 100 000 BP. His skull was large and thick-walled with a low receding forehead and massive brow ridges. He survived in Europe when glacial conditions returned, about 75 000 BP, and developed features thought to be adaptations to living in extreme cold. He was short, robust, with short limbs and a large nose. He was a hunter, used fire, and buried his dead, thus increasing the chance of their becoming fossils. He disappeared about 35–30 000 BP. People of clearly modern aspect entered Europe about 40–33 000 BP. This first type was Cro-Magnon man, tall (male 1.83 m), with a high forehead and cranial capacity about 1500 cm³ (fig. 210b). His cultural attainments surpassed those of his predecessors.

Fossils show that brain size had reached its peak by about 100 000 BP. At this stage, however, absolute size is less important than brain quality, i.e. its complexity, as a guide to intelligence in *Homo*. The range of cranial capacity in man is about 1000–2200 cm³.

Whales

A number of mammals are adapted to an aquatic life. The most specialised of these are the whales, which range from the blue whale of about 30 m length to the quite small porpoises and dolphins. While they retain mammalian characteristics like, for instance, their mode of reproduction and of temperature regulation, they have developed many fish-like features which parallel those shown by the Mesozoic ichthyosaurs (p. 318). The dolphin, perhaps, comes closest to these reptiles in appearance: it is a streamlined spindle shape, with a long head which joins the body without a neck region; the jaw is elongated beyond the nose, which is reduced to a blow-hole; and its teeth are sharp spikes, ideal for grasping slippery fish. The fore-limbs are fin-like stabilisers, (fig. 207d). When swimming the tail fluke, a horizontal fin, moves up and down instead of from side to side.

The origin of whales may lie in a carnivorous offshoot of condylarths (p. 342) which adapted to aquatic life. The earliest known fossil to show whale features is a part of a skull *Pakicetus* (L Eocene), about 33 cm in length which was found, with land-based animals, in fluvial deposits. It was not yet fully aquatic; for instance the ear was not well adapted for hearing under water. The cheek teeth, of primitive type, were sharply cusped and similar teeth are found in later whales, the archaeocetes (M Eocene–Oligocene). These show increasing specialisation for marine life: for instance the hind-limbs were reduced, and the body much elongated. Some reached over 20 m in length, e.g. *Basilosaurus* (U Eocene). Representatives of modern toothed whales occur in the upper Eocene, and of baleen whales in the middle Oligocene. The latter, such as the blue whale, lack teeth, and adapted to plankton-feeding by developing a filtering device of fringed, horny plates in the upper jaw.

Carnivores

Carnivores catch and consume living prey (other vertebrates). Several unrelated groups, including some marsupials, adapted in parallel to flesh-eating, and thus developed a range of similar characteristics. Placental forms include creodonts, an extinct order; and true carnivores, such as the highly diverse dogs, cats, weasels, bears, hyaenas and seals.

The carnivore skeleton is relatively primitive in design. It is modified so as to achieve stealthy stalking followed by a short burst of speed ending in a pounce on its prey. It attacks and holds the prey with sharp claws (retractable in cats); then, with its powerful jaws, equipped with stabbing and slicing teeth, kills and eats it. The teeth are the most specialised features. The incisors are sharply pointed for biting and tearing; the canines, longer and sharp, become, in extreme cases, flattened daggers with razor-sharp edges which may, as in *Homotherium* (Pleistocene, fig. 212b), be serrated. The cheek teeth are compressed with sharp cusps and, typically, two opposed carnassials (p. 333) are present on each side of the jaws.

Creodonts appeared in the Palaeocene and reached their peak during the Eocene. Early forms were small-brained, cat-sized animals with long tail and short limbs, and not capable of great speed. They had sharply cusped molars

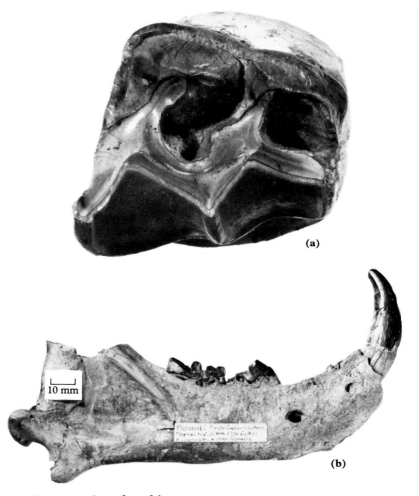

(a)

10 mm

(b)

211　A mammal tooth and jaw.
a, grinding surface of a herbivore tooth, *Palaeotherium*, a perissodactyl,
Bembridge Limestone, Oligocene, Hampshire (× 2.5). b, lateral view of the lower
jaw of a primitive carnivore, *Hyaenodon*, a creodont, lower Headon Beds,
Eocene, Hampshire (× 0.7).

and long canines suited to a diet of flesh or insects. Later forms were larger
with carnivore-type teeth including carnassials as in *Hyaenodon* (U
Eocene–Miocene, fig. 211) and, in some, 'sabre-tooth' canines. Only one
group, the hyaenodonts (Eocene–Pliocene) survived the Eocene. They
were widespread in the Miocene when some reached large size. One form,
with several pairs of carnassials, may have achieved a 30 cm gape between its
canines, and could have attacked quite large prey.

Ancestors of the true carnivores, the miacids, are first found in the mid-
Palaeocene. They were small, weasel-like and at first not diverse. Towards
the end of the Eocene an adaptive radiation resulted in new forms which
replaced the creodonts as the dominant predators. Their efficiency in this

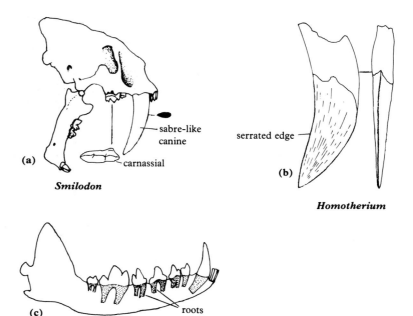

(a)

sabre-like
canine

carnassial

Smilodon

serrated edge

(b)

Homotherium

(c)

roots

Alopex

212 Carnivores.
a, sabre-tooth cat showing wide gape, U Pliocene–Pleistocene. b, razor-edged
canine adapted for slashing, Pliocene–Pleistocene. c, lower jaw of a 'dog' (arctic
fox), U Pleistocene–Recent.

role depended on the development of keen senses; on increased agility based
on longer limbs with digitigrade feet; and on the development of strongly
muscled jaws with sharp biting and slicing (carnassial) teeth (advanced
forms had one pair, fig. 212a). The main split was into two basic groups, cats
and dogs. Between the Oligocene and Pliocene diverse offshoots developed
from each of these, coinciding in time with advances in speed and size of their
herbivore prey. The most important derivatives from the cat line in the
Oligocene were: (i) lightly built 'biting' forms with quite short canines but
powerful jaws; and (ii) more heavily built sabre-toothed 'stabbing' cats, able
to attack large prey (fig. 212a). The dogs remained carnivores (fig. 212c),
but various offshoots arising from them in the Miocene developed appropriate modifications in the teeth and became either omnivores (bears),
herbivores (pandas), fully marine flesh-eaters (seals) or shellfish-eaters
(walrus). True carnivores reached their acme in the Pliocene and are both
numerous and diverse at the present day.

Herbivores
Widely differing orders of placentals are herbivores. Those considered here,
the ungulates, are plant-eaters with hoofs: the elephants (proboscideans)
with five-toed feet; and the horses (perissodactyls) and cattle (artiodactyls)
with four or fewer toes. There are also many extinct orders.

Adaptations shown by these herbivores are largely concerned with diet and locomotion. As herbivores they are usually large-bodied to accommodate and digest a large bulk of low-energy food. Their teeth are modified, the incisors being widened to facilitate cropping of vegetation; and the cheek teeth being enlarged, with a high crown and complexly ridged grinding surface, to crush plant material (fig. 208j, k). The cheek teeth may continue to grow as they are worn down. In the main, limb changes are connected with the need to escape quickly from enemies; and also to range over wide areas in search of fresh pastures. Thus, for example, the limb bones are elongated and this lengthens the stride. Also they walk on the tips of the toes only and these are protected by horny hoofs (modified claws) against jarring contacts with hard ground (fig. 207g). The side toes may be reduced or lost except in forms like the elephant, whose great weight requires the support of a broad foot (fig. 207f).

Archaic ungulates. Most, if not all, 'ungulates' are believed to have separate origins in the CONDYLARTHS (Cretaceous–Miocene) which appeared in the late Cretaceous. Early forms were long-bodied with short limbs, clawed (not hoofed) digits, a heavy tail and primitive, sharp-cusped teeth. In later members, such as *Phenacodus* (U Palaeocene–M Eocene), hoofs replaced the claws and the cheek teeth were broadened with extra and rounded cusps. These forms, and a variety of archaic derivatives, some reaching the size of a large rhinoceros, were on the decline in the latter part of the Eocene. Their place was taken by the ancestors of modern ungulates, forms which, relatively more intelligent and faster moving, were better equipped to evade carnivores.

Proboscideans. The existing elephants (*Elephas* and *Loxodonta*) are the remnants of a once varied and widespread group. They show a number of unusual and highly specialised features, for instance in the structure of their teeth. These are contained in short jaws, and at any one time there is room for only four molars, one in each half-jaw. As these are worn down they are pushed forwards and out by the next set of four (fig. 213), and this happens twice during the life of the animal so that the eventual total of molars is 12. The molars are enormous, high-crowned grinders with many transverse ridges (fig. 214b). There are no canines and only one pair of incisors, greatly elongated to form tusks (fig. 214a).

The modern elephants mark the culmination of a series of changes which can be traced in an extensive fossil record beginning in the lower Eocene in Africa. The earliest fossil to show proboscidean features is of a primitive form less than 1 m high, which had a large, high-skulled head with the nasal opening at the anterior end, and stout, weight-bearing limbs. It had a nearly complete set of teeth, lacking only one set of cheek teeth in the upper jaw, and the canines and a pair of incisors in the lower jaw. A pair of incisors in the

213 The lower jaw of a Pleistocene elephant.

On each side of this jaw is a worn-out 'grinder' (molar) which is almost displaced from the front of the jaw by the eruption of the second, relatively unworn grinder seen at the rear. *Elephas antiquus*, Pleistocene gravels, Barrington, Cambridgeshire (× 0.2).

214 Proboscidean skulls and teeth.

a–c, skull and molar of a Pleistocene mammoth; compare the short jaw and greatly elongated incisors with those of (d). d, e, a pig-sized form from the late Eocene–early Oligocene with low-crowned molars (e) in which the cusps form low ridges.

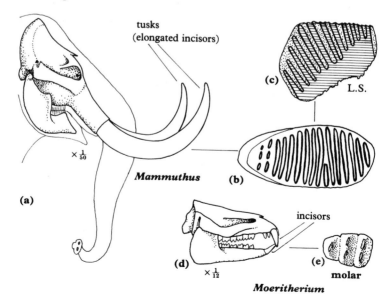

upper jaw showed a slight enlargement and perhaps represent incipient tusks.

During the late Eocene to Pliocene, proboscideans of various types developed and dispersed over much of the Old and New World. They represent divergent lines but they share a number of similar features such as a great increase in size, the largest measuring 4.5 m in height; growth of stout, pillar-like, weight-bearing legs with each toe protected by a hoof (fig. 207f); growth of a trunk; and specialised teeth (fig. 214a, b). The existing elephants belong to the same lineage and evolved in Africa. *Elephas* (Indian form) and *Mammuthus* (the ice age mammoth) appeared about 4 my ago. Later they dispersed, *Elephas* to Asia, *Mammuthus* to Europe and North America.

Perissodactyls. The living members of this group, sometimes referred to as the 'odd-toed ungulates', include the horse, zebra and rhinoceros. They are a small remnant of a once very extensive group which was at its acme in late Eocene and Oligocene times. Apart from the most primitive forms they have either three toes, in which case the axis of the foot passes through the enlarged middle toe (third), or one toe only as in the horse.

The descent of the horse, *Equus*, one of the classic examples of evolutionary radiation, is particularly well documented by fossils. Much of the material was collected in North America, to which area the main evolution was largely confined, but enough occurs elsewhere to demonstrate the patterns of dispersal to South America and the Old World.

The ancestral horse, *Hyracotherium* (U Palaeocene–L Eocene), occurred widely in both North America and Europe. It was a small animal, about 25 cm high, with arched body and long head, rather similar to *Phenacodus* (p. 342) but with longer limbs and larger brain-case. It had four toes on each fore-leg (fig. 215c) and three on each hind-leg, each toe ending in a hoof. Its teeth, relatively simple, included chisel-like incisors, small canines, and low-crowned cheek teeth with blunt cusps (fig. 215a, h). It was a woodland form which browsed on soft leaves, and its splayed-out feet were suitable for moving, or running to escape predators, over soft ground. Its diet of leaves is confirmed by a fossil, with gut contents preserved, discovered in mid-Eocene lake deposits.

The main line of descent from *Hyracotherium* leads directly to the modern horse, *Equus* (Pliocene–Recent, fig. 215b, f), and related forms like the zebra. But in detail its history is complex with various sorts of 'horses' branching off from the main lineage during the Miocene and Pliocene. Some forms retained more primitive characters, and others had a mosaic of primitive and advanced features. Distinctions between genera often rest on details of cranium and face rather than dentition alone. Some of the main progressive changes which occurred may be summarised:

(i)   There was an increase in size from about 25 cm in *Hyracotherium* to over 1.5 m in *Equus*.

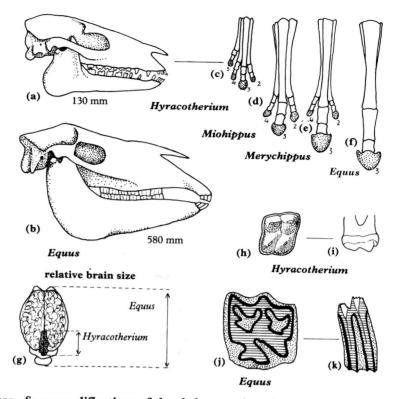

215 Some modifications of the skeleton and teeth in horses.
a, b, skulls of the earliest horse (a) and of a Pleistocene horse (b). c–f, changes
in the right fore-foot in a series of horses from c, the lower Eocene (four toes
functional); d, the Oligocene (three toes functional); e, the Miocene (lateral toes
reduced); f, Pleistocene and Recent (one toe). g, comparison of relative brain
size. h–k, upper molars: h, grinding surface; i, side view; j, transverse section;
k, longitudinal section. Ornament as in fig. 208.

(ii) The legs became long and slender as a result of the elongation of certain
    bones, while others were reduced or lost. In the foot the middle digit
    was enlarged while the lateral digits were reduced and finally
    disappeared (fig. 215c–f).

(iii) The brain-case indicates a great increase in size and complexity of the
    brain (fig. 215g).

(iv) The face in front of the eyes lengthened to accommodate the increased
    size of teeth.

(v) A marked gap developed between the incisors and the cheek teeth
    (fig. 215b); the latter became high-crowned, and the pattern of ridges
    increasingly complex (fig. 215j).

The evolution of the horse is interpreted as a response to changes in its
environment from open woodland, with relatively soft ground, to savannas
and steppes with wide expanses of firm ground. This conclusion takes other
sources of evidence, such as plant remains, into account.

The marked diversification of horses into varied habitats during the Miocene and Pliocene coincides with changes in climate towards cooler, drier conditions and also with the spread of grasses. *Merychippus* (Miocene, fig. 215e), on the main line of descent, had high-crowned, complexly ridged teeth, suitable for grazing on grass; was three-toed and about 1 m high. From it developed the first one-toed form, *Pliophippus* (Pliocene) from which, in turn, *Equus* evolved. The latter appeared in North America about 4 my ago and spread through the New and Old World, reaching Europe about 2.6 my ago. Subsequently it died out except in Eurasia.

Artiodactyls. The artiodactyls, including pigs, hippopotami, deer and cattle, as well as many extinct forms, are highly diverse. In contrast to the perissodactyls they have an even number of toes, two or four, with the axis of the foot running between the third and fourth toes. In the more primitive forms, like the pig, the cheek teeth are low-crowned with rounded cusps. In more advanced forms the upper incisors may be replaced by a hard pad for cropping grass, and the cheek teeth are high-crowned with sharp crescent-shaped ridges on the grinding surface.

The most successful grazers, such as the cattle group, are ruminants (cud-chewers) with an elaborate digestive system using a chambered stomach. They fill up with grass which they subsequently digest at leisure. The grass is regurgitated in wads, thoroughly milled and then passed back to a different stomach chamber for processing by bacteria. In this way maximum nutrition is extracted.

Artiodactyls appeared in the Eocene and quickly diverged along many lines to reach their peak in the Miocene and Pliocene, with ruminants as the major forms. They are less numerous today but remain the dominant ungulates.

15

Kingdom Plantae – plants

Plants are many-celled, mainly land-dwelling, non-motile organisms. They are autotrophs which possess CHLOROPHYLL, a green pigment using sunlight to synthesise carbohydrates (sugars and starch) from carbon dioxide and water, the process known as PHOTOSYNTHESIS. Oxygen is a byproduct and is released into the air. The carbohydrates are then processed within the cells, along with simple inorganic nutrients drawn from the soil, and built up into the more complex substances (e.g. proteins and fats) from which the protoplasm and plant tissue is formed. The basic constituent of the cell wall is cellulose, a polysaccharide. Most plants live on land but a few are aquatic.

There are two divisions, or phyla, of plants: (i) non-vascular liverworts and mosses, the BRYOPHYTA, which are small, relatively simple forms; and (ii) vascular plants, the TRACHEOPHYTA, which are the dominant forms, complex and highly diverse.

Preservation of plant remains

Dead plant tissue decays rapidly under normal subaerial conditions and, since there is no mineral matter to strengthen their structure, fossil plants occur rather sporadically and tend to be poorly preserved. But where oxygen and aerobic bacteria are excluded, as in stagnant water, plant material may be well preserved and in some cases, as in the Rhynie Chert, fine details may be found. These conditions occur mainly in habitats such as lowland swamps, lakes, flood-plains and deltas. It is inevitable, therefore, that the record of plant remains is biased in favour of plants growing in such habitats. Plants from upland areas are rare but include material washed downstream in rivers and the minute wind-blown reproductive bodies, spores and pollen.

Bodies of stagnant, oxygen-deficient water such as swamps and lakes occurred extensively throughout the northern hemisphere during parts of the Carboniferous. The coals formed then consist of variably carbonised vegetable matter in some of which only resistant cuticles (p. 351) and spore cases are identifiable. In other cases material was preserved entire and shows, under the microscope, the structure of the original woody tissue. In nodules (coal balls) occurring in some coal seams, the plant tissues are mineralised, mainly by calcite, and show cell structures. Some of the most perfectly preserved plants are those in which the tissue has been replaced by silica. The Rhynie Chert (lower Old Red Sandstone, Aberdeenshire), an outstanding example of this mode of preservation, represents a silicified peat-bog containing the remains of small animals (e.g. mites and springtails) as well as plants in which the cell structure can be distinguished in thin-section (fig. 219).

Plant fossils are usually fragments of different parts, such as roots, stems or leaves. Complete specimens showing the connection between the various parts are uncommon, and fossils have often been assigned to 'form genera' which subsequent discoveries have shown to be related: e.g. *Lepidodendron* refers to the trunk of a fossil lycopod; *Stigmaria* to roots; *Lepidophylloides* to leaves and *Lepidostrobus* to cones.

Classification

Plants are classified according to details of their structure and mode of reproduction. With advances in knowledge new names have been introduced for the main groups of plants, but many of the categories have 'common' names which are familiar through long usage and are, therefore, convenient to use here.

Non-vascular plants – Bryophyta

Liverworts and mosses (or bryophytes) are the simplest land plants. They are low growing and their tissue lacks the lignin (p. 350) which strengthens the tissue in higher plants. They have inherited from aquatic algae a life cycle involving a distinct alternation of sexual and asexual generations which is dependent on water for its completion (fig. 216c). Thus they occur characteristically in damp places.

The moss plant (fig. 216a), with which most people are familiar, is the sexual phase known as the GAMETOPHYTE. It grows from a SPORE which is HAPLOID, i.e. it has a single set of unpaired chromosomes in each nucleus. It develops aerial shoots with green leaves which can absorb water directly. It is anchored in the soil by hair-like processes, rhizoids, which absorb water and nutrients. It carries minute sexual bodies which produce either female or male gametes (i.e. eggs or spermatozoa). The male gametes are flagellate and can swim via surface water to fertilise the non-motile females, forming a DIPLOID zygote (i.e. with paired chromosomes). This completes the sexual phase of the life cycle.

The asexual, SPOROPHYTE phase follows. The fertilised egg (ZYGOTE) develops a stalked capsule (the sporophyte) which remains attached to the

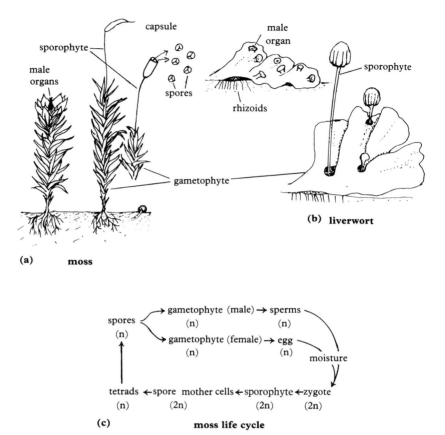

216 Bryophytes.
a, left, a male plant and, right, a female on which the sporophyte developed. b,
shows cup-shaped male structures and form of sporophyte. c, cycle of
reproduction in moss; the chromosome number is indicated by n (haploid) and
2n (diploid).

gametophyte and is nourished by it. The capsule is the organ,
SPORANGIUM, within which numerous asexual reproductive bodies,
SPORES, are produced. It contains 'mother' cells, each having paired
chromosomes (diploid). These divide, each into a TETRAD of four spores in
which the chromosomes are unpaired (haploid). Once ripe, the spores are
dispersed by wind, each one being capable of germinating into a new
gametophyte. They have thick walls impregnated with a tough waxy
substance, SPOROPOLLENIN, and can withstand desiccation.

The liverwort gametophyte is a prostrate, lobed, leafy body (fig. 216b).
Its life history is similar to that of the moss.

Bryophytes are rare as fossils though they may be beautifully preserved on
occasion. A liverwort not unlike a modern form occurs in the upper
Devonian and occasional examples are found in the Carboniferous and later
rocks. A fossil moss has been described from the Carboniferous and from a
well-preserved Permian flora. Otherwise fossil mosses are rare until the

Cainozoic, when extant genera occur, and the Quaternary where they are very abundant.

Vascular plants – Tracheophyta

In vascular plants the body is differentiated into aerial shoots, the stems, with green leaves in which photosynthesis occurs, and underground, anchoring roots which absorb water and nutrients. Food and water move through the plant between leaves and roots along a system of conducting cells, PHLOEM and XYLEM, which make up the vascular tissue. Food, made in the leaves, is moved into the plant along the phloem cells; these are not thickened and may not be preserved. Water, with minerals in solution, moves upwards from the roots along a pipe-like system of xylem cells. In these the cell wall is strengthened by deposition of woody matter, LIGNIN, a complex aromatic substance. Individual cells, known as TRACHEIDS, are elongate and tapering with perforated end walls. The lignin is laid down, over part of the walls only, as rings (annular); spirals; or a more solid ladder-like (scalariform) thickening (fig. 217b). Once formed and functioning

217   Vascular tissue.
a, simplified diagram of a cross-section of conifer wood showing two year's growth. b, longitudinal section of primary xylem in an angiosperm showing tracheids (a, annular thickening; sp, spiral thickening; sc, scalariform thickening; p, circular bordered pits). (Reproduced from W.N. Stewart).

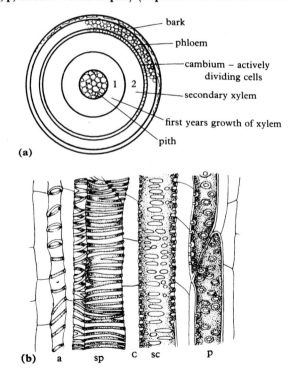

bark

phloem

cambium – actively
dividing cells

secondary xylem

first years growth of xylem

pith

(a)

(b)   a   sp   c   sc   p

xylem becomes non-living. It provides the mechanical support needed for the growth of tall plants. While some plants have only primary (first-formed) xylem others, including forest trees, grow annual increments of secondary xylem (wood) and can reach great size.

Water is conserved in the plants by an impervious layer, the CUTICLE, covering the surface of stems and leaves. The cuticle consists of CUTIN, a complex waxy hydrocarbon. Interchange of gases (oxygen and carbon dioxide) with the atmosphere takes place through breathing pores, STOMATA, in the cuticle; the pores also control the escape of water vapour from the plant.

The life cycle of vascular plants, as in mosses, involves an alternation of generations but, in contrast, the *sporophyte* is dominant, forming the major part of the plant, and the gametophyte is inconspicuous. Reproduction is by means of SPORES in the simpler and earlier forms but by SEEDS in the higher plants.

Spore-bearing vascular plants (Pteridophytes)

Spore-bearing plants with an important fossil record include the primitive, extinct rhyniopsids (Class Rhyniopsida); club-mosses or lycopods (Lycopsida); horsetails (Sphenopsida); and ferns (Filicopsida). Most are small herbaceous plants having no growth of secondary wood. Exceptional fossil forms, however, grew secondary tissue and reached great size, e.g. the giant lycopods of the Coal Measures in the Carboniferous.

Reproduction is by means of spores which develop in sporangia often borne on spore-bearing leaves, SPOROPHYLLS (fig. 218e); the spores are shed and dispersed by wind. In most cases the spores are of one size only (HOMOSPOROUS) and grow into a single gametophyte as described above. Some, however, like the modern club-moss *Selaginella* (and certain fossil forms), are HETEROSPOROUS with large MEGASPORES and small MICROSPORES (fig. 218g). After shedding, the megaspores germinate to form female gametophytes which produce egg cells, and the microspores develop into male gametophytes. The latter produce motile gametes which require the presence of water to find their way to the females for fertilisation. In some species the megaspores remain attached to the parent sporophyll, where they develop into the female gametophytes, so that fertilisation of the egg cell takes place on the sporophyll. This marks a stage towards the condition found in seed plants.

Rhyniopsids

The simplest spore-bearing plants were the extinct rhyniopsids, remains of which occur in the Silurian and Devonian rocks. Two extant, warm-climate

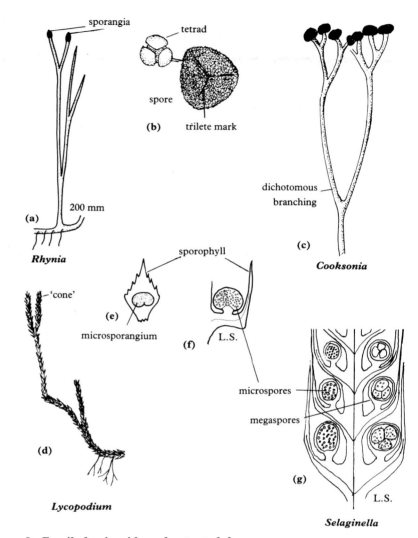

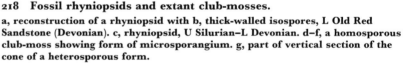

218   Fossil rhyniopsids and extant club-mosses.
a, reconstruction of a rhyniopsid with b, thick-walled isospores, L Old Red
Sandstone (Devonian). c, rhyniopsid, U Silurian–L Devonian. d–f, a homosporous
club-moss showing form of microsporangium. g, part of vertical section of the
cone of a heterosporous form.

plants show a broadly similar organisation but they have no fossil record and
are placed in a separate group, the psilopsids.

    The best-known rhyniopsids were discovered in the silicified remains of a
peat-bog in Aberdeenshire, the Rhynie Chert (lower Old Red Sandstone).
One of these, *Rhynia major* (fig. 218a), is reconstructed as a plant with leafless
stems, about 40–50 cm high and 6 mm in diameter, rising from a prostrate,
creeping stem held in the soil by rhizoids. The stems fork symmetrically
(dichotomously) and some end in elliptical sporangia containing thick-
walled spores of one size (isospores, fig. 218b). The cell structure, as seen in

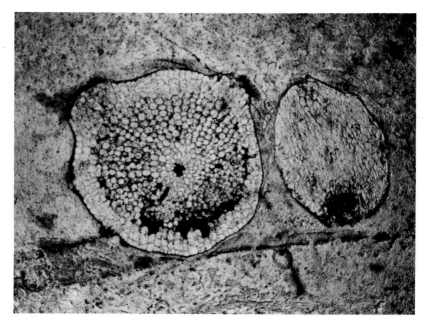

219 Plant cell structure in Rhynie Chert, L Old Red Sandstone, Aberdeenshire (× 35).
A transverse section of *Rhynia* is included; the small dark area in the centre of this represents woody tissue.

thin-section (fig. 219) includes vascular tissue, a central cylinder with xylem cells, and stomata. The earlier *Cooksonia* (U Silurian–L Devonian, fig. 218c) has stems about 6 cm high and 1.5 mm diameter with stomata; the terminal rounded sporangia have spores of one size.

Lycopsids

Extant lycopsids, the club-mosses, are small herbaceous plants with stems which sprawl over the ground and from which erect branches rise at intervals. They have a dense cover of small sharp leaves, spirally arranged and each with a single vein (i.e. a strand of vascular tissue). Some leaves are sporophylls, bearing kidney-shaped sporangia, and may be clustered into loose cones. The spores are either of one size (homosporous) or, as in *Selaginella* (U Carboniferous–Recent) of two sizes (heterosporous, fig. 218g).

The earliest lycopsids were small herbaceous forms. *Baragwanathia* (U Silurian–L Devonian), the first known, occurs in Australia in upper Silurian rocks. It resembles the extant *Lycopodium*. The stems, about 1–2 cm in diameter, branch dichotomously. The leaves are grass-like, up to 4 cm long and set in a close spiral; each has a single vein. Some are sporophylls with sporangia.

By late Devonian times a variety of lycopsids developed and the group diversified reaching its acme in the Carboniferous. The most prominent of these were tree-like (arborescent) forms for which a height of over 50 m has been estimated.

Lepidodendron (Carboniferous, Coal Measures, fig. 220a) is one of the best-known fossil lycopsids with a trunk about 30–35 m in height and over 1 m diameter at the base. The trunk had some secondary wood but the main increase in girth was due to a thick outer layer of non-woody cells. It was anchored by a symmetrical, almost horizontal system of underground

220   Arborescent lycopsids, U Carboniferous.
a, a restoration. b, microspore. c, vertical section of part of a cone. d, megaspore. e, g, part of the surface of a branch showing leaf cushions and leaf scars. f, a fragment of a root-bearing system (rhizophore) showing scars left by roots on shedding. h, leaves and leaf cushions, *Lepidodendron*. i, a microspore, *Lycospora*. (SEM courtesy of Dr I.C. Harding).

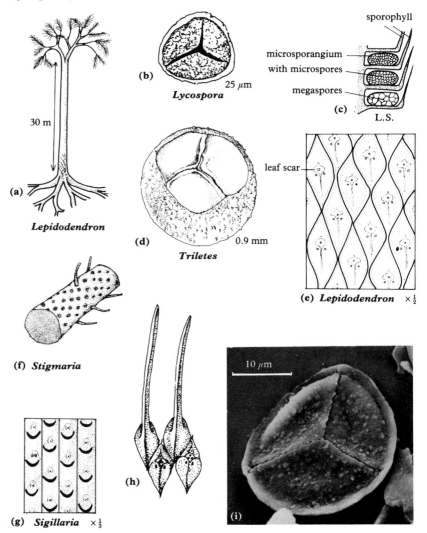

'stems' (rhizophores), four of which radiated outwards from the base of the trunk, and forked repeatedly. These bore the roots, the positions of which are marked by spirally arranged, round, pit-like scars (fig. 220f). The roots were shed as the plant grew. Such rooting systems, typical of the tree-like lycopsids, are referred to the form genus *Stigmaria*. The trunk branched profusely at the top and the stems were densely covered with elongate strap-like leaves (form genus *Lepidophylloides*). The leaves were progressively shorter towards the younger, distal branches. As they were shed they left prominent, spirally arranged leaf scars with a characteristic rhomboid shape (fig. 220e). A related genus, *Sigillaria* (Carboniferous, Coal Measures), had squarish leaf scars arranged in vertical rows (fig. 220g). The sporophylls were clustered in cones (*Lepidostrobus*) up to 35 cm long, at the ends of small branches. In some species vast quantities of both megaspores and micro-spores were produced in the same cone (fig. 220c); in others the two kinds of spore occur in separate cones. The spores are very abundant in some coals, especially in durain layers, from which they can be extracted by chemical methods (fig.220b,d). Assemblages of such spores have been used empiri-cally for correlation of coal seams over limited areas.

The tree-like lycopsids, uncommon after Coal Measure times, became extinct in the early Permian. The record of the small herbaceous forms, however, while insignificant is persistent.

Horsetails (Sphenopsids)

The small herbaceous *Equisetum* is the only surviving genus of horsetails (fig. 221a). It has jointed, aerial stems arising from an underground stem

221 Horsetails.

a, a recent example. b, a rosette of leaves, U Carboniferous. c, transverse section of a stem to show the wedges of woody tissue projecting into the pith cavity. d, internal mould of a stem. e, f, spores: e, shows appendages in coiled-up state; f shows folds caused by compression.

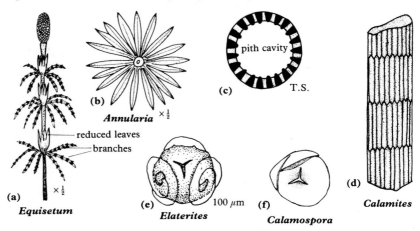

(a) Equisetum ×½

(b) Annularia ×½

reduced leaves
branches

(c) pith cavity   T.S.

(d) Calamites

(e) Elaterites   100 µm

(f) Calamospora

bearing wiry roots. The stems are ribbed vertically and each joint (node) is ensheathed by much-reduced tooth-like leaves. Whorls of slender branches, alternating with the leaves, diverge from the main stem at each node and some of these bear sporangia at their tips.

Horsetails appeared in the Devonian, were at their acme in the Carboniferous and then declined in numbers. The commonest fossil is

222 *Annularia*, **a horsetail (leaf whorls), U Carboniferous (× 1.3).**

Calamites, an arborescent form which reached a height of about 20 m. It had a much-branched stem in which, unlike *Equisetum*, there was some development of secondary wood. The axial part of the stem contained soft tissue (pith) which decayed rapidly after death leaving a cavity. Internal moulds of the pith cavity are common and may be cylindrical or flattened according to the degree of compaction of the associated sediment. The moulds show the transverse markings of the leaf nodes and longitudinal furrows and ridges (fig. 221d). The furrows are impressions left by radial wedges of wood projecting into the pith cavity (fig. 221c). The leaves of some species of *Calamites* are assigned to the genus *Annularia* (fig. 221b, 222); they are small, lance-shaped and arranged in rosette-like whorls. The sporangia were borne in compact cones at the tips of branches. The spores were mostly homosporous but some were heterosporous. They may have appendages which aided dispersal by wind (fig. 221e).

Arborescent horsetails disappeared in the mid-Permian but herbaceous forms like *Equisetites* (Permian–Pleistocene) and *Equisetum* persisted.

Ferns (Filicopsids)

A typical fern has an underground stem (rhizome) from which roots grow down into the soil, and which above ground, grows large leaves, the FRONDS, which typically are divided and subdivided into smaller leaflets, PINNAE. The veins in each pinna are forked; sporangia lie on the underside (fig. 223c).

Primitive ferns are found in the upper Devonian but the group is not generally common until the upper Carboniferous. There is a wealth of different kinds of ferns in the Coal Measures, among which tree-ferns are prominent. Tree-ferns reached an estimated height of about 8 m. They had thick, erect stems, up to 1 m in diameter at the base. The girth here was due to outgrowth of adventitious roots. The stem was crowned by pinnate leaves, some reaching about 3 m in length. Tree-ferns remained important until the Jurassic when their numbers fell. They are represented today by six genera living in the tropics.

The dominant living ferns (filicalids) are most abundant in the tropics, including rain forests, but they are also common in temperate zones in cool, shady places. *Osmunda* (Cretaceous–Recent) represents the most primitive group with a record extending back to the Permian. The group became diverse in the later Jurassic and has an extensive record since then. *Tempskya* is an extinct Cretaceous form, with a trunk-like stem, which is estimated to have been about 5 m high.

Many fern-like fossil leaves occurring in the Carboniferous are not true ferns but seed ferns (p. 361). The distinction can be made only if sporangia or seeds are present on the fossil.

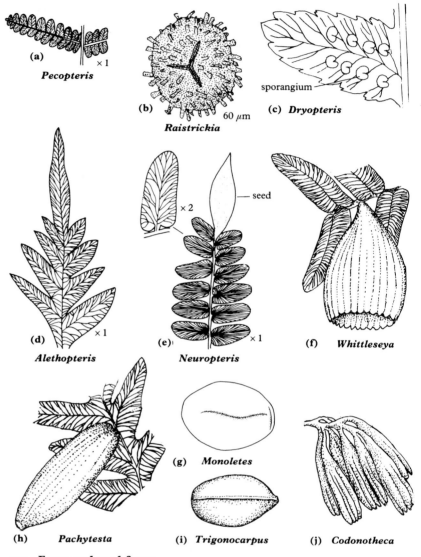

223 Ferns and seed ferns.
a, c, ferns: a, from U Carboniferous; c, part of leaf of extant form to show
sporangia on underside. b, microspore. d–j, seed ferns, U Carboniferous: d, e,
leaves; f, j, pollen-bearing organs; g, pollen; h, seed; i, internal mould of seed.
(f, g, h and j redrawn from W.N. Stwart.)

Seed-bearing vascular plants

The higher, vascular, seed-bearing plants form the dominant living
vegetation. They are heterosporous, the plant itself being the sporophyte
and the gametophyte stage being reduced to miniature size. They reproduce
by seeds, i.e. embryo plants with bud, root and seed leaves enclosed by a
protective coat. The male gamete is non-motile (except in *Ginkgo* and

cycads) and so water is not needed for its union with the egg. Fertilisation occurs on the plant. As was noted for some lycopsids (p. 351), the megaspore is not shed from the megasporangium. The latter is covered by extra layers of tissue (integument) and is now called an OVULE. A single mother cell (p. 349) is produced in the ovule. It divides into four, as in free-sporing plants, but only one megaspore survives. This is nourished within the ovule and grows into a minute female gametophyte which produces an egg. Meantime the microsporangia have produced numerous microspores, now called POLLEN GRAINS, within each of which is a tiny male gametophyte. The pollen grains are shed when ripe and transferred to the ovule either by wind or, as in some higher plants, by a more complicated process, for instance by insects. When a pollen grain makes contact with the ovule it typically grows a small process, the pollen tube, along which male gametes (spermatozoa) are passed. One of these fertilises the egg lying in the female gametophyte. The resultant zygote develops, within the ovule, into an embryo sporophyte plant, the SEED, which is then shed. The seed can remain dormant for a period until conditions are suitable for its germination.

Pollen grains are produced in vast quantities and few take part in pollination. Since the wall is highly resistant to decay the residue may accumulate in sediments. The walls may be sculptured with distinctive markings which are characteristic of the different species, and thus pollen grains extracted from rocks may be a valuable means of identifying past floras.

There are two major groups of seed plants, the gymnosperms, which are the more primitive, and the angiosperms. Differences include the nature of the reproductive organs which form cones in gymnosperms and flowers in angiosperms; and the state of the ovule at the time the pollen grain is transferred to it. In gymnosperms the ovule is exposed on the surface of the sporophyll; in angiosperms it is encased by the sporophyll, which forms a protective carpel (fig. 228).

Gymnosperms

Gymnosperms include the seed ferns (pteridosperms), glossopterids, cycadophytes, ginkgos and conifers. They are woody plants which increase in girth by annual growth of secondary wood. The basic pattern of reproduction is similar throughout the group, though in each case there is a characteristic variation in detail. The sporophylls, bearing sporangia, are usually aggregated in cones.

Gymnosperms were foreshadowed in the upper Devonian by the occurrence of primitive seeds produced by an as yet unidentified plant and also by the presence of gymnosperm-like wood in a group, the progymnosperms which, however, reproduced by spores. One of these, *Archaeopteris*, up to about 18 m in height, was crowned by spirally arranged

224   A scrambling seed fern, *Calymmatotheca*, Coal Measures, U Carboniferous, Kilwinning, Ayrshire (× 1.3).
(Courtesy of the Hunterian Museum.)

fern-like leaves. Some of the leaves were fertile, bearing micro- and megasporangia. The origin of the gymnosperms is thought to lie in the progymnosperms.

Seed ferns. The seed ferns, or pteridosperms, are extinct plants which had much-divided, fern-like leaves. These were attached in spiral arrangement to a slender aerial stem (fig. 224) or, in some which resembled tree-ferns in habit, crowned an upright trunk and grew to perhaps 10 m in height. Seeds were borne on the leaves either singly (fig. 223h), or grouped in cup-shaped structures. The pollen-bearing sporangia were aggregated in clusters or were united on the leaves (fig. 223f, j). Reference has been made (p. 357) to the fact that certain fossil fern-like leaves can only be identified as seed or true fern if the reproductive parts are present on the fossil. For instance one species of *Pecopteris* (Carboniferous) was a seed fern but other species were true ferns with sporangia on the underside of the pinnae (fig. 223a).

Seed ferns first appeared in the upper Devonian and they were abundant and diverse in the Carboniferous. They were at their acme in Coal Measure times with genera like *Neuropteris* (fig. 223e) and *Alethopteris* (fig. 223d). Such forms disappeared in the Permian. Later Mesozoic fossils, often classed as seed ferns, were widespread in the Jurassic, e.g. *Sagenopteris* (U Trias–L Cretaceous, fig. 225a).

Glossopterids. The northern hemisphere was dominated in Carboniferous times by a luxuriant flora adapted to a climate not unlike the present day tropical rain forest. The decline of this flora, which ensued from climatic changes in the late Carboniferous, saw the beginning of a contrasting flora, characterised by *Glossopteris*, in the southern continent of Gondwana (i.e. peninsular India, Africa, Australia, Antarctica and South America) where a period of glaciation was beginning. *Glossopteris* is reconstructed as a tree-like gymnosperm with a short trunk about 6 m high. The leaves, by which it is mainly known, are tongue-shaped with reticulate veins (fig. 225h). Some are fertile, bearing clusters of sporangia or groups of ovules. Its relationship to other gymnosperms is not clear. It was abundant in the Permian and Trias and disappeared in the early Jurassic.

Cycadophytes. Cycadophytes have a rough, columnar stem covered by the leaf bases of previous seasons' growth and crowned by spirally arranged pinnate leaves of tough, leathery texture. They include two groups of plants with an important fossil record in the Mesozoic and Cainozoic. One, the cycadeoids, is extinct; the other, the cycads, still survives. Superficially alike, they differ in structural details and in the reproductive organs. Their origin is thought to lie in late Palaeozoic tree-like seed ferns.

The cycadeoids had hermaphrodite cones of almost flower-like appearance growing between leaf bases on the trunk (fig. 225g). The ovules were borne on a central conical structure which was surrounded by pinnate structures carrying the pollen. The whole cone was protected on the outside by hairy, leaf-like structures.

The cycadeoids (Trias–U Cretaceous) had a world-wide distribution

225   Mesozoic gymnosperms.

a, leaf of a seed fern (Caytoniales) with part enlarged to show veining; and b, a pollen grain from the related pollen organ, *Caytonanthus* (U Trias-L Cretaceous). c, g, cycadeoid leaf (c) and reconstruction (g), Jurassic–L Cretaceous. d, cycad pollen grain (a similar type is found in *Ginkgo*, Jurassic–Recent). e, f, leaves of the recent *Ginkgo biloba* (e), and a Mesozoic form (f). h, leaf showing distinctive glossopterid veining. (a, b, d, and f redrawn from W.N. Stewart.)

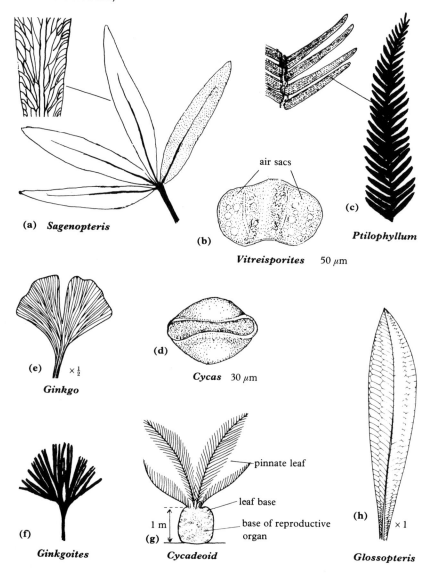

(a)   *Sagenopteris*

(b)

air sacs

Vitreisporites   50 μm

(c)

Ptilophyllum

(e)   × ½

Ginkgo

(d)

Cycas   30 μm

(f)

Ginkgoites

(g)   1 m

pinnate leaf

leaf base

base of reproductive organ

Cycadeoid

(h)   × 1

Glossopteris

226 *Ptilophyllum*, cycadeoid leaf, L Deltaic Group, M Jurassic, Yorkshire (× 1).

during the Jurassic and early Cretaceous. In Britain, for instance, they occur in the Jurassic Deltaic Group in Yorkshire (fig. 225c) and also in the Purbeck 'dirt' beds where well-preserved fossil trunks are found.

Ten genera of cycads survive today in widely separated tropical areas. Unlike the cycadeoids their cones are compact, barrel-shaped structures borne at the end of the stem; and, further, individual plants bear either pollen or ovules. They range from the Trias.

Ginkgos. Ginkgo, the maidenhair tree from South China, is a deciduous tree, reaching about 30 m, and branching repeatedly. Individual plants are either male or female, the male tree carrying its sporangia in pendulous cones (catkins), and the female tree carrying pairs of ovules on stalks. Pollen is dispersed by wind. The leaves are fan-shaped, typically bilobed and borne on short lateral spurs; they have long stalks and two veins which fork regularly (dichotomously, fig. 225e). The leaves are the most commonly preserved fossil parts, occurring for instance in the Jurassic in Yorkshire (Deltaic Group) and in the lower Tertiary leaf beds in Mull, Scotland. Some fossil leaves (occurring in rocks ranging in age back to the Jurassic) are very similar to those of the only surviving species, *Ginkgo biloba*; others, the margins of which are deeply divided, are referred to separate genera (fig. 225f).

Ginkgos appeared in the early Permian and became widespread in the Mesozoic reaching their acme in the mid-Jurassic. During this time they were present in the various parts of Gondwanaland and in the northern hemisphere. A decline set in during the later Cretaceous becoming more marked in the Tertiary and at the same time their distribution became restricted. Today only *Ginkgo biloba* remains.

Conifers. Conifers are, typically, tall trees with spirally arranged, evergreen, needle- or scale-like leaves. Their wood is compact and shows annual increments. The ovules and pollen grains are borne on separate cones, usually on the same plant but sometimes on separate plants. The ovules are borne on scales underlain by leaf-like bracts in a compact cone which, in time, becomes woody (fig. 227a). The male cone has a spiral arrangement of microsporophylls (stamens) each with two pollen sacs on the underside. The pollen grains may have balloon-like 'wings' which aid in their dispersal by wind (fig. 227c).

Conifers are diverse and flourishing at the present day, occurring widely in temperate zones where they may form dense forests. They include the familiar *Pinus* (pine), and *Larix* (larch, which is deciduous), *Araucaria* (monkey-puzzle) and the tallest of all forest trees, the redwoods (e.g. *Sequoia*, which grows to a height of 106 m). *Taxus* (yew) is sometimes classed as a conifer but differs in its reproductive organs, the plants being either male or female, and the ovules, borne singly on short shoots, sit when ripe in a red, fleshy cup.

Conifer-like plants were abundant in the upper Carboniferous. *Cordaites* (Carboniferous–Permian) is the best known. It flourished in habitats not unlike the mangrove swamps of today. It was about 5 m high with stilt-like roots and was crowned by branches with spirally arranged strap-like leaves up to 1 m long and with parallel veins, forking occasionally. It had complex reproductive organs of scales bearing either ovules or pollen sacs and representing an early stage in the evolution of the cone.

Diversification of modern conifers began in the late Trias. Early forms were broadly similar to *Araucaria*. They expanded in the Jurassic and early Cretaceous. They were, together with the cycadophytes and ginkgos, the

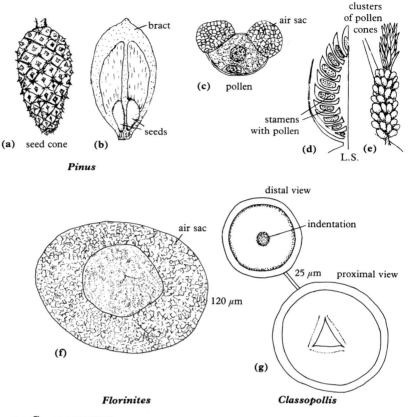

227 Gymnosperms.
a–e, reproductive structures in *Pinus sylvestris*, **Recent. f, pollen grain of**
Cordaites, **Carboniferous–Permian. g, a conifer pollen type common in U**
Trias–M Cretaceous. (f, redrawn from W.N. Stewart.)

dominant Mesozoic land plants. By late Cretaceous times, however, they
were surpassed in numbers of genera by the angiosperms.

Angiosperms

Angiosperms, or flowering plants, are the dominant plants today. They
include woody and herbaceous members which show an enormous range in
structure and habit. Typically they differ from most gymnosperms in various
ways, for instance in the presence of vessels in the wood. VESSELS (xylem
elements) form a tube-like series of cells in which the perforated end-walls
seen in tracheids (p. 350) have disappeared; they transport water more
efficiently than tracheids. But the main distinction is shown in the nature of
the reproductive organs. Typically the male and female organs occur on the
same plant and, usually, on the same structure, the flower (fig. 228). This
consists of outer parts, often brightly coloured petals, surrounding a ring of
modified microsporophylls, the STAMENS, and in the centre, the megasporo-

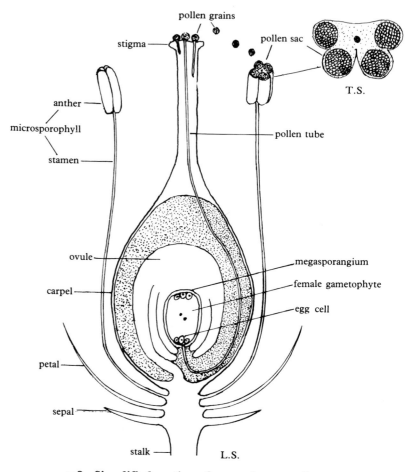

228 Simplified section of an angiosperm flower.

phyll or CARPEL. The stamens consist of stalks bearing pollen sacs (ANTHERS) inside which the pollen grains develop. The carpel comprises the OVARY, with the one or more OVULES, and STIGMAS. The stigma has a sticky surface on which pollen, once transferred to it, germinates. The pollen tube (p. 359) penetrates to the ovule where a unique double fertilisation occurs. One male gamete fuses with the egg cell to form a zygote, while a second gamete fuses with a second (female) nucleus to form nutritive tissue, the endosperm. The latter nourishes the developing embryo. When ripe, the ovary, now enclosing one or more seeds, forms a fruit.

Two classes of angiosperms are distinguished, MONOCOTYLEDONS which have one seed leaf and DICOTYLEDONS which have two seed leaves.

MONOCOTYLEDONS include palms, grasses and bulbs like onions and tulips. Typically they do not grow secondary wood; the vascular tissue occurs in scattered strands; the leaves have parallel veins; and the flower parts are grouped in threes.

DICOTYLEDONS include most of the flowering plants and all the forest trees (other than conifers) and shrubs in which secondary wood is formed each season. In transverse section the strands of vascular tissue form a ring; the leaves have branching, reticulate veins; and the flower parts are usually in fives or fours.

The ancestry of angiosperms is not known but it probably lies with the gymnosperms in which many of their features had already appeared. Their main innovation was the flower with its double fertilisation. No unequivocal fossils are known before the early Cretaceous when a few types of pollen appeared. Diversification ensued, accelerating as time passed until by the end of the Cretaceous the major groups of angiosperms had appeared; had become widespread; and had displaced the gymnosperms as the dominant land plants. Many of the fossil leaves then found resemble surviving familiar forms such as *Quercus* (oak), *Populus* (poplar), *Magnolia*, *Platanus* (sycamore)

229 *Cinnamomum*, **Miocene** (× 1.3).

and *Nymphaea* (water-lily). Pollen and leaves are the most frequent fossils but fruit, flowers and varied types of wood have also been found.

Tertiary floras are abundantly preserved in many parts of the world and, since these include an increasing proportion of modern forms, they have provided valuable information about climatic changes. In Britain the flora of the London Clay is well known. Most of the fossils are angiosperms (mainly fruit, seeds, leaves) and some of these survive today in tropical areas. One fossil closely resembles the fruit of the modern *Nipa* palm which grows in brackish water at sea-level in the Indo-Malay peninsula. Other fossils include *Ficus*, *Magnolia* and *Cinnamomum*, fig. 229).

The occurrence of husks of grass seed in lower Miocene deposits in parts of America (e.g. Nebraska) indicates the development of open prairie grass-lands replacing forest conditions. This coincided with a change in climate to cooler, drier conditions which favoured the spread of the wind-pollinated grasses. It is noteworthy that at the same time various herbivorous mammals greatly increased in number.

Another intimate relationship between angiosperms and animals is illustrated by the devices they have evolved to ensure pollination. While some are wind pollinated, many are dependent on insects and some on small animals, e.g. humming-birds and bats. Insect pollination was well es-tablished by mid-Eocene. Flies, bees, butterflies and beetles are the main insects involved. In many cases the shape of the flower has been modifed to favour particular insects: examples include bees and orchids; wasps and figs (*Ficus*). The benefits are mutual. Flowers provide food (nectar and pollen) for the insects which, in collecting it, brush pollen on to the stigmas. Self-pollination is prevented in many cases by devices which ensure that pollen from one flower is carried to the next flower visited.

Reference has already been made (p. 25) to the value of fossil plants in deciphering details of climatic changes during and since the Pleistocene. Much of the evidence is based on analysis of fossil pollen.

Spores and pollen grains

Fossil spores and pollen grains of vascular plants are widely dispersed through sedimentary rocks dating from the lower Palaeozoic onwards. Their outer wall is composed of sporopollenin (p. 349) which is highly resistant to natural decay. Thus they occur as fossils in very great numbers. In addition they can readily be extracted from rocks by chemical means. Their study is called palynology and its primary aim is to relate the spores and pollen grains to their parent plants. Distinctive characters such as shape, size, aperture for germination, and the nature of the wall and its sculpture are used for identification. Differences in these characters are connected with the mode of reproduction of the parent plant as outlined in previous pages. Frequently the dispersed fossil spores and pollen grains cannot be related to their parent plant because of the lack of direct evidence. They are then referred to form genera and species. Generally, however, the broad group of plants to which they belong can be identified.

It will be recalled from the previous section on plant macrofossils that the more primitive plants reproduce by spores, and the more advanced by seeds. The simplest spore-bearing plants (the rhyniopsids, some lycopsids and horsetails, and most ferns) are HOMOSPOROUS, producing spores of one size only, ISOSPORES. More advanced forms (e.g. *Selaginella* and giant lycopsids) are HETEROSPOROUS, producing spores of two sizes: large numbers of small, male MICROSPORES ranging in size from 5 to 20 μm; and a small number of large, female MEGASPORES mostly over 200 μm in diameter (p. 374).

These free-sporing plants are damp-loving and depend on the presence of water for the process of fertilisation. In contrast, the seed-bearing plants are adapted to drier climates and became freed from the need for water during fertilisation. The seed habit was developed by: the retention on the plant of the megaspore which developed within its protective megasporangium (now forming the ovule); the development, in parallel, of the microspores into pollen grains, each containing only a few cells (with male gametes), and which were dispersed (e.g. by air) to fertilise the egg-cells within the ovule.

Shape.   The basic shape of spores and pollen grains is broadly determined during the division of the mother cell (p. 349) into a tetrad of four daughter cells. Typically, these lie one within each corner of a tetrahedron, each cell touching the other three. They are more or less spherical but impinge on one

230   Symmetry of spores and pollen grains resulting from the two modes of mother cell division.

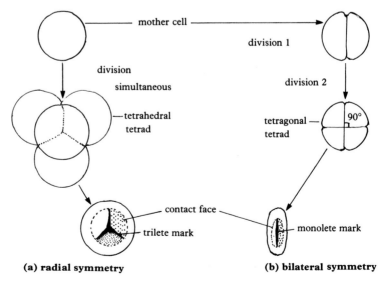

another as they grow so that each develops *three* flattish contact surfaces (fig. 230a). Less commonly the mother cell divides in two, and these in turn divide in two, the divisions occurring in two planes at right angles to each other in tetragonal arrangement. These cells are more or less ellipsoidal in

231 Orientation and wall structure of spores and pollen grains.
a–c, orientation of spores (a); aspects of trilete spores (b); aspects of monolete spores (c). d–g, orientation of pollen grains: d, tricolpate; a, triporate; f, monosulcate; g, monosulcate with air-sacs. h, i, sections of a spore (h) and a pollen grain (i) to show structure of the walls and some of the types of sculpture found. (Mainly based on West.)

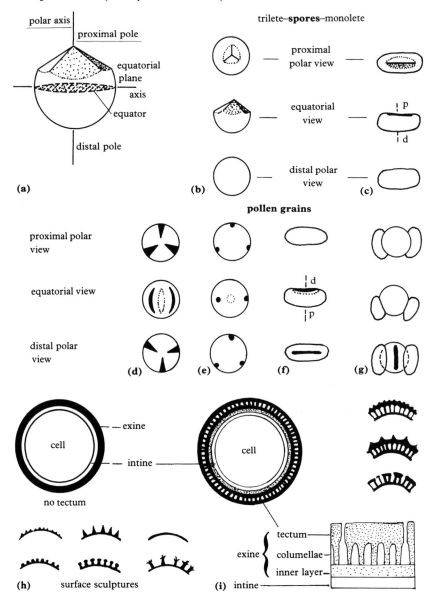

shape and each one touches two others so that *two* contact surfaces develop (fig. 230b).

A single spore or pollen grain is described with reference to the axis of an imaginary pole which runs from the centre point of the contact surfaces, called the proximal polar face, to the opposite, outer surface, the distal polar face. Between these faces lies an imaginary equator in a plane at right angles to the polar axis (fig. 231a).

Apertures. Spores and pollen grains germinate via apertures, the nature and position of which is characteristic for the type concerned. In spores the apertures lie on the *proximal* surface and are identified by a Y, or TRILETE mark; or by a I, or MONOLETE mark. In a few forms there is no aperture and these are called ALETE.

Trilete spores form when the mother cell has divided simultaneously producing the tetrahedral tetrads. The Y mark is formed by sutures separating the three contact surfaces; these meet at the proximal pole (fig. 231b).

Monolete spores arise when the tetrad has formed in two stages. The I mark is formed by one suture separating the two contact surfaces (fig. 231c).

Pollen grains are distinguished from spores by a different arrangement of apertures, marked on the surface by one or more furrows. These lie either on the *distal* surface and are known as SULCI (sing. sulcus); or across the equator and are then known as COLPI (sing. colpus). The furrows are reduced to pores in some forms. The arrangement of the apertures is related to the way in which the tetrad was formed. Pollen formed by division in two stages is MONOSULCATE with one sulcus on the distal face and is more or less ellipsoidal in shape (fig. 231f). Monosulcate pollen is typically found in gymnosperms and monocotyledon angiosperms. In pollen formed by simultaneous division, three apertures or colpi lie across the equator, the TRICOLPATE condition (fig. 231d). This type of pollen is typical of the dicotyledon angiosperms. In these there are many variants in the number and form of apertures, each distinguished by a separate term. For instance, in triporate pollen the colpi are reduced to pores (fig. 231e).

Wall structure. The wall of spores and pollen grains is designed to ensure dispersal, protection against decay and desiccation and eventually, when conditions are favourable, quick germination. It is a double structure consisting of an inner cellulose wall, (the intine), enclosing the soft contents, and an outer wall of highly resistant sporopollenin (p. 349), the EXINE. The exine may be simple or complex in character and varies in thickness. In spores and gymnosperm pollen grains it is fairly homogeneous and may have a sculptured surface (fig. 231h). In some cases it is layered so as to form air-sacs, as in winged pine pollen grains (fig. 227c). The exine of angiosperm pollen grains is more complex, consisting of an inner and an outer layer

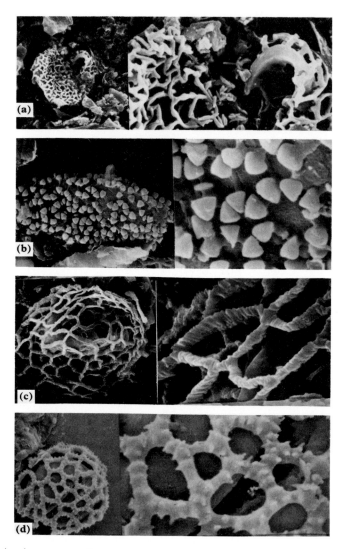

232 Angiosperm pollen grains.
A range of pollen grains showing surface sculptures; those on the left (× 1600),
on the right (× 7000). a, shows a section of the exine including columellae. b,
processes on surface of tectum. c, d, show reticulate sculpture; in (d) the inner
exine layer with aperture has rotated slightly within the outer exine layer;
columellae are absent in this form. All from L Cretaceous (Aptian/Albian),
Mersa Matruh, Egypt. (SEMs courtesy of Dr J.H.J. Penny)

separated by rodlike structures. The surface layer, the TECTUM, may be
smooth or sculptured (figs. 231i, 232, 233).

Dispersal. Dispersal of spores and many pollen grains is by wind. Such
forms are small, usually less than about 60 μm, and smooth. The air-sacs seen
in many gymnosperm pollen grains aid floating by increasing surface area.
Wind dispersal is chancy and much more pollen than is necessary for

fertilisation must be produced. This is wasteful. An alternative mechanism is found in many angiosperms in which pollen is transferred by a variety of insects and other small creatures (p. 368). Such pollen is often relatively large, up to about 200 μm, tends to have a thick exine, and is often highly sculptured and sticky. This method is efficient, and insect-pollinated flowers produce less pollen than those pollinated by wind. For this reason they may be less well represented in sediments.

Once pollination has been effected, germination occurs rapidly through an aperture.

Geological distribution of spores and pollen grains

It is clear from the fossil record that plants have evolved both in their structure and their mode of reproduction, becoming more complex in the course of time. Further, the record shows that as each of the various groups appeared it diversified, was important for a time, then went into decline and was replaced by more complex forms. The diversification of spores, and then of pollen and seeds, parallels the macrofossil record.

Spores. Spores with trilete marks occur in lower Palaeozoic rocks, initially in groups of four (tetrads) and then singly. Spores of this nature are produced by algae and bryophytes as well as by vascular plants. Accordingly, earlier spores can be authenticated only if found in organic attachment with known vascular plants. In turn, indisputable proof that these are vascular plants depends primarily on finding tracheids (p. 350) in their remains. *Cooksonia* (fig. 218c, U Silurian (Wenlock)–L Devonian) is generally accepted as the earliest vascular plant. But it has to be noted that the fossils, in which tracheids and spores were found, were separate though associated specimens. The lycopsid, *Baragwanathia* (U Silurian (Ludlow) – L Devonian), with trilete spores is, on graptolite evidence, only slightly younger. It is probable that many of the dispersed spores at this time were from vascular plants.

Only a few types of spores are found in the Silurian and these are simple and small, under 50 μm in diameter. In the course of the lower Devonian the number of spore types increased steadily. The plants at this time were free-sporing forms including, e.g. *Rhynia* (p. 352) and lycopsids like *Baragwanathia* (p. 353). They were homosporous, producing isospores, some showing surface sculpture. Sporangia on *Rhynia*, for instance, contained trilete spores, about 40–60 μm in diameter, arranged in tetrads and with short spines (fig. 218b).

Many new genera appeared during the middle and upper Devonian, typically trilete, but some (lycopsids) monolete. They show a range of surface sculpture and also an increase in size. Analysis of size distribution has shown two categories. These are arbitrarily described as microspores, under

100 μm, and megaspores, over 200 μm in diameter. By analogy with the extant *Selaginella* (p. 353) this indicates that some plants had become heterosporous. Heterospory is more clearly established in the Carboniferous with a marked divergence in size. For instance, some megaspores exceeded 2 or 3 mm in diameter. The occurrence of both microspores and megaspores within the same lycopsid cone nicely demonstrates this habit. In one species of *Lepidostrobus* (p. 355) microsporophylls are arranged at the tip, and megasporophylls at the base of the cone. These contain, respectively, microspores about 20–30 μm and megaspores about 700–1250 μm in diameter. The microspores resemble the dispersed microspore *Lycospora* (Devonian–Permian, fig. 220b, i) which is abundant in upper Carboniferous (Coal Measure) rocks. The megaspores compare with the dispersed spore *Triletes* (fig. 220d). Heterospory was general among the arborescent lycopsids but some were homosporous like the modern *Lycopodium*.

The Carboniferous horsetails were typically homosporous but some were heterosporous. In some species of *Calamostachys*, the cone associated with *Calamites*, the spores are trilete and may show some variation in size, e.g. ranging between 40 and 100 μm in diameter. In other species the cones were heterosporous, bearing microspores about 70–118 μm and megaspores about 140–275 μm. Some spores, had three coiled processes, elaters, attached. Similar processes found in the modern *Equisetum* are hygroscopic and coil or uncoil with changes in humidity. They serve to aid dispersal from the sporangium. In some cases the elaters were shed on dispersal from the cone, as in *Calamospora* (fig. 221f), but they were retained in *Elaterites* (fig. 221e).

Most ferns are homosporous; exceptions include a Carboniferous species, and the aquatic ferns (Cretaceous–Recent) which are heterosporous. Their spores are commonly trilete but some are monolete (fig. 14). They are usually small, as little as 14–20 μm in the monolete spores of one species of *Pecopteris* (Carboniferous). They show varied shapes and sculpture and may be produced in enormous quantities. For instance, a single sporangium on a Triassic fern contained an estimated 1000 isospores; and the number of sporangia on a fertile frond of *Botryopteris* (Carboniferous) was assessed at 50 000. The norm, however, is much less; e.g. about 120 isospores of the *Raistrickia* type were found in sporangia of another Carboniferous fern. *Raistrickia* (Devonian–Carboniferous) is a common dispersed spore, trilete, about 70 μm in diameter and with rod-like processes on the surface (fig. 223b).

Pollen grains. The seed habit, characteristic of the gymnosperms, evolved at some time in the Devonian, the earliest known seeds, e.g. *Archaeosperma*, being found in the upper Devonian. By early Carboniferous times the seed habit was well established in several groups of plants. It has to be assumed that these plants produced pollen as distinct from microspores. However, the 'pollen' found in the earlier seed plants, like seed ferns, cannot be distinguished morphologically from microspores because they germinated on the proximal face. Such 'pollen' is referred to as 'pre-pollen'. Pre-pollen with trilete, or monolete sutures is found *in situ* in many seed-fern pollen organs. *Monoletes* (fig. 223g) is a very common form. It is about 100–500 μm in length, with a slightly bent suture on the proximal side, and two distal grooves. In later, true pollens the point of germination moved to the distal or equatorial surface.

The pollen grains produced by cycads, cycadeoids and *Ginkgo* (all prominent in the Mesozoic) are broadly similar (fig. 225d). Typically they are monosulcate, and more or less ovate in outline, shaped almost like a rowing boat. The walls are thin and the surface usually smooth. The grains are small, around 25–30 μm, slightly larger in *Ginkgo*. One or two air-sacs are present in some forms.

The pollen grains of many conifers and conifer-like plants (occurring from the Carboniferous onwards) typically bear air-sacs, and the aperture is distal. *Cordaites*, the conifer-like tree common in the Carboniferous (p. 364), contained pollen resembling the dispersed *Florinites* (Carboniferous–Permian). *Florinites*, which has one large air-sac, with reticulated inner surface, measures about 65 by 100 μm (fig. 227f). *Classopollis* (Trias–M. Cretaceous, fig. 227g) is a common dispersed pollen produced by an extinct group of conifers of world-wide distribution. The grains, about 20–30 μm in diameter, are unusual in having both a trilete mark on the proximal surface and a dimple-like pore on the distal surface. Encircling the grain between equator and pore is a groove where the exine is thin. Perhaps the best known conifer is *Pinus* (Cretaceous–Recent), which typically carries pollen which is monosulcate, with two air-sacs giving a 'winged' appearance, and measuring about 75 by 65 μm (figs. 14b, 227c).

The development of air-sacs in conifers is varied: they may be two or three in number, they may be reduced or even lacking.

The earliest authentic angiosperm pollen grains occur in early Cretaceous rocks. They are small forms, under 30 μm diameter, the size suggesting wind pollination. Typical of these is *Clavatipollenites* (fig. 233a). It is monosulcate and oval in outline. It has been identified as an angiosperm pollen because it possesses a thin, albeit discontinuous tectum. These early forms marked the start of the major angiosperm radiation which, by the end of the Cretaceous, resulted in great diversity, including many of the modern, major plant groups and a variety of extinct forms. New types of pollen appeared in parallel with these innovations showing, among other features, elaboration of the apertures for germination. The single furrow of the monosulcate condition was succeeded by the tricolpate state as seen in *Tricolpites* (fig. 233b, c). That in turn was modified by reduction of the three furrows to pores producing, in the early upper Cretaceous, triporate pollen. This is the type of pollen produced today by *Betula* (birch, figs. 14c, 233d). Further

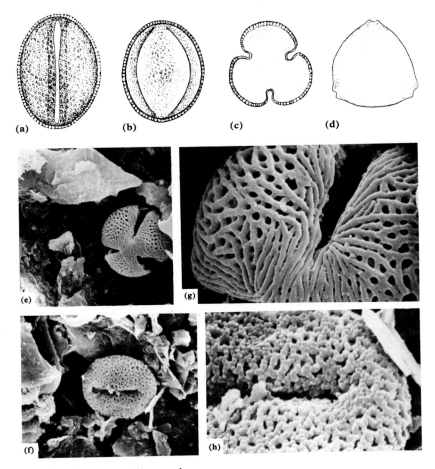

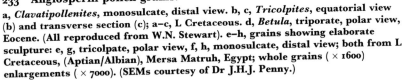

233　Angiosperm pollen grains.
a, *Clavatipollenites*, monosulcate, distal view. b, c, *Tricolpites*, equatorial view
(b) and transverse section (c); a–c, L Cretaceous. d, *Betula*, triporate, polar view,
Eocene. (All reproduced from W.N. Stewart). e–h, grains showing elaborate
sculpture: e, g, tricolpate, polar view, f, h, monosulcate, distal view; both from L
Cretaceous, (Aptian/Albian), Mersa Matruh, Egypt; whole grains (× 1600)
enlargements (× 7000). (SEMs courtesy of Dr J.H.J. Penny.)

elaboration of apertures (and also of the tectum) followed at an accelerating
rate in the later Cretaceous.

Angiosperm pollen can be distinguished in general as of monocotyledon or
dicotyledon type. Monocotyledon pollen is typically of the primitive,
monosulcate, bilateral, type, but may be modified as, for instance, in the
grasses. These have a single pore (monopore) and are more or less spherical.
The primitive, monosulcate form of pollen is retained in some dicotyledons,
for instance in *Magnolia*, one of the oldest known flowering plants.
Typically, however, the dicotyledon pollen has three or more furrows or
pores. The pollen record broadly parallels the evolution of angiosperms as
revealed by the macrofossils.

Uses of spores and pollen grains. Spores and pollen grains are valuable in biostratigraphy, particularly in rocks of continental facies which often lack other fossils. An early use in this respect was in the local correlation of Carboniferous coal seams, using the spores empirically. In younger rocks, where their biological affinities have been certainly established, they may also be used in the study of climatic change as indicated by fossil floras. Short term changes in climate during the Quaternary have been clearly demonstrated by this means.

Kingdom Fungi

Fungi (moulds and toadstools) consist of single cells or tubular filaments (hyphae). The hyphae may have cross-walls (septate) or not (aseptate). The walls contain chitin and some contain cellulose. They reproduce, sexually or asexually, by spores from toadstools. They were formerly classed as plants, but they lack chlorophyll and they now form the separate kingdom, Fungi.

Fungi are heterotrophs which live as saprophytes on organic matter or, in some cases, as parasites on live organisms. Their hyphae ramify through the host material, secreting enzymes to dissolve and digest it. In the process they release soluble nutrients which can then be absorbed by plants. Thus they play an important part in the breakdown of plant debris and humus. Certain fungi, mycorrhiza, have a mutually beneficial association with roots of a wide variety of plants, living on or within the roots. This association is important for the growth of both organisms and, for some plants, it is vital for establishment and growth, particularly for those growing in nutrient-poor or acid soils. Examples include alder, pine, heather, orchids and club-mosses (gametophytes).

Fungi, despite their fragile nature, have a long fossil record which extends back to the Silurian and possibly to the Precambrian. Fossils are rare and are generally found in association with plant remains. Both spores and hyphae may be preserved, for instance in the silicified plant tissues in the Rhynie Chert; and along with a wide variety of fossil plants in the Carboniferous Coal Measures, where mycorrhiza are reported in association with ferns and *Stigmaria*.

16

Epilogue

Summary of the fossil record

Although the fossil record is incomplete, we now have a vast amount of information about the history of life on the earth, and the sequence of change during the past 3.5 by is broadly known. It is a story of increasing complexity and diversity within the framework of a relatively small number of basic life patterns.

Some of the most important events in the fossil record are:

(i) The first record of procaryotes about 3.5 by ago.
(ii) The first record of unicellular eucaryotes by about 1.4 by ago.
(iii) The first metazoans (the Ediacaran fauna) about 630 my ago.
(iv) The Cambrian–early Ordovician radiation during which the basic designs of skeletal animals appeared over a period of about 80 my.

The first two of these events, separated by a period of more than 2 by, represent an extremely slow rate of change. In contrast to this, developments in the multicellular eucaryotes, over a period of about 150 my, seem almost rapid.

The earliest known metazoans are impressions of soft-bodied animals, known as the Ediacaran fauna, found in late Precambrian rocks in several parts of the world, from south Australia to north Russia. More than 30 different forms have been distinguished, most of which cannot, with confidence, be assigned to known phyla. Some show resemblances to jellyfish, to sea-pens and to annelid worms. They were preserved under highly unusual conditions and are, presumably, a sample of an extensive and diverse soft-bodied fauna, with all the necessary food resources, which existed during the late Precambrian about 630 my ago. This diverse fauna must have been the product of an earlier radiation, details of which are at present unknown. Body fossils are not as yet known from the rocks lying between the Ediacaran horizon and the base of the Cambrian. Trace fossils, however, reveal the presence of surface-crawling and shallow-burrowing organisms.

234 **Ranges of selected fossil groups; thicker lines show periods of greater diversity.**

Invertebrate skeletal fossils are first known in the lowest stage of the Cambrian (fig. 234). They were already clearly differentiated and must have been in existence prior to that time. They were primarily benthic organisms and include hyolithids, sponges, archaeocyathids, gastropods, inarticulate brachiopods and minute phosphatic shells of uncertain affinities.

In the succeeding two stages of the lower Cambrian, trilobites appear and become common; and are joined by the first known echinoderms, articulate brachiopods, bivalves and ostracods. The mid-Cambrian saw fewer innovations, but dendroid graptolites appeared, to be followed in the upper Cambrian by nautiloids, and in the lower Ordovician by bryozoans and tabulate corals.

During this period of about 80 my, all the major groups of skeletal invertebrates appeared and diversified to varying degree. In addition, marine 'plants' were represented by calcareous algae (lower Cambrian), and vertebrates appeared in the upper Cambrian. If one considers this important radiation as part of a larger episode including also the soft-bodied metazoans of the Ediacaran then a period of some 150 my was involved, equivalent to about one quarter of all Phanerozoic time.

The appearance of rigid skeletons marked a great advance in the invertebrate body-plan. In addition to the obvious advantage of a more or less protective covering, the skeleton served as a base for muscle attachment and allowed the development of improved mechanisms for locomotion and feeding, the basis of many observed adaptations to varying habitats.

The evolution of the major invertebrate groups established basic body-plans within which many variations appeared with time. For example, one of the major molluscan groups, the cephalopods, first found in the late Cambrian in the form of the nautiloids, later gave rise to the ammonoids and coleoids in the early Devonian. All three of these lesser groups quickly diversified, adapting to varying modes of life. As a result of these multiple adaptive radiations (which occurred in all the skeletal invertebrate groups) the marine environment was effectively occupied, presumably with displacement of many of the soft-bodied organisms which now had to compete with more efficient skeletal forms. Later, fresh-water habitats and, to a limited extent, the land were also invaded by these invertebrates.

The fossil record is incomplete, notably in the general absence of soft-bodied organisms. But unusual fossilisation of the kind found in the mid-Cambrian Burgess Shale of North America, does something to fill the gap. When we hear the words 'Cambrian arthropods' we think immediately of trilobites; yet, of the many Burgess Shale arthropods the great majority are not trilobites but other forms which are not usually fossilised. In addition there are representatives of extinct phyla not otherwise known. Worms of various kinds are also well represented.

The first vertebrates, in the form of jawless fish, appeared in the late Cambrian and the first major radiation of jawed fish started during the late Silurian. This was followed by other radiations in different groups of fish, culminating in the major diversification of the bony fish during the Cainozoic.

Amphibians appeared in the late Devonian and reptiles in the late Carboniferous. The first mammals, in the late Trias, were followed by birds in the Jurassic; both had limited diversity until the Cainozoic when massive expansion took place.

Of the vascular land plants, the earliest known appeared in late Silurian times but the great expansion in diversity took place in the Carboniferous when the pteridophytes dominated, the gymnosperms playing a lesser role (fig. 236b). During the Mesozoic, however, the roles were reversed with gymnosperms dominant. Finally, the angiosperms, first found in the early Cretaceous, rapidly assumed dominance during the Cainozoic.

In this large-scale diversification of plants, invertebrates and vertebrates, numerous examples of opportunism and interdependence are to be found: for instance between land arthropods and plants; the development of herbivory in reptiles; the evolution of horses and the spread of grasses.

Originations and extinctions

Originations and extinctions are common features of the fossil record and the rates at which they occur determine changes in diversity.

The term diversity refers to the variety of types of organisms which co-exist. It may be expressed as numbers of species, or genera, or families or orders of fossils found within a given time span, e.g. a biozone or stage or system. Many of the quoted values are for genera or families per stage. Since the fossil record is most complete for marine skeletal invertebrates, a better overall picture emerges for these than for vertebrates and plants whose fossil record is uneven, often scanty or absent completely. Even so, the diversity figures quoted for marine invertebrates are rather crude for the following reasons. Firstly, sampling is uneven and incomplete, particularly for the older rocks, most of which are subsurface and inaccessible. Secondly, some fossil groups have been studied more intensively than others. This usually results in 'splitting' into more species, genera and families, giving such groups a greater apparent diversity than others. Thirdly, the assignment of fossils to a particular category is to some extent subjective, leading to inconsistencies of treatment as between individual workers. However, in spite of these qualifications, the fossil record shows fluctuations in diversity, coupled with changes in composition of the fauna and flora, which are so striking that there can be little doubt as to their reality.

A diagram showing diversity changes with time for a major group such as the ammonoids (fig. 235, based on families) reveals considerable fluctuations. These were determined by the changing balance between originations (dominating the rising sectors) and extinctions (dominating the falling sectors). It should be noted that a period of constant diversity is rare and is not one of 'no change' but one of equality between originations and extinctions. Diversity curves for other groups show similar features though differing in detail. When the data for all skeletal marine fossil invertebrates

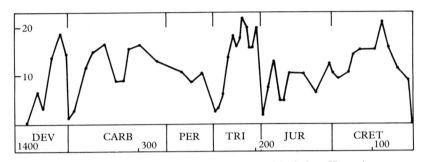

235   Family diversity of the ammonoids (after House).

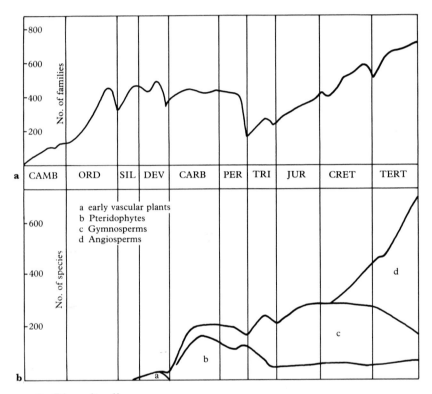

236   Diversity diagrams.
a, family diversity of marine skeletal invertebrate fossils (after Sepkoski). b,
species diversity of vascular land plants; the upper line shows total species
(after Niklas et al.).

are plotted on the one diagram (fig. 236a, based on families) certain periods
of time are marked by generally rising or falling diversity, suggesting that the
fauna as a whole was being influenced by factors favouring one or other of
these trends. The well-marked diversity falls in the late Ordovician, late
Devonian, late Permian, late Trias and late Cretaceous have been labelled
'mass extinctions'. A better term might be 'multiple extinctions' since the
significant feature is that many different groups (though not necessarily the

majority) were similarly affected during the same time period. Equally important are the periods of rising diversity. These indicate 'multiple originations', the major radiations.

Organic change, i.e. evolution, involves the appearance of new species, originations. The theory of evolution explains this in the following way. Within an inter-breeding population there is inherited genetic variability which results in numerous offspring different from each other and from their parents, and not all of which survive. Natural selection ensures that the survivors are those which are best able to exploit the range of habitat either existing or newly appearing in a changing environment. These successful variations will build up in the population, in time forming a new species. Competition, not only between individuals within a species but also between species, is an essential part of the evolutionary process.

The first metazoans, ancestors of the soft-bodied Ediacaran fauna, had for their exploitation seas occupied only by unicellular organisms. There was, presumably, considerable empty living-space into which newly emerging forms could radiate with minimum competition. That soft-bodied animals continued to evolve and flourish, presumably with further radiations, is exemplified by the diversity of the Burgess Shale fauna in the middle Cambrian. At the beginning of the Cambrian, however, skeletal invertebrates appeared and radiated dramatically, being able to compete successfully with many soft-bodied forms and also, perhaps, to occupy hitherto empty niches. This radiation continued into the lower Ordovician and obviously merits the title 'multiple'.

The precise controls (as opposed to the foregoing generalities) of a particular radiation cannot be deduced from the fossil record. It may be possible, however, to identify some of the factors most likely to have been involved. For example, the spread of gymnosperms and the decline of pteridosperms during the Permian can be explained by a general change to a drier climate to which the former were more easily adaptable (fig. 236b). Climatic change and competition are probably the main factors here. Another example is that of the mammals, which originated in the late Trias but remained inconspicuous during the long period of dinosaur domination on land. After the extinction of the latter the great radiation of the mammals occurred. Was this mere coincidence, or an example of opportunism?

Diversity of the skeletal marine invertebrates increased generally from the early Jurassic to the present day. This may, in part, be an artifact, since the younger faunas are better known, sampled and studied. But increases in ammonoids (to mid-Cretaceous), echinoids, bivalves and gastropods have all contributed to another long-term multiple radiation. The reasons for this are unclear.

The first appearance of a new fossil species in a sequence of rocks does not necessarily mark its actual time of origin. Commonly, the underlying rock is unfossiliferous, or contains fossils of a different group. There may even be a break in sedimentation, leaving a gap in the record. In such circumstances there can be no certainty about the immediate antecedents of the new form. The sudden appearance of the first metazoans is an example of this situation. Only when a new species appears within a continuously fossiliferous series of related forms can the circumstances of its origin be investigated. Such sequences can be found in oceanic deposits formed by the slow accumulation of planktonic organisms. In one sequence of this kind, several lineages of radiolarians were studied using samples ranging from late Miocene to Recent in age and representing about 6 my in time. During a 0.5 my period in the latest Miocene and early Pliocene, two new species arose, by gradual divergence and with transitional forms, from a continuing ancestral species. All three continued to show gradual variation, often irregular, with time. The new species and transitional forms appeared over the whole of the equatorial regions of the Pacific and Indian oceans during the same time interval. This is an example of new species arising by gradual change.

In other examples, involving macrofossils from various facies of sedimentary rocks, one species may follow another successively in a single varying lineage. Intermediate forms are sometimes present but frequently the record is discontinuous and the details uncertain. In many other cases, all one can say is that the new species appears 'suddenly'.

Among possible explanations for such abrupt appearances is that of extremely rapid species origination. In living forms, genetic variability is so great that it seems possible for new species to arise in a very short time: a few thousand years according to some authorities. This is most likely to happen in a very small isolated population through which the new gene could quickly spread. If, subsequently, the new species was able to migrate widely then its appearance, without transitional forms, would appear to be 'instantaneous' in the context of geological time. Once established, the new species is supposed to remain virtually unchanged until extinction. This hypothesis of long-term species stability ('stasis') and extremely rapid species formation has been termed 'punctuated equilibrium'. The extent to which it applies to the fossil record is a matter of continuing debate.

The estimated duration of fossil species is quite variable: as short as 0.25 my in some cases; as long as 10 my in others. While some show little change throughout their range, others are quite variable, both within co-existing populations and between successive populations.

The abrupt appearance of radically new features (like the invertebrate shell or the tetrapod limb) poses yet another question. Were these novelties the product of normal evolution or was some special mechanism involved? This is a problem for the molecular biologist and it has not yet been solved. It is, however, clear that genetic variability is very great. It includes the effects of mutations, which alter the genes or chromosomes, providing a mechanism for the increase in complexity necessary for radical organic change. Mutations occur naturally and can be stimulated by various types of radiation (e.g. ultraviolet and X-rays) and by certain chemical agencies. An explanation along these lines seems possible.

Extinctions are almost as common as originations in the fossil record: most fossil species are of extinct forms. Given the element of competition inherent in the evolutionary process, this fact is readily understood: successful originations invade niches at the expense of the previous occupants. Variation in the physical environment must also play an important part: changes in climate and in sea-level, mountain-building processes and natural catastrophes all affect the range and types of habitats available for organisms, to the advantage of some and the extinction of others.

It is not possible to be certain of the precise reasons for a particular fossil extinction. As with originations, however, one can make plausible suggestions in some cases. As has already been said, the pteridosperms declined during the Permian and this coincided with increasingly drier climate to which the gymnosperms were more adaptable. Another example is the increase in diversity of bivalves in the Mesozoic and Cainozoic, a period of low diversity for the brachiopods. Since both groups are filter-feeders occupying similar habitats, it may be that the bivalve feeding technique had improved to the detriment of the brachiopods.

The topic of multiple (mass) extinctions is very controversial. They have been ranked according to the percentage of invertebrate skeletal marine families which became extinct during the respective falls in diversity: late Permian 57%, late Ordovician 27%, late Trias 23%, late Devonian 19%, late Cretaceous 17%. The Permian episode is clearly the most striking. Diversity began to decline in the lower Permian and continued at an increasing rate, the most rapid fall being in the final 10 my. During all this time (some 40 my) originations continued to appear at a low level but were increasingly exceeded by extinctions. In the lower Trias, however, originations increased more than three-fold, exceeding extinctions which almost halved.

Most of the major groups of marine invertebrates were much reduced by the end of the Permian, but the reduction was selective. Fusuline foraminifera, rugose and tabulate corals all became extinct, though other foraminifera, and also hydroids and scyphozoans, survived. Ammonoids were greatly reduced but a few families carried-over into the Trias as the ancestral stock of their subsequent radiation (fig. 235). This applies also to other major groups: e.g. brachiopods, gastropods and echinoderms. The forms which became extinct had been prominent in Palaeozoic faunas; their later replacements gave the Mesozoic faunas a 'new look': e.g. the new scleractinian corals and new groups of echinoids (the euechinoids), crinoids and ammonoids.

On land, plant diversity fell throughout much of the Permian and tetrapods also declined during the earlier part of the period.

The late Cretaceous episode, lasting more than 10 my, involved extinction of only 17% of marine skeletal invertebrate families, notably the ammonoids, which declined, with minor fluctuations, from a peak of diversity to extinction during the last 30 my of the period (fig. 235). Other groups of invertebrates declined slightly and some not at all, e.g. the environmentally sensitive scleractinian corals. Diversity decline was thus much more restricted than in the Permian episode, being mainly concentrated in one group, the ammonoids. A well-known extinction event occurred suddenly at the very end of the Cretaceous: the simultaneous disappearance of about 85% of oceanic plankton species (calcareous nannoplankton and other phytoplankton; planktonic foraminifera and radiolaria). New Cainozoic forms started to appear immediately afterwards. This was an extremely short-term event.

On land, about 14% of tetrapod families became extinct, including the dinosaurs, which declined during the last 7 my or so of the Cretaceous. The fossil record is too scanty to record this decline in detail. Pterosaurs also became extinct at this time. On the other hand, plants did not show any significant change in diversity.

The cause(s) of multiple extinctions is not known. Hypotheses include climatic cooling on a global scale (which would affect particularly organisms in tropical regions); changes in area of the continental shelves caused either by sea-level fall or by plate motions eliminating or otherwise modifying continental margins (this would reduce the range of habitat available to marine invertebrates and intensify competition); changes in intensity of incoming radiation (affecting some organisms directly and with 'knock-on' effects on others); impact of a giant meteorite or comet (which would produce a global dust cloud, cutting off sunlight with immediate damage to plants and phytoplankton, again with 'knock-on' effects). Other suggestions emphasise biological factors: in the seas, diminishing food resources (e.g. lowering of phytoplankton production affecting the entire marine food chain) and direct predation (which might be associated with the emergence of new predating forms or the migration of existing ones). Additionally the possibility of coincidence of unrelated causes cannot be ruled out. Any of these mechanisms might be the cause of an individual extinction event and a combination of some of them might result in the larger scale but selective multiple extinctions. Since these are prolonged episodes it seems reasonable to suppose that long-term factors such as climatic variation and sea-level changes, which may be linked to plate movements and ocean floor spreading, are likely to have been significant. The formation of the supercontinent of Pangaea in the later Palaeozoic and its subsequent break-up starting in the Mesozoic must have had repercussions in the biosphere.

The sudden extinction of oceanic plankton at the very end of the Cretaceous has been attributed to an impact event by a giant extra-terrestrial body. Craters and fall-out deposits are known for about 100 large meteorite impacts on the earth. None of these is dated exactly at the time of the plankton extinctions and none seems to have been large enough to have produced a near-global effect. Such an impact, however might have occurred in the oceans, in which case the crater is unlikely to be found. Other evidence in support of this hypothesis comes from a thin 'boundary clay' at the Cretaceous–Tertiary boundary, marking the disappearance of the plankton. This clay is unusually rich in the element iridium which is relatively scarce in terrestrial rocks and which could have been introduced by an extra-terrestrial body. There are also minerals and mineral aggregates claimed to have been altered by high-pressure shock and by fusion. However, alternative explanations for the presence of all these features have been proposed (e.g. iridium from volcanic sources) and the matter remains controversial. Large-scale meteorite impacts have been a normal feature of the earth's history and must have affected the biosphere to varying degree. Such effects should be kept in mind when analysing the fossil record.

The major question remains: what were the controls of multiple originations and extinctions? We have to admit that though our knowledge of the fossil record is considerable this question cannot be answered convincingly. If we knew the cause(s) of each species origination and extinction in the record we could find the causes of long-term trends. Unfortunately we are very remote from this happy position. Favoured hypotheses range from those based on chance coincidence of a variety of causes, to those invoking periodic major perturbations caused by extra-terrestrial phenomena. Many hypotheses; none proven.

This final section has been concerned mainly with uncertainties and with the limits of our knowledge of the fossil record. Detail is difficult to establish and may be even more difficult to explain. But the broad picture of the history of life is plain enough: from the procaryotes of the Archaean through the whole range of fauna and flora leading to the diverse and complex biosphere of today.

ADDITIONAL READING

Abercrombie, M., Hickman, C.J. and Johnson, M.L. 1980. *Penguin Dictionary of Biology*. Penguin Books, Harmondsworth.

Barnes, R.S.K. and Hughes, R.N. 1982. *Marine Ecology*. Blackwell, Oxford.

Bignot, G. 1985. *Elements of Micropalaeontology* (English Edition). Graham and Trotman Ltd, London.

Boardman, R.S., Cheetham, A.H., and Rowell, A.J. 1987. Fossil Invertebrates. Blackwell Scientific Publications, Oxford.

Brazier, M.D. 1980. *Microfossils*. George Allen and Unwin, London.

British Museum Publications: *British Palaeozoic Fossils, British Mesozoic Fossils* and *British Cainozoic Fossils*. British Museum (Natural History), London.

Charig, A. 1983. *A New Look at Dinosaurs*. British Museum (Natural History), London.

Clarkson, E.N.K. 1986. *Invertebrate Palaeontology and Evolution*. George Allen and Unwin, London.

Glaessner, M.F. 1984. *The Dawn of Animal Life*. Cambridge University Press, Cambridge.

Haq, B.U. and Boersma, A.E. (eds.) 1978. *Introduction to Marine Micropalaeontology*. Elsevier, Amsterdam.

Lehmann, U. 1981. *The Ammonites – Their Life and World*. Cambridge University Press, Cambridge.

MacFarland, W.N., Pough, F.H., Cade, T.J. and Heimar, J.B. 1984. *Vertebrate Life*, Collier MacMillan, London.

Meadows, P.S. and Campbell, J.I. 1978. *Introduction to Marine Science*. Blackie, Glasgow and London.

Moore, R.C. (ed.), 1953– *Treatise on Invertebrate Palaeontology*. Kansas University Press. A series of volumes each dealing with a major invertebrate group on a detailed and systematic basis.

Murray, J.W. (ed.) 1985. *Atlas of Invertebrate Macrofossils*. Longman Group Ltd, Harlow.

Norman, D. 1985. *The Illustrated Encyclopedia of Dinosaurs*. Salamander Books Ltd, London.

Raup, D.M. and Stanley, S.M. 1978. *Principles of Palaeontology*. W.H. Freeman and Co., San Francisco.

Romer, A.S. 1966. *Vertebrate Palaeontology* (3rd Edition). Chicago.

Smith, A. 1984. *Echinoid Palaeobiology*. George Allen and Unwin, London.

Stewart, W.N. 1983. *Palaeobotany and the Evolution of Plants*. Cambridge University Press, Cambridge.

Taylor, T.N. 1981. *Palaeobotany*. McGraw-Hill, New York.

INDEX

Names of genera are italicised. Page numbers of illustrations are in bold type. For a simple definition of morphological terms, see also the glossary of technical terms listed under the appropriate groups.